WORKSHOPS IN COMPUTING
Series edited by C. J. van Rijsbergen

Also in this series

Michael F. McTear and Norman Creaney (Eds.)

AI and Cognitive Science '90

University of Ulster at Jordanstown
20–21 September 1990

Published in collaboration with the
British Computer Society

Springer-Verlag London Ltd.

Michael F McTear, BA, MA, PhD

Norman Creaney, BSc, MSc

Department of Information Systems
University of Ulster at Jordanstown
Newtownabbey BT37 0QB
Northern Ireland

ISBN 978-3-540-19653-2

British Library Cataloguing in Publication Data
Irish Conference on Artificial Intelligence and Cognitive Science (3rd:
University of Ulster, Newtownabbey, '90)
Irish Conference on Artificial Intelligence and Cognitive Science, University of
Ulster, Newtonabbey, 20–21 September 1990. (Workshops in computing)
1. Artificial Intelligence
I. Title   II. Creaney, Norman   III. McTear, Michael 1943–   IV. British
Computer Society   V. Series
006.3
ISBN 978-3-540-19653-2
Library of Congress Cataloging-in-Publication Data
Irish Conference on Artificial Intelligence and Cognitive Science
(3rd : 1990 : Ulster, Ireland)
AI and cognitive science '90: proceedings of the Third Irish Conference on
Artificial Intelligence and Cognitive Science, Ulster. 20–21 September 1990 /
Michael McTear and Norman Creaney [editors].
    p. cm. — (Workshops in computing)
"Published in collaboration with the British Computer Society." Includes index.
ISBN 978-3-540-19653-2   ISBN 978-1-4471-3542-5 (eBook)
DOI 10.1007/978-1-4471-3542-5
1. Artificial intelligence–Congresses.   2. Cognitive science–Congresses.   3.
Expert systems (Computer science)–Congresses.
I. McTear, Michael.   II. Creaney, Norman, 1958–   .   III. British Computer
Society.   IV. Title.   V. Series.
Q334. 163 1990   91-2722
006.3—dc20   CIP

34/3830–543210   Printed on acid-free paper

Preface

This book contains the edited versions of papers presented at the 3rd Irish Conference on Artificial Intelligence and Cognitive Science, which was held at the University of Ulster at Jordanstown, Northern Ireland on 20–21 September 1990.

The main aims of this annual conference are to promote AI research in Ireland, to provide a forum for the exchange of ideas amongst the different disciplines concerned with the study of cognition, and to provide an opportunity for industry to see what research is being carried out in Ireland and how they might benefit from the results of this research.

Although most of the participants at the conference came from universities and companies within Ireland, a positive feature of the conference was the extent of interest shown outside of Ireland, resulting in participants from USA, Canada, Austria, and England. The keynote speakers were Professor David Chin, University of Hawaii, and Professor Derek Partridge, University of Exeter, and the topics included machine learning, AI tools and methods, expert systems, speech, vision, natural language, reasoning with uncertain information, and explanation.

The sponsors of the conference were Digital Equipment Co (Galway) and the Industrial Development Board for Northern Ireland.

Michael F McTear
Norman Creaney

February 1991

Contents

List of Authors

Ambikairajah, E.
Regional Technical College, Department of Electronic Engineering,
Athlone, Co. Westmeath, Ireland

Anderson, Terry J.
University of Ulster at Jordanstown, Department of Information
Systems, Newtownabbey BT37 0QB, Northern Ireland

Ball, Jerry T.
New Mexico State University, Computing Research Laboratory, Las
Cruces, New Mexico 88003, USA

Beic-Khorasani, Reza
University of Ulster at Jordanstown, Department of Information
Systems, Newtownabbey BT37 0QB, Northern Ireland

Bell, David A.
University of Ulster at Jordanstown, Department of Information
Systems, Newtownabbey BT37 0QB, Northern Ireland

Blakley, J.
University of Ulster at Jordanstown, Department of Electrical and
Electronic Engineering, Newtownabbey BT37 0QB, Northern
Ireland

Bradley, Dearbhaile
Queens University, Department of Psychology, Belfast BT9 5AH,
Northern Ireland

Brennan, M.
Queens University, Department of Electrical and Electronic
Engineering, Belfast BT9 5AH, Northern Ireland

Byrne, Ruth M.J.
University College Dublin, Department of Computer Science,
Belfield, Dublin 4, Ireland

Chin, David N.
University of Hawaii at Manoa, Department of Information and
Computer Sciences, Honolulu, Hawaii 96822, USA

Clements, David
Queens University, Department of Psychology, Belfast BT9 5AH,
Northern Ireland

Cowie, Roddy
Queens University, Department of Psychology, Belfast BT9 5AH,
Northern Ireland

Cross, G.
University of Ulster at Jordanstown, Department of Electrical and
Electronic Engineering, Newtownabbey BT37 0QB,
Northern Ireland

Delaney, T.A.
University College Cork, Department of Computer Science, Ireland

Dunnion, John
University College Dublin, Department of Computer Science,
Belfield, Dublin 4, Ireland

Flynn, O.M.
University College Cork, Department of Computer Science, Ireland

Fusco, V.F.
Queens University, Department of Electrical and Electronic
Engineering, Belfast BT9 5AH, Northern Ireland

Gregg, J.B.
University of Ulster at Jordanstown, Department of Computing
Science, Newtownabbey BT37 0QB, Northern Ireland

Guan, Jiwen
University of Ulster at Jordanstown, Department of Information
Systems, Newtownabbey BT37 0QB, Northern Ireland

Hickey, Ray
University of Ulster at Coleraine, Department of Computing
Science, Coleraine, Northern Ireland

Hong, Jun
University of Ulster at Jordanstown, Department of Information
Systems, Newtownabbey BT37 0QB, Northern Ireland

Johnson, Peter
Queen Mary and Westfield College, University of London,
Department of Computer Science, Mile End Road, London E1 4NS

Kavanagh, Ita
University College Dublin, Department of Computer Science,
Belfield, Dublin 4, Ireland

Kenneally, W.P.
University College Cork, Department of Computer Science, Ireland

Lennon, S.
Regional Technical College, Department of Electronic Engineering,
Athlone, Co. Westmeath, Ireland

Lesser, Victor, R.
University of Massachusetts

Liu, Weiru
University of Ulster at Jordanstown, Department of Information
Systems, Newtownabbey BT37 0QB, Northern Ireland

Livingstone, Mark
Queens University, Department of Psychology, Belfast BT9 5AH,
Northern Ireland

McKevitt, Paul
University of Exeter, Department of Computer Science,
Prince of Wales Road, Exeter EX4 4PT

McTear, Michael F.
University of Ulster at Jordanstown, Department of Information
Systems, Newtownabbey BT37 0QB, Northern Ireland

Monds, Fabian C.
University of Ulster at Jordanstown, Institute of Informatics,
Newtownabbey BT37 0QB, Northern Ireland

Morrissey, J.M.
University of Windsor, Department of Computer Science,
Ontario N9B 3P4, Canada

Mulvenna, M.D.
University of Ulster at Jordanstown, Department of Computing
Science, Newtownabbey BT37 0QB, Northern Ireland

Murphy, S.J.M.
University College Cork, Department of Computer Science, Ireland

O'Flaherty, F.B.
University College Cork, Department of Computer Science, Ireland

O'Mahoney, A.B.
University College Cork, Department of Computer Science, Ireland

Oh, Jonathan C.
University of Missouri-Kansas City, Computer Science,
Kansas City, MO 64110, USA

Partridge, Derek
University of Exeter, Department of Computer Science,
Prince of Wales Road, Exeter EX4 4PT

Peschl, Markus F.
University of Vienna, Department of Epistemology and Cognitive
Science, Sensengasse 8/9, A-1090 Wien, Austria

Power, D.M.J.
University College Cork, Department of Computer Science, Ireland

Sarantinos, Efstratios
Queen Mary and Westfield College, University of London,
Department of Computer Science, Mile End Road, London E1 4NS

Sheridan, Paraic
Dublin City University, School of Computer Applications, Dublin 9,
Ireland

Shui, F.
University of Ulster at Jordanstown, Department of Electrical and
Electronic Engineering, Newtownabbey BT37 0QB,
Northern Ireland

Smeaton, Alan F.
Dublin City University, School of Computer Applications, Dublin 9,
Ireland

Sorensen, H.
University College Cork, Department of Computer Science, Ireland

Stewart, J.A.C.
Queens University, Department of Electrical and Electronic
Engineering, Belfast BT9 5AH, Northern Ireland

Sutcliffe, Richard F.E.
University of Exeter, Department of Computer Science, Prince of
Wales Road, Exeter EX4 4PT

Ward, Catriona
University College Dublin, Department of Computer Science,
Belfield, Dublin 4, Ireland

Woodham, C.
Northern Exploration Services, Waterloo, Dingwall, Ross-Shire,
Scotland

Yang, Gi-Chul
University of Missouri-Kansas City, Computer Science,
Kansas City, MO 64110, USA

Section 1:

Artificial Intelligence – Tools and Methods

What the software engineer should know about AI -- and *vice versa*

Derek Partridge *

ABSTRACT

Artificial Intelligence (AI) and Software Engineering (SE) have traditionally been largely independent disciplines. A common interest in developing computer systems is an area of overlap but fundamental differences in goals and objectives has resulted in little actual interaction. Limited success with chronic problems in both disciplines has forced practitioners in both fields to look beyond their local horizons in the hope of discovering new sources of impetus for further progress within their individual areas of interest. Software engineers are looking towards AI for both potential increase in software 'power' that AI techniques seems to promise, and exploitation of complexity-reduction strategies that AI has, of necessity, pioneered. AI practitioners, struggling with fragile demonstrations, would dearly like to import from SE tools, techniques and 'know-how' for the production of robust, reliable, and maintainable software systems. In this paper, I shall survey the range and scope of this two-way interaction between AI and SE.

INTRODUCTION

It has been the case that Artificial Intelligence (AI) and Software Engineering (SE) were largely independent subdisciplines of computer science (CS). The software engineers are preoccupied with the down-to-earth problems of actually constructing robust and reliable software systems. AI practitioners, on the other hand, are usually struggling to generate only fragile demonstrations of rather more ambitious ideas. The software engineers tend to view AI as the domain of hacked-together nebulous ideas that will never be robust and reliable software, while the AI people look at the software engineers as technicians pursuing the rather pedestrian

---

* author's address: Department of Computer Science, University of Exeter, EX4 4PT, UK. email: derek@uk.ac.exeter.cs

goal of getting some quite straightforward and uninteresting specification implemented reliably. Thus the two groups are perhaps vaguely aware of each other, but see no reason to get really acquainted. And on the face of it they seem to be right.

But times have changed (as is their tendency). The reason explosion of interest in expert systems (if not exactly robust and reliable products) has forced AI into harsh world of practical software systems which has brutally exposed many weaknesses in the technology -- for example, how do you validate software whose behaviour is not clearly either correct or incorrect, but just adequate or inadequate? And from the SE perspective, the software crisis is still with us (despite several decades of concerted attack on it) and the introduction of AI promises an increase in software 'power' but also threatens to feed the fundamental problems and exacerbate the situation. There is, however, a growing body of opinion that AI may well have something positive to offer (sophisticated support environments, for example) to the software engineer who is entangled in the complexity of the task and yet trying to keep the worst effects of the software crisis at bay.

There are then good reasons why these two groups of hitherto largely separate practitioners should get together and exchange information. We have a lot to learn about how to build robust and reliable AI software as opposed to merely fragile demonstration systems, and the software engineers most likely have some useful information in this regard. But the software engineers, who can claim no more than a steady advance against the software crisis over the last two decades, can be expected to similarly benefit from the experience of implementors who have been tackling even more complex problems (but with less stringent requirements on what counts as success), and have of necessity evolved some neat ways of enlisting the computer's help in reducing the effective complexity of their task.

In this paper I shall elaborate on a number of ways that people in AI and in software engineering can, and should, exchange information. This will lead to the argument that there is nothing special about AI software development; it just seems to be rather different from conventional software development because it embodies a number of extremes. And it is not just the case that developers of practical AI-software must learn to live with the constraints of conventional software engineering; the lesson to be learned may be more on the side of software engineers accepting (and thus addressing rather than ignoring) the fundamental problems that are illuminated by our efforts to development practical AI-software -- problems like, no complete specification, no hope of verification even in principle.

AI AND SE: WORLDS OF DIFFERENCE

If we take the view that one of the central pursuits of practitioners in both AI and SE is the design and construction of computer software, or to put it more crudely, programming, then we might plausibly infer that these two disciplines ought to have much in common. The fact of the matter is that they don't -- they typically employ very different programming languages, different system development methodologies, and the perceived sets of most-pressing problems associated with each discipline are almost disjoint. This state of affairs suggests that there has been a lamentable failure to communicate between these two computer-related disciplines. Within wishing to imply that lack of constructive dialogue is not a contributive factor, I do want to press another, alternative explanation: there are fundamental differences between AI and SE that are disguised by the use of misleadingly similar terminology.

To begin with, AI problems and typical SE problems differ in a number of important respects. The following tabulation illustrates these dimensions of difference.

AI problems	SE problems
ill-defined	well-defined
solutions adequate/inadequate	solutions correct/incorrect
poor static approximations	quite good static approximations
resistant to modular approximation	quite good modular approximation
poorly circumscribable	accurately circumscribable

There is much that is discussable about this table of differences, but I'll restrict myself to bringing just the first two features explicitly to your attention. Consider the consequences of dealing with problems that have no well-defined specification and for which there is no clear notion of a correct solution. Conventional computer science rests on two key assumptions: first, that the problem can be well-specified, and second, that potential solutions can tested for correctness if not verified in a more robust sense. Both of these assumptions are false for AI problems. This could be taken to suggest that AI problems are not programmable, and AI's oustanding track record of unrealized expectations could be taken as evidence that the suggestion is in fact more of a certainty. And I would not challenge this negative assessment too squarely given the computer science or software engineering meaning of the design and development procedure known loosely as "programming." But this is just one of those words that is important within both AI and SE and yet has a totally (but not generally recognised) different meaning within each discipline.

In CS/SE programming is:
  designing and developing a correct algorithm for a well-defined specification

In AI programming is:
  finding an adequate, tractable, approximate algorithm for an intractable 'specification'

So given these two rather different notions of programming as just one example of hidden and largely unrealised differences within the fundamentals of AI and SE, it is not at all surprising that practitioners within the two disciplines often don't seem to be speaking the same language -- they aren't.

AI AND SE: THE NEED FOR DIALOGUE

So now that we have some inkling of why practitioners in AI and SE have trouble communicating, let's see why it might be worth their while taking the trouble to talk to each other. AI holds out a promise for future software systems. It is a promise of increased software 'power' where 'power' is to be interpreted in terms of increased functionality and scope of applicability and not in terms of increased speed, etc. (hence the scare quotes).

AI promises new functionality in terms of responsiveness to operating environments -- e.g., software might adapt to each user as an individual and it might track a changing usage pattern. A rather different sort of new functionality is the ability to diagnose problems and suggest fixes *in itself*. As I've been careful to stress such new functionality is no more than a promise of future riches. The pessimist might observe that the AI world (like a communist state) is a place where all verbs are always conjugated in the future tense -- promises are abundant but the date for delivery is tomorrow, which of course never comes. But to take a more positive stance we might note that self-explanation, for example, is now commonplace (although rudimentary) within expert systems' technology.

While it remains an open question as to how much of certain AI promises will indeed be deliverable, other long-term implications of AI in practical software systems seem to be much better bets. And sadly one of these near certainties would appear to be an exacerbation of the software crisis unless we are very careful.

SE, as I've said, is not yet a robust discipline. It cannot guarantee to deliver reliable and maintainable systems. The main anchor points of practical SE are:-

1. that the problem be completely specified in sufficient detail,
and
2. that potential solutions (i.e., implemented algorithms) can be tested for correctness.

Point (1) can only be approximately satisfied when dealing with problems outside the world of abstract mathematical functions. And point (2) can deliver (at best) software that is free from known error. The anchor points are thus not as firm as we might wish them to be.

This is the basis of practical SE -- a methodology of Specify-And-Test (SAT). In formal CS, by way of contrast, we find a determination to do better than this and to formally verify the correctness of an algorithm -- Specify-And-Verify (SAV) is the basic method here.

Now we are in a position to fully appreciate the awfulness of having to incorporate AI into practical software. The nature of AI problems is such as to undermine both of these less-than-thoroughly-solid anchor points, one of which has already been weakened from verifying to testing. In short: AI in practical software systems may plunge us into a super software crisis (both parsings of this phrase give appropriate meanings).

But looking on the brighter side: what can good AI software be? The answer cannot be a correct implementation of a well-specified problem, at least not in the first instance if ever. Good AI software is much more likely to be a good-enough, tractable approximation to an impossible problem. Hence the need for an incremental or exploratory methodology for constructing such software. The basic framework for this methodology can be cast as the Run-Understand-Debug-Edit cycle, or more memorably the RUDE cycle. In its worst manifestation it is simply code hacking or the code-and-fix methodology which has little or nothing to recommend it. The important question is can we develop a disciplined version of the RUDE cycle, one that carries with it some guarantees on the final system produced as well as a reasonable assurance that the iterative activities will indeed converge and deliver a final system? As much as I'd like to answer this question with a definative 'yes,' instead I'll be honest and admit that this is an open question. We just don't really know, but there are some promising avenues of research which bolster hope that a positive answer will eventually be supportable with something more concrete than hope.

AI AND SE: THE DIALOGUE STRUCTURE

There are a number of rather different ways that potentially fruitful interaction between AI and SE can occur. Above I have concentrated on an option that is of particular personal interest but is also one of the more contentious ones -- AI in practical software systems. It is possible to separate out two other main options.

I have already mentioned, if not laboured, the notion that conventional SE is not without its problems; the phenomenon of software crisis is not of purely historical interest. Conventional SE still needs all the help it can get to produce robust and reliable software systems that satisfy user demands within acceptable time and cost constraints. Use of environments is seen as an important source of help -- system development environments, programming environments, complete life-cycle environments, etc. there's many different sorts and equally many different names for many essentially identical types of environment. What they all have in common is that they are designed to reduce the effective complexity of some (or all) stages of the SE process. And by so doing they assist the human software engineer in his or her prime task.

Using the computer to provide help, in many different ways, to the software engineer is a laudable goal -- almost everyone would agree. So the use of such environments has long been widely acknowledged as a good thing. However, what at first sight seems fairly straightforward for example, warning the programmer about the use of dubious structures, actually proves to be very difficult to realize in reality. Much of the automatic help that we can conceive of supplying to the system developer turns out to be AI-ish -- i.e., we seem to need heuristic procedures to generate best guesses on the basis of incomplete information.

So the need for interaction that arises within this thoroughly conventional SE context is a need for AI within system support environments. It is a use of AI to assist in the design, development and even maintenance of software which itself may be totally conventional -- not a heuristic or knowledge-base in sight.

There is one further general category of interaction, and it may be so obvious that it largely escapes notice. AI systems are software systems and, despite my disparaging remarks about the current state of SE, there is a considerable heap of accumulated wisdom concerning how best to go about building robust and reliable software. Clearly AI-system builders should not ignore this repository of potentially relevant expertise. But the situation is not quite this straightforward because it is only *potentially relevant* information. There are some quite sharp differences between what the AI practitioner and the traditional software engineer do -- e.g., between what they are aiming at, where they start from, and how they think they can best get there. This should be no revelation as this paper began with an expose of such differences. So clearly, there is useful lessons that the AI-system builder can learn from the (relatively) long practice of SE, but equally clearly the transfer from SE to AI must be selective.

In sum, we have three main loci of interaction: AI in practical software, AI in software support enviornments, and SE tools and techniques in AI-software. For the interested reader full details together with representative examples can be found in Partridge (1988) [1].

In reality of course, terms like AI and SE do not label mutually exclusive tools, techniques, etc. At the extremes, we can find wide agreement for many applications of these labels. But in the middle ground no such simple agreement is found, or expected. So despite the fact that a clear separation of AI phenomena from SE ones is not possible, it is both useful and instructive to examine our (software) problems from these two viewpoints because they do capture the essence of many points of contention is this general area. The following diagram illustrates the component continua: the top one is that of subject area, the middle one is of methodologies, and the bottom one is example problems (with the subfield of expert systems, ES, slotted in).

the reality is a continuum

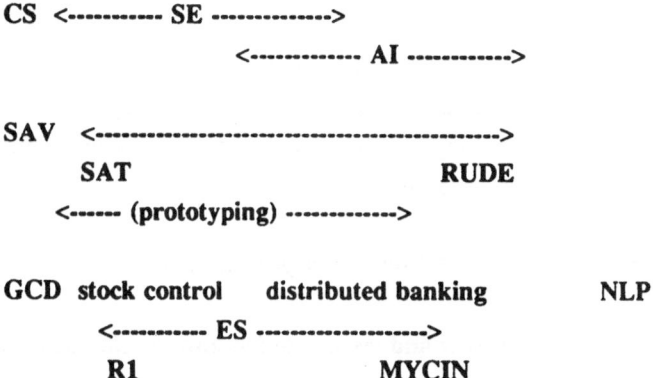

Interestingly, if there is any real discontinuity here it is not between AI and SE but between CS and SE. It may be that the leftmost column (containing CS, SAV and **GCD** -- greatest common divisor) illustrates phenomena that are really in a different world.

THE LESSONS LEARNED

Firstly,

* AI may be more like SE than SE is like CS

If this is true, then it's not AI that's a field apart but formal computer science which is quite surprising.

And what might adherents of the two extremes (AI and SE) learn as a result of dialogue? The SE person might be expected to agree to the following points:

* there is no complete specification of problems

* having a specification and knowing exactly what has (and has not) been specified are two different things

* software system design and construction can only be partially modularised

In addition, AI has pioneered the notions of exploratory programming and software support environments, and the latter (in particular) will continue to have great utility in all areas of software design, development and maintenance.

From the conventional SE perspective, the AI practitioners have learned that some disciplined version of RUDE is an essential prerequisite of widespread success with practical AI software. And pointers toward an appropriate methodology of exploratory system development may be found in the conventional wisdom of SE. I can suggest the following possibilities:

* an incremental analogue of "structured programming"

* need to explicitly represent and record design and development decisions

* divide and conquer -- separate AI and non-AI modules

In addition, AI will continue to benefit from advances in the fundamentals of computer science, advances in, for example, programming language design and implementation.

History will record the rise of separate encampments -- AI enthusiasts and practical software engineers. For many reasons this is a pity and to the detriment of both areas. I have presented an argument to this effect, I hope that it will motivate some people, at least, to look across at 'the other side' with a view that it might just be possible to learn something useful there.

References

1. Partridge D (1988) Artificial intelligence and software engineering: a survey of possibilities. Information and Software Technology 30(3): 146-154 (reprinted in Ince D and Andrews D (eds) (1990) The Software Life Cycle. Butterworths, London 375-385)

Developmental Support for Knowledge Based Systems

H. Sorensen, T.A. Delaney, W.P. Kenneally, S.J.M. Murphy, F.B. O'Flaherty,
A.B. O'Mahony, D.M.J. Power

ABSTRACT

DESKS is an integrated software package designed to support the development of advanced knowledge based systems, including expert systems and other AI applications. It provides system designers with an interacting set of sophisticated tools and facilities to assist in the development process. This paper discusses the aims and achievements of DESKS, and chronicles the design and construction of the package.

1. INTRODUCTION

In recent years, a distinction has been made between *Database Management Systems* (DBMS), which have traditionally supported data processing applications, and *Knowledge Base Management Systems* (KBMS), whose aim is to support more advanced applications of information, including those within the general area of artificial intelligence[1]. While a precise definition of a KBMS is difficult to achieve, we can informally state that such a system consists of a loose mixture of database and semantic capabilities[2]. In this paper, we describe a system which we have developed at University College Cork to support experimentation with, and development of, knowledge-based applications. In particular, we aim to provide support for users who may wish to develop expert systems which require storage and retrieval of considerable volumes of data, and who wish to experiment with various design strategies during the development process.

In the next sections, we describe the role and structure of our system - we call it DESKS (Developmental and Experimental Support for Knowledge-based Systems).

---

1. The title *Expert Database System* (EDS) has also been used in the literature in a way which is generally synonymous with KBMS [1‿2]. This 'EDS' term is meant to suggest that the purpose is to support the development of expert systems with database facilities, and corresponds broadly with the aim of the package described here.

2. We shall use the term KBMS to refer to a knowledge base management system. The term KBS (Knowledge Based System) shall then refer to applications developed around a KBMS.

Subsequently, each of the structural components is examined and its relationship to the whole is outlined. In a later section, we describe the current status of the system and cover some of its sample uses. Finally, we evaluate our system and outline some current and future developments which we consider necessary.

2. ROLE OF DESKS

In essence, DESKS is a software package which acts as a workbench for designers of knowledge-based systems. It is envisaged that it would be used by designers of advanced information systems, particularly those constructing expert systems or other AI applications requiring the storage and retrieval of large volumes information - this information may be shared with other applications, possibly being generated as a result of general data processing (DP) tasks within an enterprise. DESKS permits the integration of data requirements of AI and DP tasks, and provides a set of advanced tools for development of applications of the former type. In doing so, it aims to address several deficiencies which have been identified as pertaining to the development of realistic AI applications.

One such deficiency is the lack of database facilities present in languages and packages used for the development of AI applications. Languages such as Lisp or Prolog are based on a memory-resident model, presuming all data relevant to the program to be stored in data structures in main memory. The same approach is taken by many of the expert system building packages. While files can be consulted from disk, and while the underlying operating system may adopt a paging strategy to simulate a large virtual address space, little practical use of secondary storage is made. This has proved severely restrictive for many AI applications which, by their very nature, are likely to require access to considerable amounts of data and which may frequently generate and modify data values as a result of computation. With the adoption of Prolog for the Japanese Fifth Generation Project [3], much attention has focussed on the provision of database facilities for this language, either by extending the compiler/interpreter or by integrating it with a conventional database system. We discuss some of the research in this area later in the paper. At this point, we simply mention that we have adopted the strategy of integrating a Prolog interpreter with a conventional SQL-based database system.

Another deficiency, which is related to that mentioned above, is the fundamental mismatch between the facilities available for support of AI applications and those used in DP tasks. Many present-day DP applications use database management systems for storage and retrieval of data pertinent to the business being transacted - the use of relational databases is now quite prevalent in commercial environments. Most AI applications, on the other hand, have been developed using either high-level languages (Lisp, Prolog, etc) or specially-tailored packages, and have no facilities for accessing those databases. As a result, the AI applications do not possess the capability to utilise up-to-date data on the enterprises for which they have been designed. If this scenario were to continue, it is difficult to envisage AI techniques being applied in many commercial/industrial areas; yet, surely, these are the very areas which require the employment of such advanced mechanisms. DESKS aims to support the integration and sharing of data generated and used by both DP and AI applications.

A final deficiency which we mention at this point is the lack of good user interface facilities in most AI development systems/languages. In most Prolog implementations, for example, data can be retrieved only through the construction and execution of syntactically valid queries (or goals). Similarly, SQL queries are used for relational database access. We would claim that this one-dimensional data access mechanism ignores many of the advances which have been made in software technology, particularly within the domain of user interface design. We refer in particular to the development of natural language enquiry systems, window interfaces, browsers, icon/mouse front-ends, etc. In the design of DESKS, we attempted to provide AI designers with a sophisticated set of interface tools for access to data (and code) relevant to the application being constructed.

3. SYSTEM ARCHITECTURE

Fig. 1 contains an architectural overview of the DESKS system. The major structural components are illustrated, as are the information and control flows which take place between them. Some of these components are optional[3], and can be "switched off" by a system user via a front-end menu selection - in this way, users can tailor the system to their individual requirements and tastes. In the remainder of this section, we outline the operation of the various component modules and discuss some of the major design decisions faced during the construction of DESKS.

3.1 Prolog-DBMS Integration

Perhaps the most important decision taken was to integrate a Prolog interpreter with a conventional relational database system. We based this decision on the fact that Prolog has gained a considerable foothold in the area of AI application development. Its rule-

---

3. Optional modules are denoted in Fig. 1 by placing an asterisk in the upper right corner.

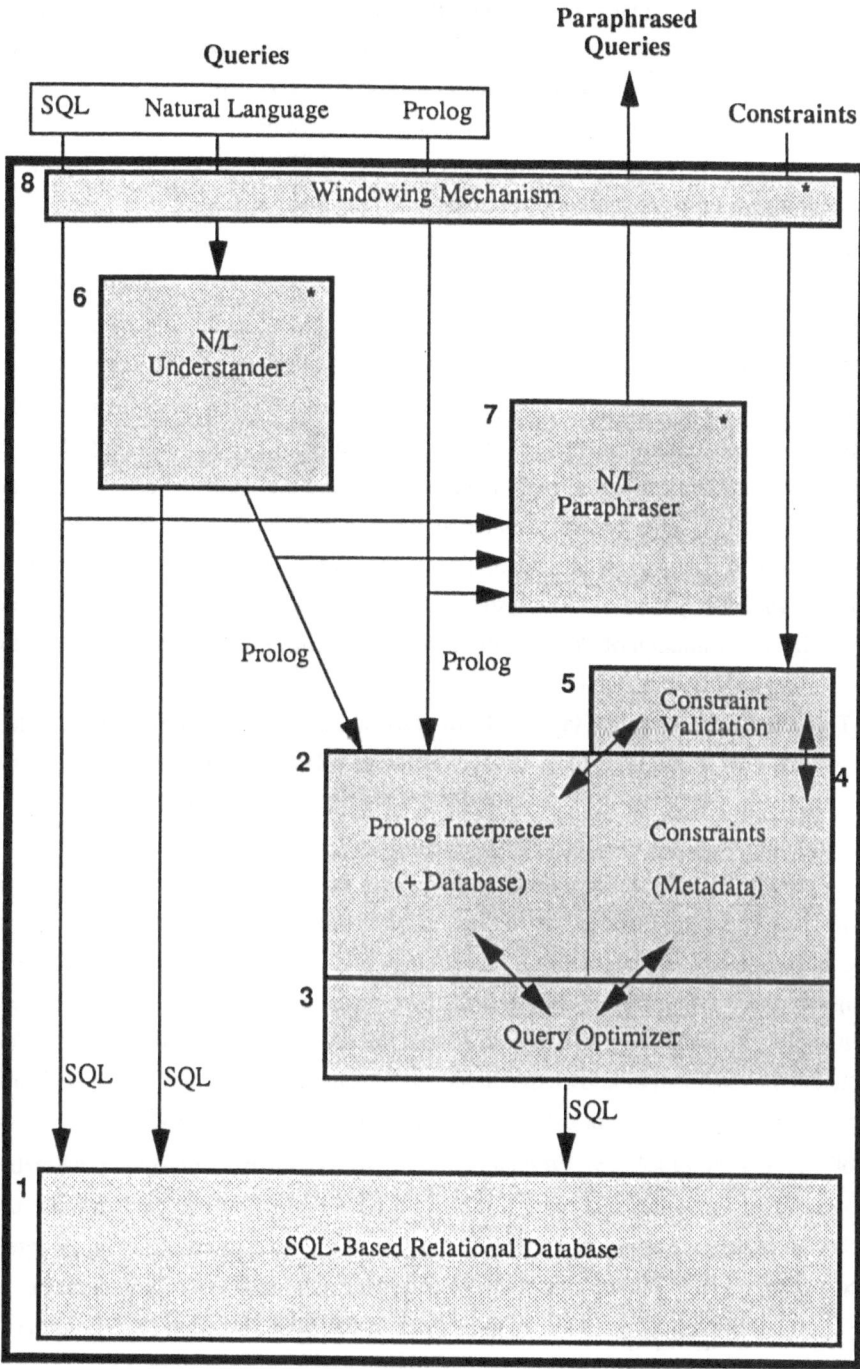

Fig. 1 - System Architecture

based paradigm and declarative programming style has resulted in its being adopted for the development of several expert systems, as well as other systems within the AI domain. Its adoption for the Fifth Generation Project has served to widen interest in the language and its applications. As many expert system design packages are themselves written in Prolog and allow the incorporation of Prolog code with the applications developed, they are unlikely to reduce the relevance of the logic programming approach to future development strategies.

Shortly after the design of the Prolog language, two issues were highlighted: firstly, its data representation methodology and the goal-based search mechanism used closely resembled the data storage and query approach of relational databases (assuming that queries are based on the *domain calculus* formalism) [4], a fact which had earlier been noted in relation to the general logic formalism [5]; and secondly, Prolog code and data are maintained in main memory only, and the language lacks any database facilities. This latter point was noted as a severe deficiency by both AI programmers and KBS developers, who recognised the importance of stored data to the applications on which they worked. The response from the database community has been to suggest the amalgamation of Prolog with database facilities [6], though there has been some dissent on this approach [7].[4]

This amalgamation might take the form of extending a Prolog interpreter/compiler with general DBMS facilities (data storage/retrieval/update, security, concurrency control, transaction management) - conversely, a DBMS might be extended with logic programming facilities to produce an equivalent result. An alternative strategy, and the one adopted in DESKS, involves the *coupling* of a Prolog front-end with a relational DBMS back-end [8]. Besides cost reduction (we could work with existing products and concentrate on their interaction), this strategy had one major attraction for us: by basing our system on a widely-used DBMS[5], we can allow sharing of information between applications and users within both AI and DP domains - this should serve to facilitate the development of knowledge based systems in industrial/commercial environments, where large volumes of data relevant to the enterprise may already exist as a result of ongoing DP processes. Of course, the *impedance mismatch* problem, whereby the operational semantics of Prolog's tuple-at-a-time evaluation mechanism must be matched to SQL's set-at-a-time approach, must be recognised in constructing an integrated package [7]. We are not alone in following this path towards integration - several other research efforts have made strides in the same direction [8, 9]. Others,

---

4. An alternative response, which might be said to have come from the programming language community, has suggested the addition of *persistence* to the Prolog language. While such persistent programming languages have their advantages, we do not discuss them further in this paper.

5. The DBMS upon which DESKS is constructed is Oracle, the most widely-used database package which employs the SQL query language.

noting that the procedural semantics of Prolog reduces its declarative power, have designed entirely new logic programming languages with inbuilt database facilities [10, 11] - this would, however, defeat our purpose of integrating KBS with current data processing applications and was not therefore pursued. There have also, of course, been attempts to integrate other high-level languages, such as Lisp, with database facilities [12], but we do not consider this matter here. What differentiates DESKS from other *deductive databases,* as they are sometimes called, is its overall design philosophy and the incorporation of other value-added components into an integrated package.

One final decision which was necessary with respect to integration related to the degree of transparency which should be built into the Prolog-DBMS interface and to whether we should adopt a *loose* or *tight* coupling mechanism between the logic and DBMS components [13]. PROSQL, developed by IBM, is an example of a non-transparent interface, whereby SQL queries can be embedded in Prolog programs, using a special predicate, and used to retrieve information from the database for processing by the program code [14]. One disadvantage of this approach is that Prolog programmers need to be familiar with the SQL language, and need to know which data is held in internal Prolog databases and which is stored in the external relational database. For this reason, we chose to avoid this approach and build a completely transparent interface. Our approach is characterised by the fact that all data, whether stored internally or externally, is treated similarly. All access to data is via Prolog queries. If this data is stored as Prolog facts, then the result can be generated without disk access. If the data is stored in database relations, then an SQL query is generated based on a syntactic transformation of the Prolog goal. This SQL query is then evaluated against the relational database contents and the retrieved data asserted as facts into the Prolog workspace for further processing. Once processed, these facts can be discarded.The advantage of our approach is that no distinction is made between memory-resident and disk-resident data, and the latter can actually be an extension of the former. The use of SQL is totally hidden from Prolog programmers - they simply see the database as an extension of the memory space available to the Prolog system. Of course, information stored on disk may still be accessed via SQL by interactive users or by other application programs - no diminution in the facilities available to DBMS users occurs. Finally, we mention that we have adopted a tight coupling mechanism, insofar as data is retrieved from the SQL database only when specifically required for evaluation of a Prolog predicate. The alternative mechanism, a loose coupling, involves the taking of a database *snapshot* and the asserting of all snapshot tuples into the Prolog workspace. This latter method has the advantage of efficiency in certain situations, but requires an intelligent strategy to determine the snapshot contents required, and is potentially demanding of memory resources.

In summary, therefore, we have augmented a conventional Prolog interpreter with a set of meta-level predicates, which serve to support a transparent tightly-coupled interface to a relational database employing the SQL language (Fig. 2). This interface allows the evaluation of Prolog goals against relations of the database by syntactic conversion of these goals into equivalent SQL queries, effectively allowing the database to appear as an extension of the Prolog workspace. As shown in Fig. 2, the interface is built in two phases: *meta-transition*, which handles the conversion of Prolog to SQL; and *communication*, involving the transfer of data from one module to the other. In the event of recursive Prolog predicates being present (since SQL has no recursion, a direct conversion is impossible), we have adopted two complementary strategies, viz. Recursive Query-Subquery (QSQR) [15] and Recursive Question Answering / Frozen Query Iteration (RQA/FQI) [16]. For a further elaboration on the meta-level predicates used and on the recursive predicate evaluation, see [17].

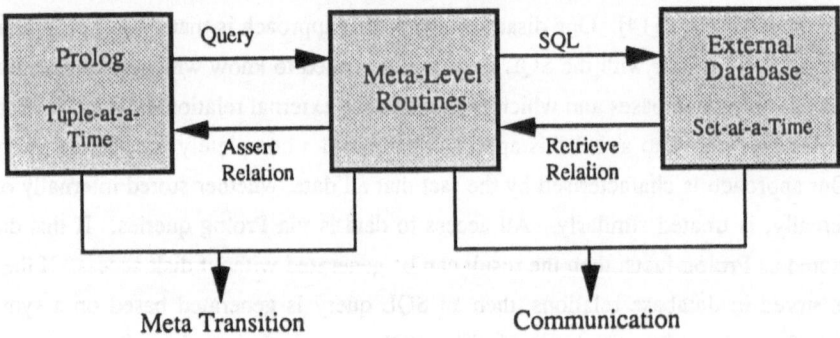

Fig. 2 - Prolog-DBMS Interface

3.2 Constraints in DESKS

The role of constraints in relational databases has long been acknowledged [18]. These include state constraints and transition constraints, and provide a mechanism for specifying valid database states. Such validity can cover issues of key integrity, referential integrity and functional dependence, as well as issues relating to the semantic validity of application-specific data (e.g. "at least half of the contributors to the AICS conference must be associated with Irish institutions" or "an employee's salary can only increase"). The logic representation mechanism used in Prolog (Horn clauses) provides a specification mechanism for a wide range of constraints which would be relevant to the development of a semantically valid knowledge-based system [19]. These constraints reside in the Prolog database (or in the associated relational database) along with the database facts, and can be used to control the transactions being processes against the facts - only transactions which convert the database from one semantically valid state to another are allowed.

In DESKS, we allow the representation and storage of constraints as logic predicates, and the monitoring of transactions to ensure that no violations result. Furthermore, we have implemented a constraint validation module using a set of meta-predicates which guarantees that constraint sets are internally consistent and that newly inserted constraints are not at variance with existing data values. For full details of this module, see [20].

3.3 Semantic Query Optimisation

A database (internal or external), when combined with a set of constraints, is often termed a *semantically constrained database*. The semantic meta-data can be used, not just to preserve validity, but also to optimise queries and general transactions against the database [21]. This is in addition to the query optimisation provided in general relational database systems, using techniques such as indexing, caching, join reduction, etc. As an example, if we know that "bonuses are not applicable to the Accounts department" (constraint), and are given the query "Find the names of employees who earned bonuses in excess of £1,000", it is clear that records relating to Accounts department staff can be ignored in evaluating the query. Several approaches have been suggested for the incorporation of semantic query optimisation in deductive databases [21], though the handling of recursive queries remains an open problem.

In DESKS, we have extended the basic Prolog interpreter with semantic optimisation meta-predicates. These meta-predicates constitute a *query reformulator* (Fig. 3), which transforms a given user query, taking into account the available integrity constraints, into a set of semantically equivalent queries. These queries can be evaluated independently, and we can use techniques such as *residue filtering* [22], to generate results in as efficient a way as possible.

3.4 Natural Language Understander

The system as described up to now constitutes an optimised, constrained deductive database, with external storage facilities provided by a relational database. Internally stored data can be accessed via Prolog, while external data may be accessed using Prolog or SQL. For many application developers (and for subsequent application users), this is a rather unproductive way of effecting data access. For this reason, we have developed a set of front-end modules to support improved user interfaces.

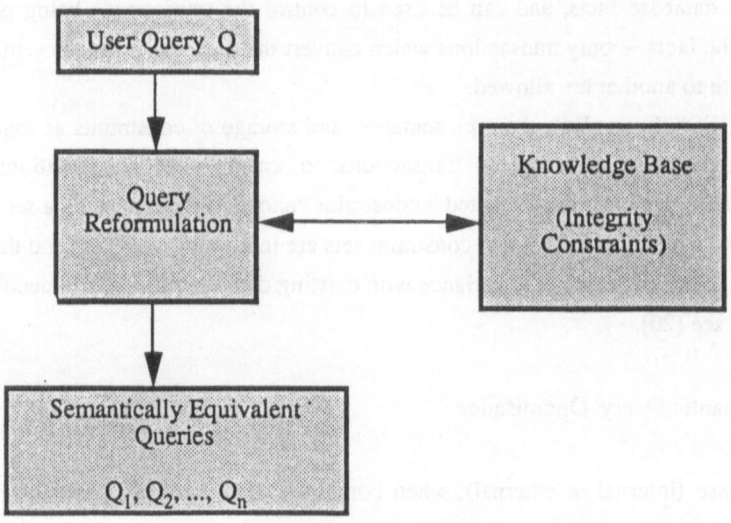

Fig. 3 - Semantic Query Optimisation

The first of these tools is a natural language understander and translator for queries against the stored data [23]. This module was designed to be *transportable* [24, 25] in two ways: it is *database independent* (or *domain independent*), in that it provides a robust interface to different sets of database relations with very little user effort; and it is *language independent*, in that it can transform natural language queries into Prolog or SQL (as shown in Fig. 1) - the user may choose which output is to be generated. Database independence is supported by the separation of domain-dependent and domain-independent information. The former can be represented by combining the database schema with a *domain lexicon*, which contains domain-specific information. By constructing different domain lexicons for different databases, it is possible to transport the N/L understander from one database to another. To support language independence for generated queries, we have employed a technique which we call a *query concept* - this is based on Schank's conceptual dependence theory [26]. This is an internal representation into which we convert N/L queries, and effectively forms a logic representation for queries. From this internal representation, we may transform to either Prolog or to SQL quite easily - [Ofl89] covers issues involving the internal representation and the transformation mechanism.

3.5 Natural Language Paraphraser

A second interface tool which we have implemented is a natural language paraphraser [27]. The motivation behind this module lay in the fact that queries submitted by a user in Prolog or SQL may not always retrieve the data as required - it may frequently occur that a query, while syntactically correct, is semantically at variance with the user's expectations. For this reason, we have allowed a user/developer to obtain an English-language paraphrase of any submitted query, which can then be used to check the user's intuitive interpretation of the query's effect. As in the REMIT system [28], which paraphrased a relational calculus-based query language (Querymaster), we found that a simple algorithmic approach could be taken towards the generation of un-ambiguous and coherent paraphrases, and that it was not necessary to resort to a so-phisticated natural language formalism. In [27], we show that a two-stage translation mechanism suffices: the first stage constitutes a *lexical gathering phase,* and also in-volves selecting a *focus* for the submitted query (i.e. determining which base relations and attributes will occur in the paraphrase) - it results in the generation of an internal syntactic query representation; and, the second stage is the *generation phase,* which produces the paraphrase from this internal representation. We have found this N/L paraphraser quite useful when debugging Prolog code used to construct different mod-ules of DESKS, and expect that it may be used in the same way by application devel-opers and other system users.

3.6 Windowing Mechanism & Browser

A final component of the user interface which we mention here is the window mecha-nism and browser which we have constructed as an optional front-end for application developers and users. Many mainframe-derived DBMS packages differ from those supplied on microcomputers in that they are entirely language-based for data lookup purposes. Thus, for example, systems such as Oracle require the use of SQL for query expression, while microcomputer-based products, such as dBase, have inbuilt browsers for data examination. Similarly, a newer generation of products have integrated the notion of windowing to achieve a visually aesthetic system interface - this is particu-larly true of workstations. While there may exist historic reasons for this development pattern (generally, mainframes and minicomputers did not possess sufficient power to handle sophisticated front-end processing), we do not believe that this situation need necessarily continue. Many approaches have been identified as promising with regard to visual interfaces [29].

We have therefore constructed a window-based front-end to DESKS, which incor-porates a browser [30]. Using this package, it is possible to inspect data and code within the Prolog workspace (and, by extension, within the external database). The ad-vantages of data browsing should be obvious - indeed, its omission is frequently cited as one of the major drawbacks in general relational systems. We have also found the code browsing facility to be useful, however. Using this mechanism, it is possible to examine the Prolog code for an application, activate a goal, and browse the data produced as a result. As with the paraphraser, we have used this facility for debugging of different system modules, and expect that it would find the same use by application developers and general system users.

4. SYSTEM STATUS AND USES

The component modules of DESKS were constructed during 1988-89. All implementation took place using standard Prolog, so that the system should be portable to different platforms with little difficulty (provided those platforms support Unix and Oracle as operating system and DBMS, respectively). At present, the integrated system is in prototype form, with further work needed to bring it up to product form. This work shall concentrate on the integration of components into a single package, with a (hopefully) simple user interface for activation and use of individual modules.

Even as it currently stands, we have found productive uses for the system. For example, as mentioned above, we used the browser and the N/L paraphraser for debugging of some of the system modules - this was possible due to the fact that different modules were implemented at different rates. We found this to be useful in producing correct and robust code, while it also served to provide a useful testbed for the browser and N/L paraphraser themselves.

We have also experimented with the development of a knowledge-based scheduler, using DESKS for development support. This scheduler (for a box manufacturing plant) involved access to considerable amounts of industrial data and specifications, the purpose being to schedule the machinery, employees and raw materials of the plant for optimal production. The data requirements suggested the use of a relational DBMS, while the rule-based and heuristic nature of the scheduling task was well suited to the Prolog language. We found DESKS to be of considerable benefit when developing the application, and felt that it improved programmer productivity - we have, as yet, no quantitative measure of the degree of improvement achieved. Nevertheless, as an example of a knowledge-based system incorporating elements of AI and of DP, we found that this type of application is suited to the type of developmental support provided by our system.

5. CONCLUSION

In this paper, we have outlined an approach towards the supporting the development of knowledge-based systems, and have described an integrated package which we are implementing towards that end. The notion of support packages exists for different fields of application - for example, the use of CASE tools and 4GLs is now spreading in DP environments. Our system, whose construction embodied extensive use of AI and KBS techniques, is aimed more at the development of applications within the domain of AI and expert systems. Our experience with the project has been very positive, and we feel that this, or similar products, are needed to improve the development and acceptance of applications of knowledge-based systems.

Besides the ongoing work in system integration, mentioned above, we have been experimenting with some new approaches. For example, an object-oriented version of Prolog using *prototyping* as a mechanism for object definition (as opposed to the general *inheritance* mechanism) is being examined [31]. If successful, this could form the foundation of an object-oriented database system on top of our relational database - we recognise, however, that this is some way down the road. Leaving this aside, it appears to us that specialised application development tools, such as DESKS, provide a very useful approach to tackling the bottleneck of software production in KBS environments.

REFERENCES

1. Kerschberg, L. (ed.). Proc. First Int. Workshop on Expert Database Systems, South Carolina, 1984.

2. Kerschberg, L. (ed.). Proc. Second Int. Workshop on Expert Database Systems, South Carolina, 1988.

3. Kunifuji, S. & Yokota, H. PROLOG and relational databases for fifth generation computer systems. Proc. of the Workshop on Logical Bases for Databases, Toulouse, 1982.

4. Gray, P.M.D. Logic, Algebra and Databases. Ellis Horwood, 1984.

5. Gallaire, H. & Minker, J. Logic and Databases. Plenum Press, 1978.

6. Li, D. A PROLOG Database System. Research Studies Press Ltd., Herts., England, 1984.

7. Maier, D. Databases in the Fifth Generation Project: Is Prolog a Database Language? in Ariav & Clifford, Eds: New Directions for Database Systems, Abtes, 1984.

24

8. Jarke, M., Clifford, J. & Vassiliou, Y. An Optimizing Prolog Front-End to a Relational Query System. Proc. ACM-SIGMOD Conf. on Management of Data, 1984.

9. Bocca, J. Educe - A Marriage of Convenience: Prolog and a Relational DBMS. Proc. Symposium on Logic Programming, 1986.

10. Zaniolo, C. Design and Implementation of a Logic Based Data Language for Data Intensive Applications. Proc. Int. Conf. on Logic Programming, Seattle, 1988.

11. Zaniolo, C. Deductive Databases - Theory Meets Practice. Proc. Second Int. Conf. on Extending Database Technology (EDBT'90), Venice, 1990.

12. Poulovassilis, A. & King. P. Extending the Functional Model to Computational Completeness. Proc. Second Int. Conf. on Extending Database Technology (EDBT'90), Venice, 1990.

13. Vassiliou, Y. Knowledge Basedand Database Systems: Enhancement, Coupling or Integration? in M.L. Brodie & J. Mylopoulos, eds: On Knowledge Base Management Systems, Springer-Verlag, 1986.

14. Chang. C.L. & Walker, A. PROSQL: A Prolog Programming Interface with SQL/DS. in [Ker86].

15. Vielle, L. Recursive Axioms in Deductive Databases: The Query/Subquery Approach . in [Ker84].

16. Nejdl, W. Recursive Strategies for Answering Recursive Queries - The RQA/FQI Strateg. Proc 13th Conf. on Very Large Databases (VLDB), 1987.

17. Power, D.M.J. Query Translation in a Tightly Coupled Logic Database System. M.Sc. Thesis; Computer Science Dept., University College, Cork, 1989.

18. Kowalski, R.A., Sadri, F. & Soper, P. Integrity Checking in Deductive Databases. Proc 13th Conf. on Very Large Databases (VLDB), 1987.

19. Ling, T-K. Integrity Constraint Checking in Deductive Databases using the Prolog NOT-Predicate. Data & Knowledge Engineering, 2, 2, 1987.

20. Kenneally, W.P. The Design of an Integrity Subsystem for a Logic Database. M.Sc. Thesis; Computer Science Dept., University College, Cork, 1989.

21. Chakravarthy, U.S., Grant, J. & Minker, J. A Logic Based Approach to Semantic Query Optimization. Tech. Report 1921, University of Maryland, College Park, 1987.

22. O'Mahony, A.B. A Semantic Based Approach to Query Optimisation in Deductive Databases. M.Sc. Thesis; Computer Science Dept., University College, Cork, 1989.

23. O'Flaherty, F.B. Design of a Transportable Natural Language Database Interface. M.Sc. Thesis; Computer Science Dept., University College, Cork, 1989.

24. Grosz, B.J., Appelt, D.E., Martin, P.A. & Pereira, F.C.N. TEAM: An Experiment in the Design of Transportable Natural Language Interfaces. Artificial Intelligence 32(2), 1987.

25. McFetridge, P., Hall, G., Cercone, N. & Luk, W.S. System X: A Portable Natural Language Interface. Proc. 7th Biennial Conf. of the Canadian Society for Computer and Statistical Software, June 1988.

26. Schank, R.C. Conceptual Information Processing. North-Holland, Amsterdam, 1975.

27. Murphy, S.M.J. A Natural Language Paraphraser for Queries to a Logic Database System. M.Sc. Thesis; Computer Science Dept., University College, Cork, 1989.

28. Lowden, B.G.T. & DeRoeck, A. The REMIT System for Paraphrasing Relational Query Expressions into Natural Language. Proc. 12th onf. on Very Large Databases (VLDB), Kyoto, Japan, 1986.

29. Rowe, L.A., Danzig, P. & Choi, W. A Visual Shell Interface to a Database. Software - Practice & Experience, 19, 6, 1989.

30. Delaney, T.A. A Windowing Interface for a Logic Database System. M.Sc. Thesis; Computer Science Dept., University College, Cork, 1989.

31. Kissane, J. M.Sc. Thesis; Computer Science Dept., University College, Cork, to be presented 1990.

Problem description and hypotheses testing in Artificial Intelligence[1]

Paul Mc Kevitt and Derek Partridge[2]

ABSTRACT

There are two central problems concerning the methodology and foundations of Artificial Intelligence (AI). One is to find a technique for defining problems in AI. The other is to find a technique for testing hypotheses in AI. There are, as of now, no solutions to these two problems. The former problem has been neglected because researchers have found it difficult to define AI problems in a traditional manner. The second problem has not been tackled seriously, with the result that supposedly competing AI hypotheses are typically non-comparable. If we are to argue that our AI hypotheses do have merit, we must be shown to have tested them in a scientific manner. The problems, and why they are particularly difficult for AI are discussed. We provide a software engineering methodology called EDIT (Expériment Design Implement Test) which may help in solving both problems.

1   INTRODUCTION

One of the most serious problems in Artificial Intelligence (AI) research is describing and circumscribing AI problems. Unlike traditional computer science problems, problems in AI are very hard to specify, never mind solve (see [1], p. 53-87). Take, for example, the problem of natural language processing

---

[1]This research has been funded in part by U S *WEST* Advanced Technologies, Denver, Colorado, under their Sponsored Research Program.
[2]The authors' address is Department of Computer Science, University of Exeter, GB Exeter EX4 4PT, EC. E-mail: {pmc,derek}@cs.exeter.ac.uk

in AI. The problem of building a program, P, to understand any natural language utterance which is passed to P is very difficult. There will always be some word, or utterance structure, or relationship between utterances, which the program will not understand. In any event, it is hard to circumscribe the problem, as there are so many words and sentence structures, and it would take a long time to specify all of these. This point is elaborated with more detail by Mc Kevitt in [2]. Also, Wilks in [3] (p. 131) says, "However, if there is any validity at all in what I called the circumscription argument, then this is not so, for a natural language cannot be viewed usefully as a set of sentences in any sense of those words. The reason for this, stated briefly and without detailed treatment given in [4] and [5], is that for no sequence of words can we know that it cannot be included in the supposed set of meaningful sentences that make up a natural language."

However, it is possible to limit the problem of natural language processing by taking a specific domain and have a system which only accepts utterances within a prescribed domain. Take the problem of building a natural language interface which answers questions about the UNIX operating system. Such a program, called OSCON, is proposed by Mc Kevitt in [6]. Yet, even in this simplified case the difficulty of describing the problem, and testing solutions to it, remains. It is hard to define what sort of questions people will ask as there are so many of them. AI researchers tend to try and predict what sort of questions people "might" ask and develop a system to cater for these. The question we would like to ask here is, "Is there any better method of defining AI problems rather than just, a priori, deciding on a few of the specific components of the problem?"

A second problem arises out of the first, when we do not limit our goal to a few prespecified examples, and that is how to test a solution to an AI problem if such a potential solution is found. Today, the test of solutions to AI problems is the implementability test; i.e. if a program can be implemented and demonstrated to solve some selected examples of the problem then that program is considered a good one. This is, of course, no test at all. If AI is a science then there needs to be some method for testing or experimenting with programmed solutions rather than just implementing and demonstrating them. We take a quote from Partridge in [1] (Preface) which emphasizes the point, "AI is a behavioral science; it is based on the behaviour of programs. But it has not yet come to grips with the complexity of this medium in a way that can effectively support criticism, discussion, and rational argument - the requisites of scientific 'progress' are largely missing. Argument there is, to be sure, but it is all too often emotionally driven because rational bases are hard to find." This leads us to ask the question, "Is there any better method than just implementation and demonstration which tells us if an AI hypothesis or theory is correct?

We believe that the questions we ask here are central to the methodology and foundations of AI, and answers to the questions need to be found. Otherwise, there will be little disciplined progress in the development of AI systems and the science of AI.

2 DEVELOPING AI PROGRAMS

Computer science has already developed techniques for describing problems. Programs have been developed to aid programmers in developing and testing software for specific domains. Practical software engineering tools such as CASE (Computer Assisted Software Engineering) products are being used today (see [7]). CASE tools are very useful for building computer software for limited domains, yet not very useful for tackling problems outside their scope. For example, there are very few CASE tools which would be useful for tackling the problem of building a machine translation system for translating Swahili into Chinese.

Methodologies for the development of computer software have also been defined. Partridge in [1] (p. 91-96) argues for the a Run-Debug-Edit methodology which is later modified to RUDE (Run-Understand-Debug-Edit) in [8]. RUDE is a software development methodology for the design of, mainly,

AI programs. It is also pointed out by Partridge in [1] that the RUDE methodology may be useful for traditional computer science problems too. This methodology calls for a discipline of incremental program development where programs are run and, if they fail on input, are edited and rerun. Partridge and Wilks in [8] (p. 369) say, "Essentially what we shall propose is a disciplined development of the 'hacking' methodology of classical AI. We believe that the basic idea is correct but the paradigm is in need of substantial development before it will yield robust and reliable AI software." The problem with RUDE is that it is tedious, and takes a long time, as the programmer is just hacking piecemeal at solving the problem without really knowing what the problem is. There is no specific goal toward which the program or programmer converges.

Another methodology called SAV (Specify-And-Verify), coined by Partridge in [9], calls for formal specification of problems, and formal verification of the subsequent algorithm. The SAV approach is advocated by Dijkstra, Gries, and Hoare in [10], [11], and [12] respectively. The use of formal techniques in proving programs correct for real world complex problems in computer science has proven difficult. One of the problems with Artificial Intelligence (AI) programs is that, as we've said, they are very difficult to specify. The application of proof logics to the intricacies of complex programs is too tedious and too complex. The technique has only become useful for small, simple programs.

Both RUDE and SAV only ensure that a program is developed for a particular specification. There is no guarantee that the specification is correct or solves the real world problem at hand for which it is intended.

3   THE EDIT METHODOLOGY

EDIT (Experiment-Design-Implement-Test) is a software development methodology which attempts to integrate elements of the SAV and RUDE methodologies. EDIT incorporates experimentation as an integral component. This is particularly useful in the AI problem domain which incorporates the added difficulty of the researcher not knowing how to describe a problem while trying to solve it. Briefly, EDIT has the following stages:

1. **Experiment:** Experiment(s) (E) are conducted to collect empirical data on the problem. This data can be stored in log file(s) (L).

2. **Design:** A design[3] (D), or specification, can be developed from L together with relevant theories of the program domain.

3. **Implement:** The description, or specification, is implemented (I) as a computer program (P).

4. **Test:** P is sent around the cycle and tested by placing it through E again. However, this time E involves P whereas initially E did not involve a program. The cycle is iterated until a satisfactory P is found.

The system developer(s) initially use(s) E to help define the problem, and successively use(s) E to develop and test P. In the initial stage E does not involve a program. However, each subsequent E involves a partially implemented P until the final P is decided upon. EDIT will always terminate after E and before I in the cycle. The EDIT cycle is shown in Figure 1 below.

At the experiment stage (E) an experiment is conducted to gather data on the problem or the quality of the current potential solution, P_n. Say, for example, the problem is to develop a natural language program which answers questions about computer operating systems like UNIX. Then, valid experimentation software would be a program which enables a number of subjects (S) to ask questions about UNIX, and an expert to answer these questions. An example setup for this experiment would be the Wizard-of-Oz[4] paradigm. In fact, an augmented Wizard-of-Oz technique is used where

---

[3] By "design" we mean any reasonable description whether it be in English, Hindi, Gaelic, logic, algorithmic form, or assembly code.

[4] A Wizard-of-Oz experiment is one where subjects interact with a computer through typed dialogue at a monitor and are led to believe that they are conversing with the

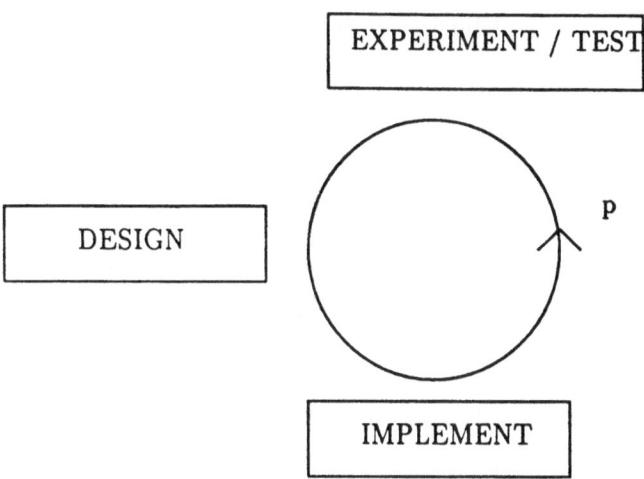

Figure 1: EDIT (Experiment Design Implement Test) Cycle.

a program (P) can be inserted and used in the interaction with the subject. A number of S and Wizard(s) (W) from varying backgrounds may be used in the experiment. Of course, the greater the number of S and W the more comprehensive the data collected will be. Also, there may be groups of S and W rather than just a single S and single W. Information on exchanges between S and W is logged in a log file (L) for later inspection. S and W operations are flagged in the file. Such an experiment is described with greater detail by Mc Kevitt and Ogden in [13] and the implications of that experiment are described by Mc Kevitt in [2].

At the design stage (D), L from E is analysed and inspected. In the initial stage the data here gives a snapshot description of the problem and how it is characterised. Further stages of the cycle will give snapshots of how well the problem is characterised in the current P. An analysis of L will give a pic-

---

computer. For example, in the case of a Wizard-of-Oz test for a natural language interface, a subject's utterances are sent to another monitor where a "Wizard", or expert, sends back a reply to the subject monitor.

ture of the information needed in various components of a program, such as knowledge representation, user modeling, and reasoning components. Many researchers and domain experts from various backgrounds may be called in to analyse L and determine what aspects of the software need to be developed. In fact, the type of researcher brought in will determine the type of program eventually developed and the best of all worlds would be to have a wide span of researchers/experts from different backgrounds. The job of the researchers is to develop algorithms with the help of the data and to specify these algorithms in some manner.

At the implementation (I) stage the algorithms or designs in D are implemented. These designs may be implemented in any programming language (P) that the implementers find most appropriate. Finally, the implemented program is sent back to E again and tested. Then, a new cycle begins.

During the initial state of the EDIT manifestation described here, S and W interact over the problem and the data is logged in L. Data may be collected for a specific task within a domain, or the whole domain itself. Each successive run of E involves the incorporation of P, which tries to answer questions first and if it fails W steps in while P restarts. The cycle may be operated in real time, or batch mode. In batch mode the experimentation component would involve a number of batched questions which are collected from S, and processed by P, with W interrupting where P fails.

The EDIT cycle continues until the program performs satisfactorily to the requirements of the designers. The designer(s) may wish P to perform satisfactorily only 50% of the time, or 80% of the time. The success or failure of P will be determined at design time, D, when the L is analysed. L will show where P has failed and where it has passed the test. W entries will show up why P did not work and will indicate what components of P need to be updated. In the case of natural language question answering W entries might show up the fact that certain types of question are not being answered very well, or at all. Therefore, W entries would indicate how P needs to be augmented in principle to solve a recurring pattern of failure. W entries could be analysed for such recurring patterns. In effect, what is happening here is that P is "learning" by being investigated and augmented in the same way as a mother might teach her child noticing the child's failure to complete certain tasks[5].

The success of P is measured by the number and type of answers P can give, and the number of answers P gets correct. The measure of capability and correctness is determined by inspection of L. During the development of P the initial coding may need to be recoded in some manner as data collected later may affect P's design.

There are many forms in which the EDIT cycle may be manifested. The experimentation stage may involve experiments other than the Wizard-of-Oz type. Another experiment might involve an observer sitting beside the subject during testing and helping the subject with the program as he/she uses it and also restarting the program itself.

The design stage and inspection of L may involve only one, or a number of designers. These designers may know much about the domain, and little about design, or vice versa. Experts and good designers may both be used at the design stage. Also, E could consist of a set of experts with different points of view and different backgrounds. E is in the spirit of Partridge and Wilks in [8] (p. 370), "Rather than the implementation of an abstract specification, we propose exploration of the problem space in a quest for an adequate approximation to the NLP problem." Hence, EDIT may consist of many manifestations of the methodology, yet, the basic methodology involves developing and testing through experimentation.

EDIT is a useful technique in that it allows the iterative development of systems and gives feedback on how to design an AI system as it is being developed. Sharkey and Brown in [14] (p. 282) point out that the belief that an AI system can be constructed first, and then tested later, as argued by Mc Kevitt in [2], is not the way to go. Sharkey and Brown show that (1) an

---

[5]This analogy was provided in personal communication by Brendan Nolan of University College Dublin (UCD).

AI system takes a long time to build, and it may be wrong at the beginning, and (2) an AI theory, and its implementation in the final state, may not be configured in a way that allows psychological testing.

It is important to point out here that the idea of using a Wizard for testing AI programs has a parallel in standard software engineering (see, for example, [15]). In the testing of standard software, top-down testing schemes use dummy modules or "program stubbs". The modules can be implemented with the following constraints: (1) Exit immediately if the function to be performed is not critical, (2) Provide a constant output, (3) Provide a random output, (4) Print a debugging message so that the programmer knows the module is entered, (5) Provide a primitive version of the final form of the module. Yourdon in [15] points out that in theory top-down testing could be done with only the main program and with all the lower-level modules implemented as stubs. However, he notes that in practice this would be clumsy. The same holds true for the Wizard-of-Oz experiment: an initial analysis of data might be one where most of the answers are given by the Wizard and only a few are handled by the system. This would be clumsy in practice as the Wizard would end up answering most of the time and the system would only respond to a few utterances here and there.

Pressman in [16] (p. 508) discusses the use of stubs and points out that subordinate stubs can be replaced one at a time with actual modules. Tests are conducted as each module is integrated. On the completion of each test another stub is replaced with a real module. Also regression testing may be conducted where all, or some, of the previous tests are rerun to ensure that new errors have not been introduced.

4   THE SCIENCE OF AI

Now that we have described our position on a methodology for developing good algorithms we shall move on to the problem of testing them. There needs to be some technique for testing if the algorithm works. Today, the test of AI theories is one where the programs, embodying those theories[6], are implemented, and demonstrated to work, over a few selected examples. This, however, is not a test at all, as any AI theory, or hypothesis, can be implemented. At most, researchers tackle the problem of testing AI systems in a weak sense by showing that they work for a few examples.

One of the problems with AI today is that it is not appreciated as a science and has no scientific test methodology. Narayanan in [19] (p. 46-47) points out, "It can be argued that the criterion of implementability is vacuous at the level of the Church-Turing thesis". The thesis basically says that any process which can be described by an algorithm can be implemented on a computer. Thus, any AI theory which can be described by an algorithm can be implemented on a computer, and hence all AI theories are valid no matter what they say. Sharkey and Brown in [14] (p. 278) also point out this problem: "To say that a theory is implementable is simply to say that it can be expressed in the form of a computer program which will run successfully", and suggest that a solution needs to be found (p. 280), "Another question we would like to raise here is this: At what point in implementation do we decide that there are too many patches to accept that the running program is actually a test of a theory." Sutcliffe in [20] argues for more empiricism and says, "I see the use of norming studies and other techniques from psychology as being relevant to AI." EDIT calls for not just implementability but also for the implementation to work on experimentation over real data. Also, EDIT moves forward on helping to solve the problem of how to check whether an AI theory is valid. Narayanan in [19] (p. 48) points out, "In any case, even if a criterion of complexity for AI programs (theories) can be found, there still remains the suspicion that no criterion exists for determining whether an AI theory is true or accurate." EDIT provides a criterion for the testing of theories embodied in programs by the inspection of log files.

There have been many arguments as to whether AI is, or is not, a science

---

[6]We do not make any strong claims here as to the relationship between programs and theories. However, this issue is discussed in [17], [18], and [3].

(see [21], [22]). Schank in [23] brings up the question as to whether AI is a technology of applications or a science. He points out that researchers have taken two directions, the scientists interested in working on problems like the brain or more neat logic problems, and the applications people working on building real practical systems.

Bundy in [24] calls AI an "engineering science". He says that it consists of the development of computational techniques and the discovery of their properties. He argues that AI is exploratory programming where one chooses a task that has not been modelled before and writes a program to implement it. On the other hand Dietrich in [25] (p. 224) argues that AI is a science and says, "Then, I will suggest a new theoretical foundation, and argue that adopting it would provide a clear, unequivocal role for programs: they would be controlled experiments, and AI would become a science." He points out that such experiments can be operated over natural systems such as ecosystems, and populations such as ant colonies. He says (p. 231), "In the science of intelligent systems, therefore, computer programs would have a definite role: they would allow scientists to experiment with hypotheses about the nature of intelligence".

Sparck Jones in [26] (p. 274) says that AI is engineering and points out that "... AI experiments are engineering experiments serving the designs of task systems, i.e. of artefacts." However, although we would agree with Sparck Jones in the sense that AI programs can be tested and redesigned by such experiments we would argue that AI hypotheses can also be tested with experiments such as those argued for in EDIT. Such experiments might be scientific ones, rather than engineering ones.

Let's assume that AI is a science. One of the problems is then to decide what the methodology of this science is. Narayanan in [21] (p. 164) brings up the point nicely, "The relationship between AI and cognitive psychology is strong. Does that mean that AI theories must conform to the same methodological rigour as psychological theories? If not, then a clear methodology must be provided for constructing and testing AI theories, otherwise AI might end up being a completely speculative subject, more akin to science fiction than science We must also ask the question of how to test hypotheses in AI if it is a science.

EDIT acts as methodology for testing hypotheses in AI where such hypotheses may be solutions to parts of problems. The advantage of the Wizard-of-Oz technique incorporating W, is that if P fails for reasons other than the hypothesis then the W can step in, keeping P alive, so to speak. Meanwhile, no data is lost in the current experiment dialogue. Of course, the log file marks where W interrupts. During testing as far as S is concerned the program has never failed as S does not necessarily know that W has intervened. The data from the testing phase can be logged in a file and system developers can then observe where the system failed, and where the W interfered. This information will be used in updating the system and any theory which the system represents.

EDIT addresses the problem brought out by Narayanan in [19] (p. 44) where he says, "The aim of this paper, apart from trying to steer well clear of terminological issues, such as the distinction between 'science' and 'study', is to demonstrate that unless AI is provided with a proper theoretical basis and an appropriate methodology, one can say just about anything one wants to about intelligence and not be contradicted; unless AI is provided with some reasonable goals and objectives little of current AI research can be said to be progressing." It is believed that EDIT might be the methodology that Narayanan asks for.

We would like to point out that EDIT is compatible with the methodologies of three philosophies of science: (1) the narrow inductivist, (2) the Hempel approach, and (3) the Popper approach. The 'narrow inductivist conception of scientific inquiry' (see [27]) is one which follows: (a) observation and reasoning of all the facts, (b) analysis and clarification of these facts, (c) inductive derivation of generalisations from them, and (d) further testing of the generalisations. This is in accord with EDIT when used in the sequence, Experiment-Design-Implement where log files are observed, and a program is developed from them.

However, Hempel in [27] argues that this type of scientific inquiry is not useful, and his approach is to develop hypotheses as tentative answers to a problem under study, and then subject them to empirical test. Hempel argues that the hypothesis must be testable empirically and that even if implications of the hypothesis are false under testing, the hypothesis can still be considered true. Hypotheses can be modified under experimentation until a limit is reached whereby the theory has become too complex and a simpler theory should be sought. Again, EDIT is compatible with Hempel's test ([27]) of AI systems as scientific hypotheses. EDIT can allow the AI scientist to have an hypothesis, a priori, (D) and place it into the cycle at I where it will be passed to E. If the program P fails at E then the log file must be analysed to see why it failed. If the hypothesis has failed then P can be modified and tested again.

Popper in [28] argues that scientific hypotheses must be developed and if they fail a scientific test then they must be thrown away and a new hypothesis developed. There is no room for hypothesis modification. Also, there must be a test which can show the hypothesis to be false. EDIT can allow scientific testing in the Popperian framework where again an hypothesis is formulated at D, is implemented at I, and is then tested at E. If the hypothesis fails then a new one must be developed and placed into the cycle at D again.

Marr in [29] describes two types of theory. The first type (type I) of theory is one where one uses some technique to describe the problem under analysis. Marr refers to Chomsky's notion of "competence" theory for English syntax as following this approach. The point is that one should describe a problem before devising algorithms to solve the problem. The second type (type II) of theory is one where a problem is described through the interaction of a large number of processes. He points out that the problem for AI is that it is hard to find a description in terms of a type I theory. Most AI programs have been type II theories. The EDIT technique enables the development of both types of theory in Marr's terms. A type I theory can be developed in terms of developing an initial complete description (starting at D) and then implementing it (at I) and testing it (at E). Also, a type II theory can be developed by starting at E stage and iteratively developing the description D of the complete complex process.

Marr in [30] defines a three-level framework within which any machine carrying out an information processing task is to be understood:

Computational theory (Level 1): The goal of the computation and the logic of how it can be carried out.

Representation and algorithm (Level 2): The implementation of the computational theory and the representation for the input and output. Level 2 also involves the algorithm for the transformation.

Hardware implementation (Level 3): The physical realisation of the representation and algorithm.

EDIT can be described in terms of Marr's framework where level 1 is the level at which D is completed, although D does not necessarily ask for a logic. Level 2 is also conducted at D. Level 3 is conducted at the I stage. Marr does not discuss experimentation in his three level framework.

EDIT is an attempt to address the problem brought forth by Narayanan in [21] (p. 179) where he says, "What we need here is a clear categorization of which edits lead to 'theory edits', as opposed to being program edits only. It is currently not clear in the AI literature, how such a categorization might be achieved. AI does not have the sort of complexity measure which would help identify when the theory, as opposed to the program, should be jettisoned in favour of another theory." Using EDIT an inspection of L should show up, in many cases, where a program has failed because of an hypothesis failure, or because of other reasons, and hence there will be distinct implications for the theory and the program. Also, Narayanan in [21] (p. 181) says, "But given the above comments, it appears that there can, currently at least be no scientific claims for claiming that one AI theory is better than another and that AI is making progress, simply because the conceptual tools for measuring one theory against another, and so for measuring the progress of AI are missing." We believe that EDIT may be a step along the road to such conceptual tools. It may be the case that EDIT has a lot to say in the development of foundations for AI as a science rather than a technology (see [21], [22]).

5 COMPARING EDIT TO RUDE

EDIT is not just a rearrangement and renaming of RUDE. The difference is that EDIT offers a means of convergence on a solution. EDIT is a significant refinement which we expect will be widely (although not universally) applicable in AI. The difference between EDIT and RUDE is that the algorithms are developed in conjunction with data describing the problem rather than from what the problem "might" be. Too often in the field of AI there are attempts at deciding, a priori, what a problem is without any attempt to analyse the problem in depth. As was pointed out earlier one of the problems with developing AI programs is that it is very difficult to specify the problem. One solution to that might be to collect data on the problem, rather than algorithms being concocted from hopes, wishes and intuition. The second major difference is that experimentation involves testing software over real data in the domain. Also, by using the Wizard-of-Oz technique the testing phase breaks down less as the wizard keeps the system going. We argue that this is important because if a test fails then data can be lost due to temporal continuity effects. Failure happens a lot while testing AI programs. For example, if one is testing a natural language interface, with an hypothesis for solving reference in natural language dialogue, then if the test fails the continuation of that dialogue may never happen, and data will be lost.

The problem with RUDE is that it does not include any goal as part of the process of development; only the update of a program. We argue here that E must be included to produce log files which measure how close P is to the goal that needs to be achieved. EDIT can be considered a more "tied down" version of RUDE where it is clearer what the problem is, and how well P is solving the problem. In fact Partridge and Wilks in [8] (p. 370) say, "What is needed are proper foundations for RUDE, and not a drift towards a neighbouring paradigm." Also, Partridge and Wilks in [8] (p. 370) point out a recognition of the need for convergence, "The key developments that are needed are methodological constituents that can guide the exploration — since a random search is unlikely to succeed."

The EDIT cycle is conducted until the implementation performs satisfactorily over a number of tests. The EDIT cycle enables the iterative development of a system through using the problem description itself as part of the solution process. EDIT is not just an hypothesis test method, but is also a method by which the *reason* for failure of software is logged and a method where that reason does not cause data loss. EDIT is useful for the development of software in an evolutionary way and is similar to those techniques described in [31]. Again, 100% reliability is very difficult to guarantee but we believe that problem description and implementation through experimentation will lead to better implementations than either RUDE or SPIV on their own.

EDIT is like the general methodology schemes proposed by researchers who are developing expert systems. The stages for the proper evolution of an expert system are described by Hayes-Roth et al. in [32]:

- IDENTIFICATION: determining problem characteristics

- CONCEPTUALIZATION: finding concepts to represent knowledge

- FORMALIZATION: designing structures to organize knowledge

- IMPLEMENTATION: formulating rules that embody knowledge

- TESTING: validating rules that embody knowledge

This is in the spirit of EDIT where, of course, identification is similar to E and conceptualization and formalization to D, and implementation to I. However, with EDIT, E is involved in both identification and testing and we argue that this is the way to go about testing if P is to meet the problem head on.

EDIT is currently being used in the development of AI software which answers natural language questions about computer operating systems. An initial computer program was developed called OSCON (see [33], [6], [34], and [35]) which answers simple English questions about computer operating systems. To enhance this research it was decided that an experiment should be conducted to discover the types of queries that people actually ask. An experiment has been conducted to acquire data on the problem. More details on the experiment and its implications are given in [2].

There are probably AI domains where EDIT will fit nicely and other domains which will not — i.e. are not open to simple data collection. For example, theories of knowledge representation could only be developed with rather indirect inferences from data collection. We do not wish to stress that EDIT will be used for all AI domains but that it may be useful in some.

6   CONCLUSION

It is pointed out here that the EDI methodology can provide a solution to the development and testing of programs in Artificial Intelligence (AI), a field where there are no sound foundations yet for either development, or testing. EDIT is compatible with scientific test philosophies at each end of a scale, and if AI is to be a science, then a technique like EDIT needs to be used to test scientific hypotheses. A sound methodology will reduce problems of how to compare results in the field.

EDIT may help in the endeavour of transforming AI from an ad-hoc endeavour to a more well-formed science. EDIT provides a methodology whereby AI can be used to develop programs in different domains and experts from those domains can be incorporated within the design and testing of such programs.

We leave you with a quote from Partridge and Wilks in [8] (p. 370), "A RUDE-based methodology that also yields programs with the desiderata of practical software — reliability, robustness, comprehensibility, and hence maintainability — is not close at hand. But, if the alternative to developing such a methodology is the nonexistence of AI software then the search is well motivated." EDIT is part of such a search.

7 ACKNOWLEDGEMENTS

We would like to thank Simon Morgan, Richard Sutcliffe and Ajit Narayanan of the Computer Science Department at the University of Exeter and Brendan Nolan from University College Dublin for providing comments on this work.

8 REFERENCES

1. Partridge, Derek. *Artificial Intelligence: applications in the future of software engineering.* Halsted Press, Chichester: Ellis Horwood Limited, 1986.

2. Mc Kevitt, Paul. *Data acquisition for natural language interfaces.* Memoranda in Computer and Cognitive Science, MCCS-90-178, Computing Research Laboratory, Dept. 3CRL, Box 30001, New Mexico State University, Las Cruces, NM 88003-0001, 1990b.

3. Wilks, Yorick. *One small head: models and theories.* In "The foundations of Artificial Intelligence: a sourcebook", Partridge, Derek and Yorick Wilks (Eds.), pp. 121-134. Cambridge, United Kingdom: Cambridge University Press, 1990.

4. Wilks, Yorick. *Decidability and Natural Language.* Mind, N.S. 80, 497-516, 1971.

5. Wilks, Yorick. *Grammar, meaning and the machine analysis of language.*

London: Routledge and Keegan Paul, 1972a.

6. Mc Kevitt, Paul. *The OSCON operating system consultant.* In "Intelligent Help Systems for UNIX – Case Studies in Artificial Intelligence", Springer-Verlag Symbolic Computation Series, Peter Norvig, Wolfgang Wahlster and Robert Wilensky (Eds.), Berlin, Heidelberg: Springer-Verlag, 1990a. (Forthcoming)

7. The Byte Staff. *Product Focus: Making a case for CASE.* In Byte, December 1989, Vol. 14, No. 13, 154-179, 1989.

8. Partridge, Derek and Yorick Wilks. *Does AI have a methodology different from software engineering?.* In "The foundations of Artificial Intelligence: a sourcebook", Partridge, Derek and Yorick Wilks (Eds.), pp. 363-372. Cambridge, United Kingdom: Cambridge University Press. Also as, *Does AI have a methodology which is different from software engineering?* in Artificial Intelligence Review, 1, 111-120, 1990b.

9. Partridge, Derek. *What the computer scientist should know about AI — and vice versa.* In "Artificial Intelligence and Cognitive Science '90", (this volume) Springer-Verlag British Computer Society Workshop, Mc Tear, Michael and Creaney, Norman (Eds.), Berlin, Heidelberg: Springer-Verlag, 1990.

10. Dijkstra, E.W.. *The humble programmer.* Communications of the ACM, 15, 10, 859-866, 1972.

11. Gries, D.. *The science of programming.* Springer-Verlag, NY, 1981.

12. Hoare, C.A.R.. *The emperor's old clothes.* Communications of the ACM, 24, 2, 75-83, 1981.

13. Mc Kevitt, Paul and William C. Ogden. *Wizard-of-Oz dialogues in the computer operating systems domain.* Memoranda in Computer and Cognitive Science, MCCS-89-167, Computing Research Laboratory, Dept. 3CRL, Box 30001, New Mexico State University, Las Cruces, NM 88003-0001, 1989.

14. Sharkey, Noel E. and G.D.A. Brown. *Why AI needs an empirical foundation.* In "AI: Principles and applications", M. Yazdani (Ed.), 267-293. London, UK: Chapman-Hall, 1986.

15. Yourdon, Edward. *Techniques of program structure and design.* Engelwood Cliffs, New Jersey: Prentice-Hall, Inc., 1975.

16. Pressman, Roger S. *Software engineering: a practitioner's approach.* McGraw-Hill: New York (Second Edition), 1987.

17. Bundy, Alan and Stellan Ohlsson. *The nature of AI principles: a debate in the AISB Quarterly.* In "The foundations of Artificial Intelligence: a sourcebook", Partridge, Derek and Yorick Wilks (Eds.), pp. 135-154. Cambridge, United Kingdom: Cambridge University Press, 1990.

18. Simon, Thomas W. *Artificial methodology meets philosophy.* In "The foundations of Artificial Intelligence: a sourcebook", Partridge, Derek and Yorick Wilks (Eds.), pp. 155-164. Cambridge, United Kingdom: Cambridge University Press, 1990.

19. Narayanan, Ajit. *Why AI cannot be wrong.* In Artificial Intelligence for Society, 43-53, K.S. Gill (Ed.). Chichester, UK: John Wiley and Sons, 1986.

20. Sutcliffe, Richard. *Representing meaning using microfeatures.* In "Connectionist approaches to natural language processing", R. Reilly and N.E. Sharkey (Eds.). Hillsdale, NJ: Earlbaum, 1990.

21. Narayanan, Ajit. *On being a machine.* Volume 2, Philosophy of Artificial Intelligence. Ellis Horwood Series in Artificial Intelligence Foundations and Concepts. Sussex, England: Ellis Horwood Limited, 1990.

22. Partridge, Derek and Yorick Wilks. *The foundations of Artificial Intelligence: a sourcebook.* Cambridge, United Kingdom: Cambridge University Press, 1990a.

23. Schank, Roger. *What is AI anyway?.* In "The foundations of Artificial Intelligence: a sourcebook", Partridge, Derek and Yorick Wilks (Eds.), pp. 1-13. Cambridge, United Kingdom: Cambridge University Press, 1990.

24. Bundy, Alan. *What kind of field is AI?.* In "The foundations of Artificial Intelligence: a sourcebook", Derek Partridge and Yorick Wilks (Eds.), p. 215-222. Cambridge, United Kingdom: Cambridge University Press, 1990.

25. Dietrich, E. *Programs in the search for intelligent machines: the mistaken foundations of AI.* In "The foundations of Artificial Intelligence: a

sourcebook", Derek Partridge and Yorick Wilks (Eds.), 223-233. Cambridge, United Kingdom: Cambridge University Press, 1990.

26. Sparck Jones, Karen. *What sort of thing is an AI experiment.* In "The foundations of Artificial Intelligence: a sourcebook", Partridge, Derek and Yorick Wilks (Eds.), pp. 274-285. Cambridge, United Kingdom: Cambridge University Press, 1990.

27. Hempel, C.. *Philosophy of natural science.* Prentice Hall, 1966.

28. Popper, K. R.. *Objective knowledge.* Claredon Press, 1972.

29. Marr, David. *AI: a personal view.* In "The foundations of Artificial Intelligence: a sourcebook", Derek Partridge and Yorick Wilks (Eds.), 99-107. Cambridge, United Kingdom: Cambridge University Press, 1990.

30. Marr, David. *Vision.* Freeman, 1982.

31. Connell, John L. and Linda Brice Shaffer *Structured rapid prototyping: an evolutionary approach to software development.* Engelwood Cliffs, New Jersey:Yourdon-Press Computing Series, 1989.

32. Hayes-Roth, F., D.A. Waterman and D.B. Lenat *Building expert systems.* Reading, MA: Addison-Wesley. 1983.

33. Mc Kevitt, Paul. *Formalization in an English interface to a UNIX database.* Memoranda in Computer and Cognitive Science, MCCS-86-73, Computing Research Laboratory, Dept. 3CRL, Box 30001, New Mexico State University, Las Cruces, NM 88003-0001, 1986.

34. Mc Kevitt, Paul and Yorick Wilks. *Transfer Semantics in an Operating System Consultant: the formalization of actions involving object transfer.* In Proceedings of the Tenth International Joint Conference on Artificial Intelligence (IJCAI-87), Vol. 1, 569-575, Milan, Italy, August, 1987.

35. Mc Kevitt, Paul and Zhaoxin Pan. *A general effect representation for Operating System Commands.* In Proceedings of the Second Irish National Conference on Artificial Intelligence and Cognitive Science (AI/CS-89), School of Computer Applications, Dublin City University (DCU), Dublin, Ireland, European Community (EC), September. Also, in "Artificial Intelligence and Cognitive Science '89", Springer-Verlag British Computer Society Workshop

Series, Smeaton, Alan and Gabriel McDermott (Eds.), 68-85, Berlin, Heidel-berg: Springer-Verlag, 1989. .

Section 2:

Learning

Machine Learning in Subject Classification

Catriona Ward Ita Kavanagh John Dunnion

ABSTRACT

This paper describes the application of Machine Learning (ML) techniques to the problem of Information Retrieval. Specifically, it presents a system which incorporates machine learning techniques in determining the subject(s) of a piece of text. This system is part of a much larger information management system which provides software support for the creation, management and querying of very large information bases. The information stored in these information bases is typically technical manuals, technical reports or other *full-text* documents. This paper gives a brief description of the overall system, followed by an overview of Machine Learning and a summary of a number of ML systems. We then describe the classification algorithm used in the system. Finally, the learning module, which will be incorporated into the classification algorithm, is described.

INTRODUCTION

SIMPR (Structured Information Management: Processing and Retrieval) is a project in the CEC's ESPRIT II programme. The SIMPR system provides software support for the creation, management and querying of very large information bases on CD-ROM. The information stored will typically be technical manuals, libraries of technical reports or other *full-text* documents. A full-text document is one with no pre-requisites on its content or format. Each of these documents is composed of a number of *texts*. Each text is processed in two stages. It is first indexed to extract words and phrases with a high meaning content. Secondly, the subject(s) of the text are identified and appropriate *classificators* are attributed to the text.

The classification of texts is (initially) performed manually by a human editor. This classification work is supported by the Subject Classification Management System which creates and maintains the classification scheme. Our contribution to the project, and the work described in this paper, is the design of an intelligent agent to help classify texts. It is expected that as texts are processed, the SIMPR system (or more specifically, the Subject classification Intelligent System (SCIS)) will suggest suitable classificators, and as the number of processed texts increases, the suggested classificators will become more accurate.

This paper describes UCD's work on the role of learning in subject classification. The next section describes Project SIMPR in more detail. The third section introduces the topic of Machine Learning. Section 4 outlines a number of ML systems, comparing them and discussing their applicability to the subject classification of texts in SIMPR. Section 5 presents the classification algorithm and describes the learning module which will be incorporated into this algorithm. The paper concludes by suggesting directions in which the research might usefully develop.

PROJECT SIMPR

The aim of SIMPR is to develop techniques for the management of large information bases. The documents which constitute the information base are first morphologically and syntactically analysed. They are then indexed by the indexing module of SIMPR, which extracts *analytics* from texts. An analytic is a word or sequence of words that accurately represents the information content of a text [1].

The SIMPR system is intended to help in the management of very large documents, prepared by teams of authors working under the supervision of editors. It will help the editors to use, index, validate and store information so that a reader can search the resultant information base to find the answer to a query. The SIMPR information base is built by adding texts to it until all the texts have been processed. A text in this case is defined to be the information between two headings, no matter at what level the headings occur [2].

UCD's role in the project is to take the result of the indexing, i.e. the analytics, observe a human editor assign classificators to the text based on the analytics extracted from the text, and try to learn how to do this using machine learning techniques from artificial intelligence. These classificators form the basis of the retrieval mechanism of the SIMPR project. Unlike traditional Information Systems, the retrieval mechanism does not depend on pattern matching; rather retrieval is based on matching the user's query, which is also processed and indexed, against classificators previously assigned to the document. Our area of interest in the project is the application of Machine Learning techniques to subject classification.

In SIMPR, texts are classified using a **faceted classification scheme**. A faceted classification scheme is a type of systematic classification using the analytico-synthetic principles introduced by S.R. Ranganathan in 1933 [3]. All knowledge is divided into classes, which are then divided into sub-classes. Ranganathan imposes for most main classes the citation order PMEST:

- Personality: key facet depending on the subject
- Matter: relates to materials
- Energy: relates to processes, activities or problems
- Space: represents geographical areas
- Time: represents time periods [4]

These terms can appear several times in a subject classification if necessary, e.g. if two materials are used both may be represented.

In **subject classification** subjects are analyzed into their component concepts. Compound subjects may be built up synthetically from elementary concepts. The concepts are arranged in groups called facets. Each concept is allocated a notational symbol derived from the facets which make up the concept. These concepts can then be combined or synthesised, using their notations, to form more complex subjects [5].

In order to subject-classify a document, an indexer examines a document, selects appropriate index terms and decides how they are related [3]. These relationships can be represented using the notation mentioned above. For example, if the indexer selected the terms *music* and *instrument* as the index terms for a text, and music is represented by $[P_1]$ and instrument by $[M_1]$, then the text would be represented by the term $[P_1 M_1]$.

In SIMPR the number and type of basic facet groups will depend on the domain in question. For example, in experimentation with texts from car manuals, index terms were selected from six facet groups, car-type, model, date, component, operation and fault. Other domains will yield different facet groups.

The next section introduces the topic of Machine Learning, describing and comparing different Machine Learning techniques.

MACHINE LEARNING

Learning is an inherent part of human intelligence. The manner in which a person acquires new knowledge, learns new skills and, in particular, his ability to improve with practice is all part of the human learning process. Therefore, the ability to learn must be part of any system that would exhibit general intelligence. Various researchers in artificial intelligence (AI) argue that one of the main roles of current research in AI is understanding the very nature of learning and implementing learning capabilities in machines [6], [7]. In recent years *Machine Learning* (ML) has developed into a separate subfield within AI.

Different Types of Learning

Originally, learning systems were classified depending on the kind of knowledge represented and manipulated by the system. Three distinct criteria developed:

- neural modelling and decision theoretic techniques
- Symbolic Concept Acquisition (SCA)
- Knowledge intensive, Domain specific Learning (KDL)

The different strategies differ in the amount of *a priori* knowledge initially built into the system, how this knowledge is represented and the way this knowledge is modified [8]. Learning in neural networks tends to be numerical in nature, in contrast with the more conceptual nature of SCA and KDL. For a more detailed discussion see [9].

In the early eighties two distinct approaches to ML developed: Similarity Based Learning and Explanation Based Learning. More recently hybrid systems have emerged which combine different aspects of each.

One of the first areas explored in Machine Learning was that of learning from examples, i.e. learning by discovery or Similarity Based Learning (SBL). Given an adequate description language, a training set of examples, generalisation and specialisation rules and other background knowledge, an SBL algorithm induces a list of concept descriptions or class membership predicates which cover all the examples in the given set of examples.

An alternative approach to SBL is Explanation Based Learning (EBL) [10]. Unlike SBL, the system can learn new concept descriptions from a single training example. Given an adequate description language, a single positive example of a concept, generalisation and specialisation rules and an extensive domain theory, EBL adopts a two step approach. An explanation is first derived for the given training example and this "explanation" is then generalised to obtain a more general and more useful description of the object or set of objects.

Comparison of SBL and EBL Methodologies

Each technique has inherent advantages and disadvantages over the other [11]. The major differences between the two techniques are as follows:

1. Size of the training set. Use of SBL techniques necessitates a large training set, preferably containing a number of "near misses". The SBL approach depends largely on similarities and differences between a large number of examples to determine a concept description. In contrast, the EBL approach requires only a single training example to discover a concept description.

2. Amount of background knowledge required by each approach. SBL uses background knowledge to constrain the infinite number of possible concept descriptions which can be deduced from the initial input set. EBL techniques depend heavily on an extensive background knowledge, in particular to build the explanation to explain what is an example.

3. The SBL approach usually considers both negative and positive examples of the concept to be learned. In contrast an EBL system usually only considers a single positive example.

4. The main theoretical difference between the two approaches is that EBL is said to be *truth-preserving* while SBL is only *falsity-preserving*. This means applying EBL techniques ensures that if a property is true in an input example, then this property will be true in the output concept description. Applying a SBL approach to the same problem does not ensure that the property will still hold. All negative assertions in the input example are guaranteed to remain so in the output example.

5. The two approaches are complementary, SBL techniques exploit the inter-example relationship, while EBL techniques exploit the intra-example relationships [9].

6. In general the descriptions inferred using induction techniques are more comprehensible. This is important especially in cases where human experts may be asked to validate the induced inferences. In general, if the user is not asked to validate the inferred assertions, the system does so using deductive techniques. This is an example of the interplay between the two techniques.

7. Generalisation in EBL is somewhat different from generalisation in SBL. In SBL the process of generalisation does not guarantee that the result will be useful or semantically well-formed when interpreted or applied. This is entirely due to its dependence on the "aptness" of the training set of examples (for determining feature significance). EBL, on the other hand, guarantees that the generalisation specifies a meaningful concept as its construction is governed by the system's world knowledge.

Hybrid Systems

Many ML researchers are currently investigating methods of combining the analytical techniques of EBL with the more empirical techniques of SBL. Hybrid systems are usually categorised as one of the following:

- Systems which use explanations to process the results of empirical learning. In such systems, SBL methods are first applied to the training set to derive concept descriptions, which are then refined or discarded by the EBL process.

- Systems which use empirical methods to process the results of an explanation phase. Here the results of the EBL process are subjected to SBL to derive a general concept definition.

- Systems which integrate explanation-based methods with empirical methods. An example of this method is where SBL and domain-independent heuristics are used to extend the domain theory whenever EBL methods fail.

All learning systems must address problems of *clustering, i.e.* given a set of objects, how does one identify attributes of the objects in such a manner that the set can be divided into subgroups or clusters, such that items in the same cluster are similar to each other. Previous research has concentrated on numerical methods to establish a measure of similarity between objects and concepts. Recently this emphasis has changed, and the emphasis is now on conceptual clustering with particular importance on the development of goal-oriented classifications. The main advocates of this approach are Stepp and Michalski [12].

Classification is an inherent part of any learning system, in that any form of learning involves the invention of a meaningful classification of given objects or events. It is clear that developments in Machine Learning will have a knock-on effect on automatic classification research, and vice versa.

SURVEY OF SELECTED ML SYSTEMS

As part of our background work to SIMPR we examined other machine learning systems to investigate their techniques, advantages and disadvantages and decide their applicability to the SIMPR domain. In this section we discuss three of these systems, ID3, STAR and SOAR. ID3 and STAR are inductive methods and SOAR is deductive.

ID3
The ID3 algorithm [13] [14] is used to form efficient classification procedures from large numbers of examples. It iteratively forms a succession of decision trees of increasing accuracy.

Method
The method used is to select at random a subset (*window*) of instances from the large number of examples and to perform the following loop:
DO
    form a rule to explain the current window
    find the exceptions to this rule in the remaining instances
    form a new window from the current window and the exceptions to the rule generated from it
UNTIL
    there are no exceptions to the rule.

There are two ways to enlarge the subset, either add to the current subset or identify 'key' objects and replace the rest with exceptions, thus keeping the window size constant. The algorithm works on a large mass of low-grade data and uses a process of inductive inference. The objects are described in terms of a fixed collection of properties and the system produces a decision tree which correctly classifies all the objects.

The computation time increases linearly with:

- no. of objects
- no. of attributes
- complexity of the concept

Almost all of the effort must be devoted to finding attributes that are adequate for the classification problem being tackled.

The STAR Methodology

The main concept representation employed in Michalski's INDUCE system is that of a *star* [15]. Michalski defines the concept of a star as follows: a star of an event E under constraints F, is the set of all possible alternative non-redundant descriptions of event E that do not violate constraints F. This concept is useful as it reduces the problem of finding a complete description of a concept, to sub-problems of finding consistent descriptions of single examples of the concept.

SOAR

SOAR was developed as a system which combines learning and problem solving [16]. SOAR is a learning system based on chunking [17]. When a subgoal is processed a chunk is created which can be substituted for the processing of the same subgoal the next time the subgoal is generated.

Of the three systems mentioned above both ID3 and STAR are inductive methods and have the same disadvantage for the SIMPR domain, namely that they require a list of attributes associated with each text, and the possible values for each attribute. Because of the Natural Language domain of SIMPR the range of values for the possible attributes is too large.

Currently we do not see ID3 or STAR as being suitable to the needs of SIMPR. Because ID3 builds up a classification tree from its training set, and STAR a star, and then uses this tree or star to classify new data, it can not be used by the SIMPR learning module which has to learn on an on-going basis.

However, it may be possible to apply the chunking mechanism of SOAR to our problem domain. Thus if a sequence of rules used to suggest a particular type of classificator is detected this sequence of rules may be "chunked" to form a single new rule.

THE SUBJECT CLASSIFICATION INTELLIGENT SYSTEM (SCIS)

The SIMPR Subject Classification System comprises two parts, the Subject Classification Maintenance System (SCMS) and the Subject Classification Intelligent System (SCIS). The SCMS creates and maintains the subject classification system. Among its tasks are the maintenance of the facet scheme and the *heading hierarchy*, described below. The SCIS consists of a classification algorithm incorporating a learning module. In this section we describe the inputs, outputs and processing involved in the classification algorithm and the learning module.

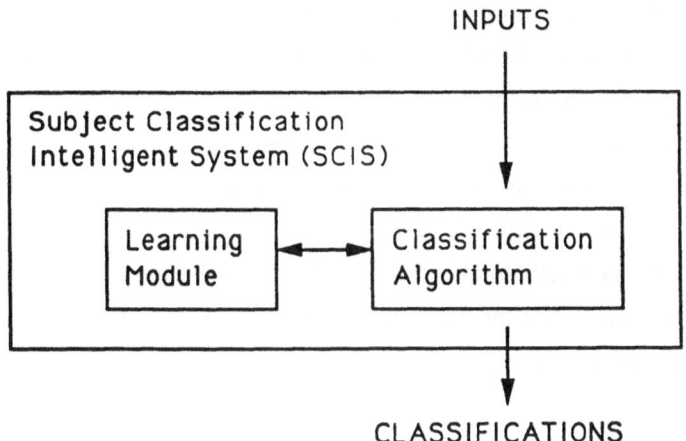

Fig 1 Overview of SCIS

INPUTS

There are five major inputs to the learning system: the processed text, the facets, background knowledge, suggested classificators and the original text. Listed in decreasing order of their contribution to the learning module these are:

Processed text: This is generated by the morphological and syntactical analysis of the original document followed by its indexing. The parts of the processed text of most use to the SCIS are the generated analytics and various statistics on the document, including word frequency and co-occurrence analyses.

Facets: The facets used are those applicable in a particular domain. These are generated by a domain expert. Each facet is represented by a frame and connections can exist between facets via PARENT and IS-ALSO-A relationships. One slot of the frame which is very important to the learning system is the EVIDENCE slot. This contains information about when that facet should be attributed to a text. This slot is not necessarily filled at the inception of the learning process. As the system learns it may place information in this slot which can then be used at a later learning stage. Hierarchical information about the organisation of the facets is also stored, e.g. details about how different facets are related to each other.

Background Knowledge: The background knowledge for the learning system is extensive and consists of information either about the document or about classification. The system also uses statistical information gleaned from previously processed texts and can generate such information about the current text. The analytics derived for the current text are also available. Here, the background knowledge will also contain information about how the analytics were generated, i.e. information about them and how they were selected as analytics.

In order to discern patterns of facet attribution we keep a record of facet attribution in the **Facet Attribution File (FAF)**. We define the Facet Attribution File to be a file which stores the list of texts processed so far and the facets which have been attributed to each text e.g.

    text1: faceta, facetf, facetr

    text2: facete, (faceti, facetj), facetq   etc.

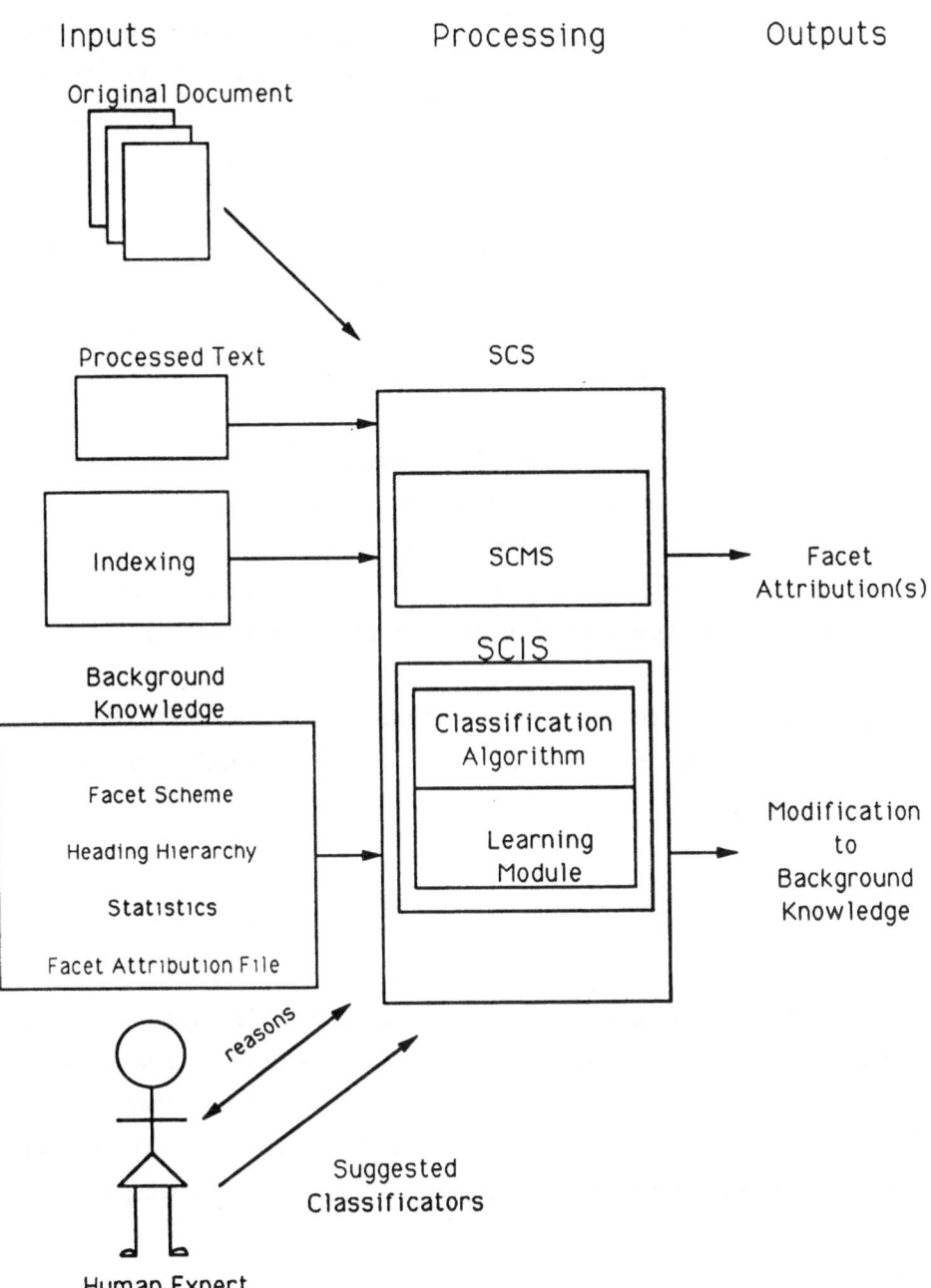

Fig 2 Overview of SIMPR System

A library of operations on the FAF is available, e.g. matching operations on the FAF, comparing the sets of facets attributed to related texts, etc. As texts are entered into the system they become part of what is known as the **Heading Hierarchy**. This is similar to a table of contents but it is dynamically constructed by the editor who decides the relation of the current text to those previously entered. The heading hierarchy also forms part of the background knowledge in that the human editor states the relation of the current text to previously entered texts. The background knowledge is a very important input to the system and the information in it is constantly updated as the system develops.

List of suggested classificators: As well as the classification algorithm suggesting classificators, which the Human Editor accepts, rejects or modifies, the Human Editor is able to add any classificators s/he wishes. However as the system progresses and the overall efficiency and effectiveness of the SCIS (in particular the Learning Module) increase the role of the Human Editor will decrease.

Original Text: The original text is always available for reference or for any other reason (as are any of the intervening files created by any module of the SIMPR system).

OUTPUTS

The outputs from the SCIS may be divided into two, the suggested classificators for the text and any modifications to either the background knowledge or to the evidence slot. The evidence slot may be filled with new rules or guidelines as to when a facet should or should not be attributed to a text. If the learning module suggests a classificator but the human editor rejects it, the human editor will be presented with a list of reasons why a suggested classificator is not acceptable and if any of these reasons is selected they can be used to modify the evidence slot of the frame representing the facet.

The background knowledge of the learning module may be modified or added to in many ways, e.g. relations between word frequencies, analytics and facets may be discerned or changed if they already exist; the importance of words which occur in the text may be noted relative to the length of the text. If the word occurs often in a paragraph, and nowhere else in the text, the system will try to decide if this signifies its importance in this paragraph.

PROCESSING

The aim of our research is to incorporate learning techniques into an algorithm for subject classification. However, before we can incorporate learning into any classification process we must first have a classification algorithm. At present, we are working on an algorithm for attributing facets to a text. This algorithm will be refined to incorporate a more detailed learning strategy.

The classification algorithm we are developing involves a number of strategies. These are:

Strategy One
For a set of texts derive the Normalised Evidence Word Tokens (NEWTs) [1] and discern if there is any connection between these and the normalised analytics produced by the indexing system for the same texts. NEWTs are the result of a simple word-count technique. To generate the NEWTs for a text, the original text is first normalised, the frequency of occurrence of the normalised words is calculated and words with a frequency value above a certain threshold value form the NEWTs.

Strategy Two
Examination of the FAF entries of texts which are related in the Heading Hierarchy to identify discriminatory (and possibly common) facets between the related texts. In particular we are investigating facet attributions for:

- parent and child texts;
- sibling texts.

Strategy Three
Examination of words which occur in the heading of a text and other words in the text. From investigations [18] with sample texts we have discovered that words

which occur in the heading of a text and which also occur in the text itself usually provide good guidelines when determining possible subject classificators for a text.

Strategy Four

Examination of the FAF to identify if a pattern of facet attribution exists. These patterns may exist in texts which are unrelated in the Heading Hierarchy.

Strategy Five

The *type* of a text, for example diagnostic, explanatory, etc. can sometimes be discerned from the structure of the sentences. If a lot of sentences in the text begin with the word "If" then the text is probably diagnostic. If many sentences start with a verb then the text is probably instructive. The type of the text may also be discerned from the presence of key phrases such as "This is how to ..." or "We now describe ...". This strategy will be concerned with exploiting information about the type of a text in suggesting classificators for the text.

Strategy Six

Examination of the occurrence of synonyms of words in the text, in particular synonyms of words which occur in the heading of the text.

Strategy Seven

From previous research [18] we have discerned that emphasised text, i.e. bolded, underlined and/or italicised, is important in deciding the subject matter of a text.

THE CLASSIFICATION ALGORITHM

At present the classification algorithm consists of four steps. These are:
1. generate for a text
   - statistics and NEWTs
     Note: the statistics here include frequency
        and co-occurrence statistics along with any
        other statistics needed.
     We count
        - the number of different occurrences of

each "type" of facet in a text. For
example, the number of components mentioned in
a text, the number of operations, etc.
(There will probably be a maximum of 20
"types" of facet groups).
- the number of imperatives in the text. This
information may indicate the "type" of text.

- analytics

Note the position of that text in the Heading Hierarchy (HH).

2. Determine if there exists a connection between the NEWTs and the analytics, i.e. if there are words which are ranked high in both the list of analytics and the list of NEWTs. Check if these "important words" part of the classification scheme. If so, using the analytics and the predefined classification ordering as a guide, combine these "important words" to form a list of possible classificators.

3. Retrieve facets attributed to associated texts (parent and/or sibling texts) from the FAF using the Heading Hierarchy as a guide.
Compare/Match/Expand the list of possible classificators and classificators previously attributed to associated texts. Modify the list of possible classificators accordingly.

4. Use of Evidence Slot. The classification scheme should contain entries for "other components" such as plastigage, hammer etc. (Perhaps these will be one of the "types of facet group" in the domain, e.g. in a domain with a "tool" facet group). The evidence slot of frames representing such facets should contain rules suggesting highly relevant classificators. For example, the evidence slot of a frame representing *Plastigage* would contain a rule suggesting *"Adjusting the Big-End"* as a likely classificator.[1]

---

[1]This example is taken from a Car domain.

The evidence slot of all facets mentioned in a text are examined and if new classificators or modifications to existing classificators are suggested then the text is re-examined to confirm such classificator attribution.

LEARNING AND THE CLASSIFICATION ALGORITHM

"Learning denotes changes in the system that are adaptive in the sense that they enable the system to do the same task or tasks drawn from the same population more efficiently and more effectively the next time" [19].

Incorporating learning into our classification algorithm will result in modifications to the background knowledge. These modifications will improve future facet attribution, either by improving the efficiency for the same facet attribution or suggesting improved "alternative" classificators for the same text. By "alternative" classificators we mean classificators which would not have been suggested by the system unless learning had occurred.

The learning algorithm will consist of a number of strategies which we call "learns". At present we have identified a number of possible learns. These are:

Learn One
If a text contains a number of different entries from the same facet group, e.g. ten different car components are cited in the same text, the facet group is examined in the classification scheme to see if generalisation is possible.

Learn Two
The analytics suggested by the indexing system but rejected by the Human Editor (HE) may be a used as negative input into the Learning algorithm.

Learn Three
The HE, if s/he chooses to reject a particular classificator is asked to "explain" any reasons for doing so. This is done by presenting him/her with a set of reasons. The system will then take this into account, most likely by altering the background knowledge accordingly.

Learn Four

We expect to be able to apply rules of generalisation to the rules contained in the evidence slot representing each facet, or any other rules comprising the background knowledge.

A number of standard generalisation rules ([20], [15], [21]) exist which we may incorporate into our learning algorithm. These are:

1. Dropping condition rule
2. Adding disjuncts rule
3. Turning constants into variables rule
4. Adding names of variables
5. Extending the domain rule
6. Climbing generalisation tree (structured domain)
7. Suppressing the antecedent of implications
8. Turning conjunction into disjunction rule
9. Extending the quantification domain rule
10. The inductive resolution rule
11. "Extension against" rule
12. "Counting arguments" rule
13. Chunking-generation chain property rule
14. Detecting descriptor interdependence.

CONCLUSIONS AND FUTURE WORK

In this paper we have described the SIMPR information management and retrieval system, the subject classification component of this system and the learning module of this classification component, in which we are applying machine learning techniques.

Our Learning Module contains ML techniques which we are applying to the problem of identifying the subject of a text, given "analytics" describing that text. The principal output of this learning module is a list of suggested classificators to describe the subject of the text. This work is being programmed in Lisp on a Sun Workstation.

Comparing our system to other ML systems reveals novel aspects of our work. Using ML as a tool in natural language processing is a new and relatively untested technique, but the area is certainly ripe with research possibilities. Similarly, the application of ML to the problem of subject classification of texts and its use in the more general domain of Information Retrieval is also new; again, the worth of this technique in this domain and for this particular application is unproven, and conclusions on its ultimate usefulness is one of the aims of this work.

A number of existing ML systems have been examined to evaluate their applicability to the problem at hand, and much work has been done on the identification of potentially relevant information. Library experts and domain experts have been interviewed to determine the techniques used by human classifiers when classifying texts. Currently we are concentrating on the development of the classification algorithm and incorporating a learning strategy into this algorithm. Afterwards, it is hoped to investigate the further applicability of the algorithm developed to other areas of Information Retrieval and Natural Language Processing.

Acknowledgements

This work was carried out as part of project SIMPR, project 2083 of the ESPRIT II programme. We would like to thank our partners for helpful comments and suggestions.

REFERENCES

1. Smart, G., March 1990, Year One: The Results so far, CRI/AS Bregnerodvej 144, DK-3460, Birkerod, Denmark.

2. Smart, G., March 1989, SIMPR: An Introductory Description, CRI/AS Bregnerodvej 144, DK-3460, Birkerod, Denmark.

3. The Subject Approach to Information, 3rd Ed., Bingley Ltd., 1977.

4. Aitchison, J., Gilchrist, A., Thesaurus Construction. ASLIB, London, 1987.

5. Sharif, C., Subject Classification Research, University of Strathclyde, September 1989.

6. McCarthy, J., President's Quarterly Message: AI needs more Emphasis on Basic Research, AI Magazine, Vol 4, 1983.

7. Schank, R.C., The Current State of AI: One Man's Opinion, AI Magazine, Vol 4, 1983.

8. Luger, G.F., Stubblefield W.A., Artificial Intelligence and the Design of Expert Systems, Benjamin/Cummings Publishing Company, 1989.

9. Michalski, R.S., Understanding the Nature of Learning. In Michalski, R.S., Carbonell, J.G., Mitchell, T.M., eds. Chap 1, Machine Learning, An AI Approach, Vol 2, Morgan Kaufmann, 1986.

10. DeJong, G., An Approach to Learning from Observation, Machine Learning, Vol 1, Morgan Kaufmann 1983.

11. Kavanagh, I., Machine Learning: A Survey, University College Dublin, SIMPR-UCD-1989-8.03, October 1989.

12. Stepp, R.E., Michalski, R.S., Conceptual Clustering: Inventing Goal Oriented Classifications of Structured Objects, In Michalski, R.S., Carbonell, J.G., Mitchell, T.M., eds. Chap 1, Machine Learning, An AI Approach, Vol 2, Morgan Kaufmann, 1986.1986.

13. Quinlan, J.R., Learning Efficient Classification Procedures and their Application to Chess End Games. In Michalski, R.S., Carbonell, J.G. and Mitchell, T.M., eds. Machine Learning, Chapter 15, pages 463-482, Morgan Kaufmann 1983.

14. Quinlan, J.R., Simplifying Decision trees, International Journal of Man-Machine Studies 27, 1987.

15. Michalski, R.S., A theory and methodology of inductive learning. In Michalski, R.S., Carbonell, J.G. and Mitchell, T.M., eds. Machine Learning, Chapter 4, pages 83-134, Morgan Kaufmann, 1983.

16. Laird, J.E., Rosenbloom, P.S., Newell, A., Chunking in SOAR: the anatomy of a general learning mechanism. Machine Learning, 1(1):11-46, 1986.

17. Rosenbloom, P.S., Newell A., The Chunking of Goal Hierarchies: A Generalized Model of Practice, In Machine Learning: An Artificial Intelligence Approach, Volume 2, 1986.

18. Ward, C. et al, The Heading Exercise, Department of Computer Science, University College Dublin, SIMPR-UCD-1990-16.4.4., April 1990.

19. Simon, Herbert A., Why should machines learn? In Ryszard S. Michalski, Jamie G. Carbonell and Tom Mitchell, editors, Machine Learning, chapter 2, pages 24-38, Morgan Kaufmann, 1983.

20. Kodratoff, Y., Introduction to Machine Learning, Pitman, 1988.

21. Ellman, T., Explanation-Based Learning: A Survey of Programs and Perspectives, ACM Computing Surveys, Vol 21, No 2, June 1989.

A Test Bed for Some Machine Learning Algorithms

Ray Hickey

ABSTRACT

A theory of noise is proposed for use in the development and testing of machine learning algorithms which induce rule sets from examples. A universe is defined to be a probabilistic model of a domain complete with noise. Examples randomly generated from the universe are used to induce rules which may then be compared to those specified by the universe. Attributes are defined to be pure noise, partially informative or fully informative. A good induction algorithm should be able to detect a pure noise attribute and exclude reference to it in the induced rule set. Some experimental results from an evaluation of the algorithm CN2 are reported.

1.   INTRODUCTION

Machine Learning is currently a very active and promising area of research in Artificial Intelligence. Within this field a number of different approaches and characterisations of learning have been adopted. By far the most successful development up to now has been attribute based machine learning in which general rules appropriate to a domain are induced from examples using an algorithm. The testing of such algorithms is the subject of this paper.

2.   ATTRIBUTE BASED MACHINE LEARNING

The acquisition of rules for use by a knowledge based system, perhaps an expert system, presents a formidable problem to a knowledge engineer and constitutes the well-known 'knowledge engineering bottleneck'. Because human experts find it easier to ply their trade than to articulate their knowledge, an algorithm which can induce the true but essentially hidden rules from instances or examples of the application of that expertise (e.g. sample diagnoses in a medical domain) can offer enormous benefits in terms of productivity to the knowledge engineer.

In this setting, an example is itself actually a rule albeit a very specific one. It has the form:

if <condition> then <conclusion>

where condition can be a logical combination of sub-conditions. The sub-conditions and the conclusion can usually be expressed as attribute-value pairs where each attribute can have a finite number of values. The attribute in the conclusion is called the class attribute and its value the class.

Given a set of such examples, an attribute based machine learning algorithm offers a recipe for producing either:

(a) a compressed representation of the knowledge involved, i.e. which is declaratively equivalent to the examples set but briefer and more comprehensible, or

(b) a generalised rule set (or decison tree) from which conclusions can be inferred for hitherto unseen conditions.

Henceforth, (a) will be referred to as the compression problem and (b) as the induction problem. This paper is concerned with the induction problem.

Thus from, say, 2000 examples a rule set of perhaps 100 rules might be produced. The rule set obtained is often referred to as a classifier. The algorithm is usually called an induction algorithm although in case (a) this is probably misleading as the term "induction" has a connotation of uncertain inference, whereas in that case the knowledge has merely been transformed. Using attribute based machine learning algorithms, considerable success in terms of both speed of development and subsequent performance has been achieved for expert systems in a wide variety of domains from oil exploration to financial services; see, for example, [1].

2.1 The ID3 Algorithm

Probably the best known attribute based algorithm is ID3 developed by Quinlan [2]. In the first instance ID3 builds a decision tree rather than a rule set. It is possible to obtain a rule set from this tree as shown in [3]). The ID3 algorithm is

To construct a decision tree for examples, S:
If all the examples have the same class, c, then result is c.
else
amongst all attributes, select most informative attribute, A;
partition S into subsets $S_1, S_2, ...$ (one subset for each possible value of A) such that in subset i all the examples have the ith value of A;
construct, recursively, sub-trees for each S_i from amongst remaining attributes B,C... .

To explicate the notion of "most informative attribute", Quinlan used the entropy function,

$$H(Q) = \Sigma\, q_i * ln\, q_i \tag{2.1.1}$$

where $Q=(q_1,...q_n)$ is a relative frequency distribution of classes, as the building block of an evaluation function, i.e. a heuristic which selects the best attribute at each stage in the tree construction. If each S_i has proportionate size p_i in the example set and has relative frequency distribution, Q_i, of classes associated with it then the evaluation function is just average of the entropies for each of these.

Thus the information content of attribute, A, is

$$I(A) = \Sigma\, H(Q_i) * p_i \tag{2.1.2}$$

(in information theory, this is the expected conditional information). ID3 has the feature that it will correctly classify all examples in the original example set. It can thus be used as in (a) above to compress knowledge.

An algorithm, CART, bearing some similarlity to ID3 was developed by Breiman et al. (see [4]). CART also builds decision trees and uses an evaluation function based on entropy. The major difference is that CART does not branch on a whole attribute. Instead, at each node of the tree the best attribute value to split on is selected and the node is expanded in "yes" (= has the attribute value) and "no"(= does not have the attribute value) branches, so that only binary splits are possible at each node.

Mingers [5] has compared entropy as a heuristic, with several alternatives; its appears that there is little to choose between entropy and the others.

2.2   The AQ Algorithm

In contrast to ID3, the AQ algorithm builds rules directly by specialising a very general condition until the examples that it matches or "covers" are all of the same class, c, thus obtaining the rule:

if <specialised condition> then c.

The covered examples are then removed from the example set and the specialisation process is repeated until all examples have been processed. AQ retains a star of best conditions so far (where the star size is a user-defined parameter) and uses a beam search technique to specialise all members of the star.

Members of the star are ranked according to an evaluation function. Different evaluation functions are used, one popular choice being "number of positive examples included plus number of negative examples excluded". Like ID3, AQ also guarantees to correctly classify all examples in the example set.

3. COPING WITH NOISE IN THE EXAMPLES

3.1 The Noise Problem

The example set from which rules are generated may contain apparent contradictions, sometimes called clashes - the same condition may appear in more than one example yet with different conclusions. It is clear that algorithms such as ID3 and AQ cannot produce a set of definite rules from such examples. Instead, at best some conclusions will be indefinite, that is an 'ored' conclusion is obtained with, perhaps, probabilities given for each of the possible conclusions. This is an instance of noise in an example set. Such noise may be the result of errors creeping in during the recording of data and thus may be genuine mistakes, others however just reflect the true state of affairs - that when the particular condition involved holds there can be different possible conclusions.

As was noted in [6] the sources of noise can be divided into the following categories:

1. Insufficiency of Attributes, i.e. if more attributes were available it might be possible to state exactly the class associated with each condition.
2. Representational limitations of the attributes used, i.e. the attribute values do not accurately capture the property in question; for example integer values might be used where the property is real-valued or descriptions of pain as attribute values may not correspond to what a patient feels.
3. Inaccuracy in the recording of attribute values in examples.

Noise in data has a severe impact on the induction problem. Algorithms such as ID3 and AQ fit the examples exactly including all the idiosyncracies that a particular example set might have even though some of these may be due purely to noise. This may be sufficient if compression is all that is required. In the induction problem, however, what is needed is an algorithm which looks beyond the noise and attempts to find the true underlying set of rules. Michie [7] refers to such algorithms as 'noise proofed'. Inevitably a rule set induced from such an algorithm will have to sacrifice classification accuracy on the examples given to it.

3.2 Pruning to Eliminate the Effects of Noise

Since algorithms grow either a decison tree or rules, a way of smoothing the effects of noise is to prune the induced tree or rule set. This can be done at run time of the induction (run time or construction time pruning) or after the induction has been completed (post pruning).

In the case of ID3 there has been a considerable amount of research carried out into pruning (particularly post pruning) the aim of which is to cut back the tree from its leaves just sufficiently far to remove the noise contribution of the data whilst leaving enough of the tree to successfully classify new examples. The resulting tree also has the bonus of being much simpler to comprehend. Post pruning algorithms are themselves quite complex; see [3] and [8]. Quinlan [9] describes an algorithm, C4, a descendant of ID3, which makes extensive use of pruning techniques. Another descendant of ID3 which copes with noise is ASSISTANT [10].

3.3 The CN2 Algorithm

Clark and Niblett [11] have recently developed an interesting algorithm called CN2. This is a noise-proofed adaptation of AQR, a variant of AQ which they also developed.
In describing CN2, Clark and Niblett use the concepts of selectors and complexes (also employed for AQR). A selector is an assignment of a value or a disjunction of values to an attribute, e.g.

[Employment = unemployed]

[MaritalStatus = single v separated]

A complex is a selector or conjunction of selectors, involving several different attributes. Effectively it is the condition part of a rule. A rule has a complex as its condition and a selector (with a single value assignment) involving the class attribute as its conclusion:

[Advice=refer_loan_request] <- [Employment = unemployed] &
[MaritalStatus = single v separated]

A complex may be regarded as a template for a condition in that it "covers" specialisations of itself. The examples whose conditions are covered by a complex in this way are called the covered examples for that complex. The empty complex covers the whole example set. Covered examples have a frequency distribution of classes; for the empty complex this is the distribution of classes over all examples. In effect CN2 searches for complexes which have good class distributions where "good" means that the distribution is concentrated, substantially, at one particular class. A rule can then be formed associating this majority class with that complex.

CN2 grows rules by progressively specialising a general complex until it, as nearly as is possible, covers a single class. A specialisation of a condition is accepted, however, only if it passes a statistical test which assesses whether it offers a significant improvement in classification accuracy over the empty complex (i.e. run time pruning is being carried out).

The heuristic used to rank complexes (through the covered class distribution) is entropy. Significance of a specialisation is assessed using a standard likelihood ratio test to compare the original class distribution for all examples to that for the covered examples. At each stage in the growing of a rule a size-limited set of best complexes is retained and all of these are further specialised, if possible, and compared to the current best complex, i.e. a beam search technique is used through the space of possible specialisations.

Clark and Niblett have shown in [11] that CN2 compares well to other induction algorithms and also a Bayesian classifier.

4. UNIVERSES AND THE EVALUATION OF ALGORITHMS

Usually an algorithm is tested on several standard example sets and perhaps one or two artificial domains; different settings of parameter values such as significance level (or even choice of an evaluation function) will typically be tried. Often a large example set is divided into a subset for induction (about 2/3) and a test set (the remainder) on which to try the induced tree or rule set.

The major difficulty with the evaluation of these trials is the assessment of quality of the induced rules with regard to both declarative simplicity and reliability when used with new examples. With real examples from natural domains there is no true set of rules with which to compare the induced rules. With artificial domains, noise is sometimes introduced in an unsystematic way. Often, as a result, limited insight is gained from the experiments.

In this paper an approach to evaluation of algorithms is proposed which uses artificial domains, or universes as they are called. These universes provide a specification of the true rules, including, in general, an element of noise. Examples are then generated randomly from the universe and an algorithm induces rules from these. Different types of universes can then be employed to explore the strengths and weaknesses of various algorithms and parameter settings within a particular algorithm.

4.1 Definition of a Universe

A universe is a full probabilistic model for how the examples arise. A number of attributes and their possible values are specified. Complexes are defined in terms of these attributes. A full complex is one with a value for every attribute; these are the most specialised complexes. An (unconditional) probability of occurrence is stated for each full complex; some such probabilities may be zero . For each full complex a class probability distribution is given. This will be denoted $Q = (q_1, \ldots q_n)$ where n is the number of values taken by the class attribute and q_i is the probability of ith class.

Noise is modelled by the class distribution. The actual sources of the noise can be any of those noted in section 3.1. Thus a universe is a collection of statements of the form

if <complex> then <class distribution>

The cases of (i) a degenerate distribution (all probability concentrated at a single class) and (ii) probabilities spread equally amongst the possible classes represent the two extreme situations of minimum and maximum noise respectively. In case (i), knowledge of the complex determines completely the class which will occur, whereas in case (ii), given the complex, we are as unsure as we can be about which class will actually occur. Distributions between these two extremes model an intermediate degree of noise.

The degree of noise in a class distribution can be assessed using a measure of information such as entropy (a larger value of entropy corresponds to a more noisy situation).
To model a situation where the class (the majority class) is fairly certain, place the largest probability at that class with the remaining probability spread over the remaining classes. To increase the level of noise, transfer some of the probability at the majority class to another class.

For example suppose there are five classes and it is required that the second class is the most likely with probability 0.7 then a possible distribution is $Q=(0.1,0.7,0,0.2,0)$. To increase the noise level replace Q by a distribution such as $Q'=(0.15,0.6,0.05,0.1,0.1)$. Here the largest probability has been degraded (although the majority class is unchanged), however the largest probability can be retained and the remaining probabilities made more equal if required. In fact for a given largest probability, a least noisy distribution is, intuitively, one with the remaining probability concentrated at a single class and the noisiest distribution is that with the remaining probability split equally amongst the other classes.

4.2   Information Content of a Universe

Class distributions give the probabilities for the possible classes given a particular complex, i.e. they are conditional distributions. From these and the probabilities of the complexes may be obtained the unconditional class distribution describing the probabilities of the classes in the absence of knowledge of the attribute values.

Suppose in a universe that there are k full complexes with probabilities $p_1,...,p_k$ and the class distributions conditional on the complexes are $P_1,...,P_k$. Then the unconditional class distribution is:

$$P^* = \Sigma\ p_i^*P_i \qquad\qquad (4.2.1)$$

so that P^* is a weighted average (i.e. a convex combination) of the P_i.

The amount of information provided by the universe about which class will occur can then be defined as the average gain in information about the class given by the full complexes, i.e.

$$\text{Information Gain} = H(P^*) - \Sigma\ p_i * H(P_i) \qquad\qquad (4.2.2)$$

It is possible that each class distribution, P_i, is more informative than P, for example each P_i can be degenerate at a particular class. Typically, though, only some class distributions will be more informative. It cannot be the case that all class distributions are less informative than P^* (this would contradict the concavity of entropy); thus the information gain in (4.2.2) is non-negative.

If the P_i are all identical then P^* is this common distribution and thus no complex provides any additional information about class over and above the unconditional class distribution. The attributes are thus totally uninformative. Such a universe will be called non-informative. If the class distributions P_i are not all identical (and assuming that all $p_i>0$) then there is some information in the attributes.

4.3   Information in an Attribute

Individual attributes may differ considerably in the extent to which they provide useful information about the class. It was noted in the previous section that, unless all the conditional class distributions are identical, some complexes are informative. This information may derive from particular attributes only or may be contributed to by all attributes.

One particularly important case is that of a totally uninformative attribute also called a pure noise attribute. To illustrate this idea suppose a universe has three attributes A,B and C.

If, in every complex in which B appears, the conditional class distribution does not vary when the value of B is varied and the values of A and C are fixed then B is totally uninformative. Equivalently, specialising a complex by adding a value of B does not alter the conditional class distribution. As a special case of this it follows that if a complex consists of just a value of B then its class distribution is the unconditional class distribution. This is just the property of conditional independence described by Pearl [12]: the class distribution is conditionally independent of B given A and C. To create a pure noise attribute it is sufficient to ensure that all full complexes have class distributions that do not depend on its values.

Attributes other than pure noise attributes are informative in the sense that there are some complexes for which the class distribution is altered by a change in the value of that attribute only. Some values of an attribute may be contributing information, however, and others not. Thus if A can take values a_i, i=1,...,4 it might be that a_1 and a_2 are informative (they influence class distribution) but that a_3 and a_4 do not. Such an attribute is said to be partially informative; if an attribute has no uninformative values it is said to be fully informative.

There is a spectrum of possibilities, therefore, for an attribute from being pure noise at one extreme to having all its values informative (to some degree) at the other extreme. The degree of informativeness of an attribute is called its status; status must be one of pure noise, partially informative or fully informative.

It is possible that an attribute is merely an alias for another attribute. For this to be the case the two attributes involved must have the same number of values. There will then be a one-to-one function which maps the values of one attribute to values of the other; note that either attribute may be regarded as the alias of the other. For example suppose a universe has attributes A, B and C where C is an alias for A. Let the mapping, for simplicity, be:

$$a_i -> c_i , i=1,2,3$$

It follows that complexes a_i & b_j & c_k can occur only when i=k while, for each i and j, complexes a_i & b_j & c_i, a_i & b_j and b_j & c_i must have the same class distributions to reflect the fact that in the presence of a_i, c_i contributes no further information and vice versa. An attribute and its alias must have the same status.

There are a number of ways of making an overall assessment of the importance of an attribute. One method involves omitting the attribute from the universe and noting the resulting degradation in information gain as defined in (4.2.2); the greater the degradation, the more important the attribute.

The degradation can also be assessed by comparing the classification rate (see the next section) for the reduced universe with that of the original. An alternative approach is to rank attributes in terms of average information content over its possible values in the manner used (on data) in the algorithm ID3, namely to evaluate the conditional expected information in the class distributions given an attribute. A drawback with this approach is that an attribute is deemed important if it is informative on average even though, as noted above, it may have several uninformative values.

4.4   Best Classifier associated with a universe

The task of an induction algorithm is to produce a set of rules (or a tree) which is declaratively simple and which classifies new examples to a high degree of accuracy. Here classification means the assignment of a particular class when presented with a condition (although a distribution over classes could be given, this being an estimate of the corresponding class distribution in the universe, provided the induction was made from sufficiently many examples to justify this).

Every induced rule set or tree will have a classification rate - a rate of success in classifying unseen examples. An upper limit to this is provided by the corresponding rate for the universe. This is the rate for the best classifier, i.e. the collection of rules of the form:

if <complex> then <majority class>.

Its classification rate is

$$\Sigma \; \max(P_i) \; {}^*p_i \qquad\qquad\qquad (4.4.1)$$

where $\max(P_i)$ is the probability of the majority class in class distribution P_i. If there is little noise in the universe (high information content) then (4.4.1) will be close to 1. In fact, because the max function is convex it can be used to provide an information measure as noted in Hickey [13]; (4.4.1) is a conditional expected information function analogous to (2.1.2). The best classifier does not depend on the complex probabilities only on the class distributions. Altering complex probabilities, however, may change the best classification rate; it can be improved by concentrating probability at those complexes with little noise in their class distributions.

Two universes which differ only in their full complex probability distributions, then, have the same best classifier (the determination of which is the goal of rule induction), but have, possibly, different best classification rates. They present problems of differing severity to an algorithm. If the complex distribution has a concentration of probability at a few complexes then these complexes will occur frequently in a set of examples randomly generated from the universe and finding rules for them will be comparatively easy; rarely obtained complexes will provide a more difficult problem for the algorithm.

5.   SOME SIMPLE UNIVERSES

To illustrate the ideas in section 4, three universes are now described. The first of these will be used to experiment with CN2.

5.1   Universe 1: A Universe with some Pure Noise Attributes

A is a totally informative attribute having values a_i, $i=1,...,4$ and B and C are pure noise variables with values b_i, $i=1,2$ and c_i, $i=1,2,3$ respectively. There are, therefore 24 full complexes. These are assigned equal probabilities.

There are four classes: $class_i$, $i=1,...,4$. Since B and C are pure noise variables it is only necessary to provide conditional class distributions for the four values of A. These are defined as follows:

a1	(0.8,0.05,0.1,0.05)
a2	(0.05,0.9,0.025,0.025)
a3	(0.05,0.1,0.75,0.1)
a4	(0.05,0.05,0.1,0.8)

The majority class given a_i is $class_i$ with a small degree of noise in each case. A full complex has the class distribution determined by its value of a_i. Thus the universe states, for example that:

> if a2&b1&c3 occurs (with probability 1/24) then the probabilities
> of the classes are 0.05,0.9,0.025,0.025 respectively.

The unconditional class distribution as given in (4.2.1) is:

$$(0.2375,0.275,0.24375,0.24375).$$

This is a very uniform distribution in contrast to the individual class distributions indicating that the information content of the universe is high, i.e. the attributes provide a great deal of knowledge about the class (in fact all this knowledge comes from A with B and C contributing nothing). The best classifier consists of rules such as

> if a_3 & b_2 & c_1 then $class_3$

Since B and C are pure noise variables, the 24 rules thus generated can be compressed to a set of just 4:

$$\text{if } a_1 \text{ then class}_1$$
$$\text{if } a_2 \text{ then class}_2$$
$$\text{if } a_3 \text{ then class}_3$$
$$\text{if } a_4 \text{ then class}_4$$

5.2   Universe 2: A Universe with Fully Informative Attributes

In some universes there may be particular complexes, specialisations of which do not alter the class distribution. A universe in which this never happens can be created by requiring the class distribution to depend on full complexes. For example, if all the attributes are binary taking values 0 and 1 and the class distribution depends on the number of 1's in the complex then this will be the case. Such a universe with four binary attributes has 16 full complexes. These are taken to be equally likely. The classes are then 0,1,2,3,4.

To introduce a degree of noise, the appropriate class (corrresponding to the number of 1's) is assigned a probability of 0.8 with immediate neighbours to this being assigned 0.1 each. In the case of 0, the immediate neighbours are 1 and 2.; for 4 they are 2 and 3.

5.3   Universe 3: Quinlan's Probability Disjunction Example

Quinlan [3] describes an artificial domain involving 10 binary attributes one of which is pure noise with the remaining 9 being fully informative. All 2^{10} full complexes are equally likely. The class, which is 0 or 1, is determined with probability 0.9 to be 0 for complexes with particular combinations of 1's and to be 1 with probability 0.9 otherwise.

6.   A PROLOG IMPLEMENTATION OF THE TEST BED

6.1   The Universe Declaration

A universe is first described in a file called the declaration file which saves the labour of specifying a rule for every full complex. First a header gives details of attributes and their values and states whether full complexes are equally likely or not. Then a highly compressed statement of the universe is given. In the case of universe1, described in section 5.1, it is sufficient to declare that for any valid full complex, the class distribution is that of the value of A. If the full complex probabilities are unequal they must all be specified.

6.2   The Full Specification of the Universe

From the (sometimes comparatively brief) description of the universe in the declaration file, a full description of the universe is made, using a Prolog procedure, for use by the example generator and held in a file called the specification file. In this file the declarations of attributes, their values and the classes are represented as in the declaration file but the description of complexes, their probabilities and associated class distributions is expanded so that each full complex is represented explicitly. The full set of universe rules is declared in rules/1 whose argument is a list of individual rules of the form:

r(Complex-Probability,ClassDistribution)

For many universes, of course, the specification file will be quite large. The specification file for universe 1 is shown in appendix 1.

6.3   The Example Generator

The example generator, Exgen, which has been implemented in Prolog, draws a random sample of examples from a specification file for a universe. In the random sampling process, complexes are drawn with replacement in accordance with the complex probability distribution. When a complex is drawn, a class is selected using the conditional class distribution for that complex.

An example is recorded in a structure $e/(m+1)$ where m is the number of attributes; the attribute values for the example are given in the first m arguments and the final argument gives the class. The whole example set is then stored in examples/1 whose argument is a list of the individual examples. An example set is shown in appendix 1.

7.   SOME EXPERIMENTS WITH CN2

CN2 was used to induce rule sets for several example sets generated from universe 1. The sizes of the example sets were 30, 100 and 300. A Prolog implementation of CN2 written by Clark and Niblett was used. The experiments were run on a Sun 3. The result of an induction involving a set of 100 examples is given in appendix 2. Note that the rules are ordered, i.e. are induced in an "if then ... else" format. These are also shown after conversion to an unordered, i.e. a modular form.

7.1   Choice of Parameters

There are two parameters to be set in CN2 - the significance level for the likelihood ratio test of a specialised complex and the size (MaxStar) of the Star of complexes retained for further specialisation. After some preliminary investigation it was decided that a setting MaxStar = 4 was sufficient for the domain of universe 1; increasing MaxStar did not have any major impact on the results whilst slowing the induction time. The significance level was set at 0.95 as this is something of a standard in the statistical world.

7.2   Discussion of results

The induced rule set in appendix 2 is typical of those obtained from many inductions. Examination of it shows that the rules are largely correct in that the majority class is associated with the appropriate value of A. The rules are, however, grossly over specialised. In universe 1, only attribute A is informative with B and C offering only pure noise. The ideal induced rule set is thus:

if A=a1 then Class = class1
if A=a2 then Class= class2
if A=a3 then Class= class3
if A=a4 then Class = class4

Instead most induced rules make reference to irrelevant values of B and C.
Increasing the size of the example set did not appear to improve matters substantially: the algorithm still tends to overspecialise.

7.3   A Suggested Improvement to CN2

The problem of over-specialisation is caused by the significance testing procedure adopting specialisations which it should reject. This problem does not appear to diminish when the significance level is increased, as might be expected. Difficulties with the testing procedure have also been noted by Chan [14].

A possible source of the problem is that the notion of significant improvement is not cumulative: if complex X is specialised to Y which is significantly different from it and Y is further specialised to Z where Z offers only a marginal improvement over Y, then Z will register as significantly better than X and will be adopted by the algorithm. There are also problems of non-transitivity of significance: if Y is significantly different from X and Z is likewise from Y it does not follow that Z will be significantly different from X. In fact X and

Z may be very similiar.

In an attempt to overcome these difficulties the algorithm was altered to require a proposed specialisation of a complex to be both significantly better than the original complex
(i.e. nil) and its immediate predecessor. The experiment was re-run and the results for the
same example set are given in appendix 3. It can be seen that there is substantial compression of the induced rule set as compared to those in appendix 2, especially after conversion
to an unordered representation.

In conclusion, therefore, while CN2 appears to be able to cope with noise corresponding to minor misclassification, it has difficulty with whole attributes which are noisy, i.e.
contribute nothing to the classification. Further investigation of over-specialisation is
currently being carried out.

APPENDIX 1:   SPECIFICATION FILE FOR UNIVERSE 1 AND AN EXAMPLE SET

Table 1.   Specification file for universe 1

rules([
           r([a1,b1,c1]-0.0416666,[0.8,0.05,0.1,0.05]),
           r([a1,b1,c2]-0.0416666,[0.8,0.05,0.1,0.05]),
           r([a1,b1,c3]-0.0416666,[0.8,0.05,0.1,0.05]),
           r([a1,b2,c1]-0.0416666,[0.8,0.05,0.1,0.05]),
           r([a1,b2,c2]-0.0416666,[0.8,0.05,0.1,0.05]),
           r([a1,b2,c3]-0.0416666,[0.8,0.05,0.1,0.05]),
           r([a2,b1,c1]-0.0416666,[0.05,0.9,0.025,0.025]),
           r([a2,b1,c2]-0.0416666,[0.05,0.9,0.025,0.025]),
           r([a2,b1,c3],0.0416666,[0.05,0.9,0.025,0.025]),
           r([a2,b2,c1]-0.0416666,[0.05,0.9,0.025,0.025]),
           r([a2,b2,c2]-0.0416666,[0.05,0.9,0.025,0.025]),
           r([a2,b2,c3]-0.0416666,[0.05,0.9,0.025,0.025]),
           r([a3,b1,c1]-0.0416666,[0.05,0.1,0.75,0.1]),
           r([a3,b1,c2]-0.0416666,[0.05,0.1,0.75,0.1]),
           r([a3,b1,c3]-0.0416666,[0.05,0.1,0.75,0.1]),
           r([a3,b2,c1]-0.0416666,[0.05,0.1,0.75,0.1]),
           r([a3,b2,c2]-0.0416666,[0.05,0.1,0.75,0.1]),
           r([a3,b2,c3]-0.0416666,[0.05,0.1,0.75,0.1]),
           r([a4,b1,c1]-0.0416666,[0.05,0.05,0.1,0.8]),
           r([a4,b1,c2]-0.0416666,[0.05,0.05,0.1,0.8]),
           r([a4,b1,c3]-0.0416666,[0.05,0.05,0.1,0.8]),
           r([a4,b2,c1]-0.0416666,[0.05,0.05,0.1,0.8]),
           r([a4,b2,c2]-0.0416666,[0.05,0.05,0.1,0.8]),
           r([a4,b2,c3]-0.0416666,[0.05,0.05,0.1,0.8])]).

Table 2.   Set of 30 examples from universe 1 produced by Exgen

attributes([a,b,c]).
classes([class1,class2,class3,class4]).
att_values(a,[a1,a2,a3,a4]).
att_values(b,[b1,b2]).
att_values(c,[c1,c2,c3]).

examples([

e(a1,b2,c3,class1),e(a3,b2,c1,class3),e(a4,b1,c1,class4),e(a2,b2,c1,class1),e(a1,b2,c3,class3
e(a2,b1,c2,class2)e(a4,b1,c3,class4),e(a3,b2,c2,class4),e(a2,b1,c1,class2),e(a4,b1,c2,class4),
e(a2,b1,c1,class2),e(a4,b2,c3,class4),e(a2,b1,c1,class2),e(a3,b2,c2,class3),e(a1,b1,c1,class1),
e(a1,b2,c3,class1),e(a3,b2,c1,class3),e(a2,b1,c1,class1),e(a2,b2,c2,class2),e(a4,b2,c1,class3),
e(a2,b1,c2,class2),e(a1,b2,c1,class3),e(a2,b2,c2,class2),e(a4,b1,c1,class4),e(a4,b2,c1,class4),
e(a4,b2,c3,class4),e(a3,b1,c3,class3),e(a2,b2,c3,class2),e(a4,b2,c1,class4),e(a1,b2,c1,class1)]).

APPENDIX2:   RULES INDUCED WITH CN2 USING 100 EXAMPLES FROM
UNIVERSE 1

Table 3:   Ordered rules

logged_cutoff(95).
logged_maxstar(4).
logged_number_exs(100).   % cov : class freqs for covered exs.

if a = a4 and c = c2 then class = class4	% cov: [0,0,0,7]
else if a = a1 and b = b1 then class = class1	% cov: [7,0,0,0]
else if a = a2 and b = b2 and c = c3 then class = class2	% cov: [0,6,0,0]
else if a = a1 and c = c2 then class = class1	% cov: [4,0,0,0]
else if a = a2 and b = b2 and c = c2 then class = class2	% cov: [0,4,0,0]
else if a = a2 and b = b1 and c = c1 then class = class2	% cov: [0,4,0,0]
else if b = b2 and c = c2 then class = class3	% cov: [0,0,7,1]
else if a = a1 and c = c3 then class = class1	% cov: [5,0,1,0]
else if a = a3 and c = c1 then class = class3	% cov: [0,0,6,2]
else if a = a3 and c = c2 then class = class3	% cov: [0,1,4,0]
else if a = a4 and c = c3 then class = class4	% cov: [2,0,0,6]
else if b = b1 and c = c1 then class = class4	% cov: [0,0,1,5]
else if a = a4 then class = class4	% cov: [2,0,0,3]
else class=class3.	% end default rule

Table 4. Induced rule set converted to yield unordered rules

if a = a4 and c = c2 then class = class4.
if a = a1 and b = b1 then class = class1.
if a = a2 and b = b2 and c = c3 then class = class2.
if a = a1 and b = b2 c = c2 then class = class1.
if a = a2 and b = b2 and c = c2 then class = class2.
if a = a2 and b = b1 and c = c1 then class = class2.
if a = a3 and b = b2 and c = c2 then class = class3.
if a = a1 and b = b2 and c = c3 then class = class1.
if a = a3 and c = c1 then class = class3.
if a = a3 and b = b1 and c = c2 then class = class3.
if a = a4 and c = c3 then class = class4.
if a = a4 and b = b1 and c = c1 then class = class4.
if a = a4 and b =b2 and c = c1 then class = class4.
if (a =a1 or a = a2) and b = b2 and c= c1 then class=class3.
if a = a2 and b = b1 and not(c = c1) then class = class3.
if a = a3 and c = c3 then class = class3.

Table 5. Rules rearranged by class

if a = a1 and b = b1 then class = class1.
if a = a1 and b = b2 and c = c3 then class = class1.
if a = a1 and b = b2 c = c2 then class = class1.

if a = a2 and b = b2 and c = c3 then class = class2.
if a = a2 and b = b2 and c = c2 then class = class2.
if a = a2 and b = b1 and c = c1 then class = class2.

if a = a3 and b = b2 and c = c2 then class = class3.
if a = a3 and c = c1 then class = class3.
if a = a3 and b = b1 and c = c2 then class = class3.
if (a =a1 or a = a2) and b = b2 and c= c1 then class=class3.
if a = a2 and b = b1 and not(c = c1) then class = class3.
if a = a3 and c = c3 then class = class3.

if a = a4 and c = c2 then class = class4.
if a = a4 and c = c3 then class = class4.
if a = a4 and b = b1 and c = c1 then class = class4.
if a = a4 and b =b2 and c = c1 then class = class4.

APPENDIX 3:   INDUCED RULES USING ALTERED CN2

Table 6.   Ordered rules

% Rule Set from 100 examples
% Mode of Induction: Ordered; Significance test: New

logged_cutoff(95).
logged_maxstar(4).
logged_number_exs(100).   % cov: class freqs for covered exs.

if a=a1 then class = class1		% cov: [17,0,2,0]
else	if a=a4 and c=c2 then class = class4	% cov: [0,0,0,7]
else	if a=a2 and c=c2 then class = class2	% cov: [0,7,1,0]
else	if a=a2 then class = class2	% cov: [1,16,0,1]
else	if c=c2 then class = class3	% cov: [0,1,11,1]
else	if a=a4 then class = class4	% cov: [4,0,1,14]
else class = class3.		% end default rule

Table 7.   Induced rule set converted to yield unordered rules and rearranged by class

if a = a1 then class = class1.

if a= a2 and c = c2 then class =class2.
if a = a2 and not(c = c2) then class = class2.

if a = a3 and c =c2 then class = class3.
if a = a3 and not(c = c2) then class = class3.

if a =a4 and not(c = c2) then class = class4.
if a = a4 and c = c2 then class = class4.

References

1. Michie D. Current developments in expert systems. In: Quinlan JR (ed) Applications of expert systems. Addison Wesley/Turing Institute Press, Wokingham, UK, 1987, pp 137-156
2. Quinlan JR. Discovering rules by induction from large collection of examples. In: Michie D (ed) Introductory readings in expert systems. Gordon and Breach, London, 1979, pp 33-46
3. Quinlan JR. Simplifying decision trees. Int. J. Man-Machine Studies 1987; 27: 221-234
4. Breiman L, Friedman J, Olshen R, Stone C. Classification and regression trees. Wadsworth, Belmont, CA, 1984
5. Mingers J. An empirical comparison of selection measures for decision tree induction. Machine Learning 1989; 3: 319-342
6. Hickey RJ. An experimental framework for the investigation of attribute based machine learning. Turing Institute/DTI Journeyman Report, Turing Institute, Glasgow, 1989
7. Michie, D. Machine Learning in the next five years. In: Sleeman D (ed) Proceedings of the third European working session on learning. Pitman, London, 1988, pp 107-122
8. Niblett T. Constructing trees in noisy domains. In: Bratko I, Lavrac N (eds) Progress in machine learning - Proceedings of the 2nd European workshop on learning. Sigma, Wilmslow, UK, 1987, pp 67-78
9. Quinlan JR. Induction of Decision Trees. Machine Learning 1986; 1:81-106
10. Cestnik B, Kononenko I, Bratko I. ASSISTANT 86: A knowledge elicitation tool for sophisticated users. In: Bratko I, Lavrac N (eds) Progress in machine learning - Proceedings of the 2nd European workshop on learning. Sigma, Wilmslow, UK, 1987, pp 31-45
11. Clark P, Niblett T. The CN2 algorithm. Machine Learning 1989; 3:261-283
12. Pearl J. Probabilistic reasoning in intelligent systems: Networks of plausible inference. Morgan Kaufmann, CA, 1988
13. Hickey, RJ. Majorisation, randomness and some discrete distributions. J Appl Prob 1983; 20:897-902
14. Chan, PK. A critical review of CN2: A polythetic classifier system. Technical report CS88-09. Vanderbilt University, Nashville, 1988

Section 3:

Expert Systems

Application of Expert Systems in Electrical Engineering

G. CROSS

J. BLAKLEY

F. SHUI

ABSTRACT

The extent to which computer-aided engineering has been
applied to heavy current electrical engineering,
embracing mainly the areas of electric power systems
and electrical machines and drives, is used to
demonstrate the potential for the application of
artificial intelligence in these areas. Examples of a
number of expert system applications are given.

1. INTRODUCTION

The definitions of an expert system are generally well
understood [1]. Such a system may incorporate one or
more of the following features:

(i)   contain intelligent software which embodies the
knowledge of a human expert in ways which make possible
the solution of problems based on the expert's
experience;

(ii)   have the ability to make modifications to
software or data as a result of computations, such that
the system learns by experience and can therefore
respond to developing situations;

(iii) have the facility to simulate human thought
processes when dealing with problems which are
incompletely structured and which would therefore
otherwise require solution by an experienced
professional.

Thus expert systems software is built extensively
on human empirical knowledge and uses logic and rules
of thumb in the role of high-level, decision-support
systems which offer a solution which has previously
been shown to work. They are not intended as
a complete substitute for human decision-making but, in
making basic decisions, can release the expert from
routine operations to permit work on more demanding
tasks.

In the field of electrical engineering, which is
generally regarded as embracing mainly the specialist
areas of electrical power systems engineering and the
design and control of electrical machines and drives,
the desirability of implementing expert systems may be
considered to assume a special importance beyond that
of implementation in the wider areas of electronic
engineering.

Within the UK, and in many other industrial
economies, the attractions of electronic engineering
have lead to a general lack of interest on the part of
many students in the study of electrical engineering.
Consequently the numbers of engineers at both
professional and technician level entering electrical
engineering employment has continued to fall. This
will lead to a projected manpower shortage in the next
decade.

While the implementation of expert systems is not a
complete solution to this problem, the facility of such
systems to assist in decision-making processes will
clearly help to alleviate some of the anticipated manpower difficulties by archiving existing expertise.

Since many of the applications of expert systems
inevitably derive from existing computer-aided-
engineering (CAE) techniques, this paper first reviews
the extent of CAE applications in electric power
systems and electrical machines and drives before
discussing present and potential developments in expert
systems applications in these areas. Power systems
applications which are considered include generation,

transmission and distribution and protection. Electrical machines and drives applications include the design of variable-speed electric motors and work currently being undertaken at the University of Ulster on an integrated design and test expert system for variable-speed dc motors.

2 CAE APPLICATIONS IN ELECTRICAL ENGINEERING
2.1 Electrical Power Engineering

CAD applications to power systems have been a source of interest for some years and have centred around the design, simulation and operation of both large and small systems ranging from those on board ships, oil rigs and aircraft to large interconnected national grid systems. Such systems are inherently large and complex and have demonstrated some of the advanced applications of CAE for both off-line and on-line tasks.

Off-line applications have included power flow studies, stability and transient calculations, network fault studies and protection relay co-ordination and settings.

Power flow software calculates voltage magnitudes, phase angles and transmission line real and reactive power flows and transformer tap settings for networks operating in the steady-state mode. Computation of solutions for networks having more than 2500 transmission lines and 2000 busbars can be carried out in less than one minute. Interactive versions of these programs are available in which power flows are displayed in single-line diagram form on the VDU. These may be used by the engineer to modify the network with a light pen or from the keyboard and permit the engineer to examine the many different sets of operating conditions necessary to analyse system expansion options.

Stability programs investigate system performance under disturbance conditions due to sudden loss of generators or transmission lines, sudden increases or decreases in load, and sudden short circuits. They determine whether generators and motors remain in synchronism with the system under such conditions by using machine dynamic equations and power flow equations to compute angular swings of machines and critical network fault clearing times.

Transient study software determines the transient overvoltages and currents which arise from line-switching operations or from lightning-strike surges and facilitates the design of plant insulation requirements and surge diverters.

Fault analysis programs calculate network inter-phase and line-to-earth fault currents to determine circuit breaker ratings and circuit breaker control relay design at each circuit breaker location in a network. Fault levels can be computed for any combination of system operating conditions such as generators or lines out of service.

An extension of this type of software is its application to protective relay setting and co-ordination studies to ensure protection discrimination between circuit breakers and fuses on different sections of a network such that only the faulty sections are disconnected. Discrimination on a radial feeder is shown in Fig. 1. The features of a typical protection program are shown in Fig. 2. The engineer is required to specify the type and setting for each relay, circuit breaker and fuse as input data for the data check block. The program then determines the effectiveness of the device types and settings and ideally recommends alternative devices if necessary. In the co-ordination study mode, the engineer specifies device types with no settings or permits the program to select the devices. Devices and settings are chosen to provide optimum co-ordination throughout the network. The final-co-ordination study indicates how the overall protection scheme will perform with the selected settings.

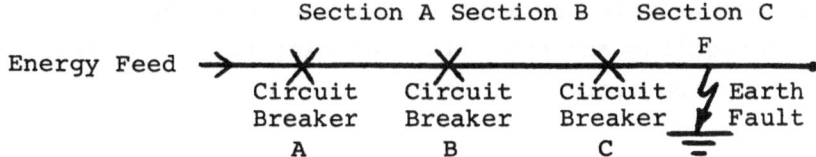

Section A Section B Section C
F
Energy Feed
Circuit Breaker A
Circuit Breaker B
Circuit Breaker C
Earth Fault

Only circuit breaker C should trip for Fault F with correct co-ordination of control relays.

Fig 1

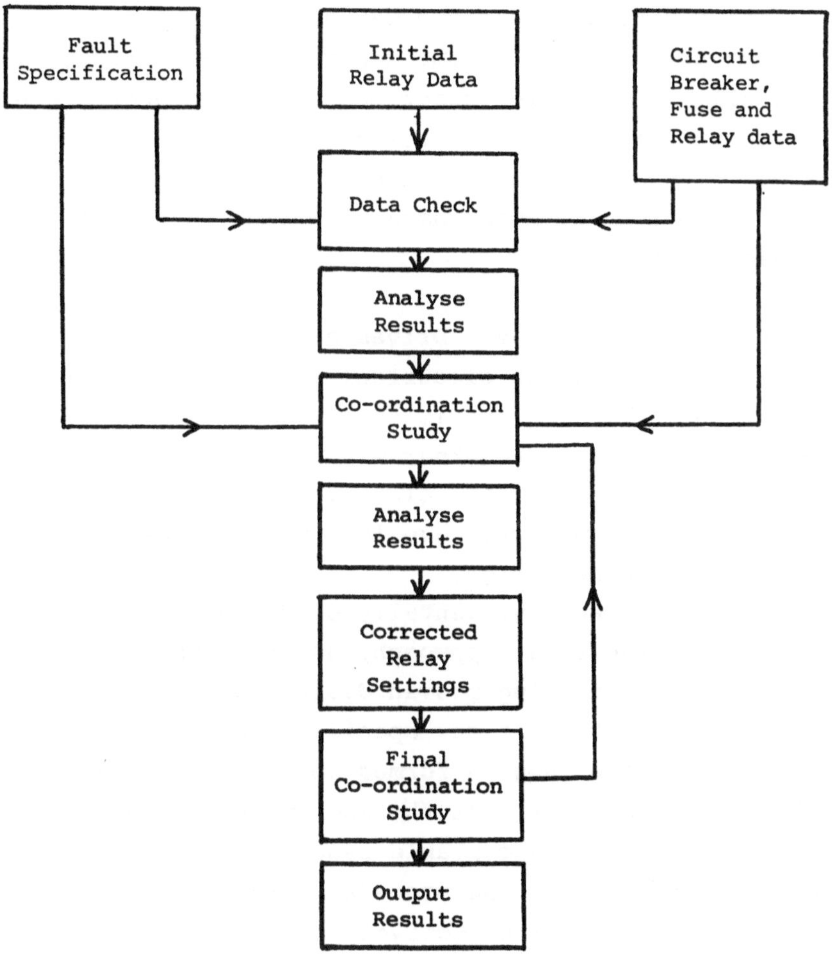

Fault Specification
Initial Relay Data
Circuit Breaker, Fuse and Relay data
Data Check
Analyse Results
Co-ordination Study
Analyse Results
Corrected Relay Settings
Final Co-ordination Study
Output Results

Fig 2

Other software applications include generating
plant scheduling and economic load dispatch to ensure
optimum system operating economy, load frequency
control, data acquisition for supervisory control, and
sequence-of-events monitoring for both confirmation of
the consequences of switching operations and for post-
mortem investigations following severe system outages.
At the consumer end of the system, load and energy
management schemes have received considerable
attention.

While many of the above CAE applications may be
thought to incorporate some expert-system features,
they all, in the forms described, fail to incorporate
the essential characteristics of providing solutions
based on human experience although their potential for
the development of such characteristics is clearly
considerable.

2.2 Electrical Machines, Drives and Utilisation

The widespread use of electrical machines, including
transformers, in the provision of energy and motive
power for industry, together with the extremely
competitive nature of the electrical machines
manufacturing industry, have lead to the widespread use
of CAE in machine design.

There are two fundamentally different approaches to
design. In the first approach, an incremental change
is made to an existing proven design in the expectation
that it will satisfy a new specification. Routine
calculations are then performed to ascertain whether or
not this is the case. If the incremental changes are
found to produce unacceptable results, then further
changes are made and the process is repeated. In this
process the designer effectively uses the computer as a
fast and powerful calculator.

The second approach to design is a purely con-
ceptional one in which no formal design methodology is
used. This approach may be applied to innovative
products and often involves the production of a number

of different prototypes which are then successively eliminated by engineering experts. The process is expensive and time-consuming. It is intriguing to speculate on the full potential for the application of artificial intelligence to this type of design process.

In the incremental-change design process applied to electrical machines, CAD packages are commonly used to check aspects of performance though these are usually discipline-specific.

The area of electromagnetics design has received considerable attention though the software which is applied depends to a large extent on the type of machine. A designer of synchronous motors or generators, or of dc motors or stepper motors is likely to use finite elements rather more than the designer of induction motors. The induction motor presents complex analysis problems and therefore present design methods are based upon the equivalent circuit model approach in which the equivalent circuit parameters are determined by finite element methods. Incremental changes to the design result in circuit parameter changes being made according to well-established rules with occasional recourse to finite element analysis as a check on the accuracy of these changes.

A number of electromagnetics CAD packages are commercially available, almost all based on finite elements techniques and all able to handle realistic geometries and model non-linear materials. Packages for the analysis of two-dimensional field problems include MagNet, FLUX, PE2D, GE2D and MAXWELL. Three dimensional field solution packages include TOSCA, CARMEN and 3-D Mag Net.

To use these packages requires a considerable knowledge of numerical methods and at the present time they would seem to lack the ability to be readily integrated into a comprehensive analysis/design environment.

A closely-related area is that of thermal-CAD. The useful life of all electrical equipment is closely

linked with accurate temperature rise analysis at the
design stage and it is perhaps surprising therefore
that although some application - specific packages are
available, much of the design is based on empirical
approaches and even on guesswork. Available packages
tend to use finite-difference rather than
finite-element solutions.

.   Printed circuit board thermal analysis packages,
which include Scitherm, Q-Board and PCH Thermal Planes,
are in some cases able to display the results as
coloured isotherms and may be interactive in permitting
the designer to modify the circuit layout or its
thermal environment. Freeman [2] suggests that
electrical devices which do not move should be capable
of thermal analysis using pcb packages though there is
no evidence of this having been attempted. Electro-
mechanical equipment with cooling air flows produced by
fans presents more complex problems and machines
manufacturers tend to employ empirical design methods.

    A further extensive area of electrical engineering
is that of building services and site-contract work.
Such work includes the provision of cabling, transformers, switchgear, and motors and control gear for
lifts, pumps and air conditioning and heating together
with lighting, power-factor correction capacitors and
standby generators. The electrical installations must
comply with relevant wiring regulations and the IEE
15th Edition Wiring Regulations are commonly employed
in many parts of the world. A wide range of software
packages are available including IEE-15, and design
packages from FACET and CYMAP. Such packages include
illumination design, acoustic design, cable, fuse and
battery sizing, and lift and pump system design. Many
aspects of this type of work clearly extend into the
areas of electric power systems and electrical
machines.

    The extensive use of empirical design methods again
indicates considerable potential for the application of
expert systems.

3 EXPERT SYSTEMS APPLICATIONS

3.1 Range of Applications

The previous sections have shown a wide range of CAE
techniques to have been applied in electrical engineering. During the last five to six years, expert systems
have found engineering application on a large scale,
especially in the USA and Canada. In electrical
engineering the areas which have received particular
attention range from diagnostics, through control and
planning, to design. While the development of power
and control system self diagnosis using linked computer
networks is readily envisaged, it is likely that
complete design processes using knowledge-based-systems
are unattainable. Designers generally use large
amounts of data from a number of knowledge-domains and
it is generally not practicable to integrate all the
necessary knowledge and the creative thought process of
the designer. However, recent publications demonstrate
that fairly high levels of knowledge-based expertise
are being used to assist in design processes.

The following sections describe a number of
applications which indicate the present state of expert
system usage in electrical power systems and machines.
The applications described do not represent an
exhaustive list but serve to demonstrate the general
trends in present development of systems.

The implementation languages used are mainly LISP,
OPS5 and Prolog. The widespread use of the former two
languages is probably due to their being firmly
established in the United States where the largest
number of users are located. However Prolog appears to
be an attractive alternative in engineering design
applications [3].

3.2 Power Systems Applications

Applications of expert systems in power engineering can
be classified into two groups, namely those which

operate on-line for control and diagnosis and those
which operate off-line for planning and design.

An example of the former in assisting with the
diagnosis of power system faults is described by
Talukdar, Cardozo, and Perry [4]. The architecture of
the programs is shown in Fig. 3. For an initial
configuration of a power system network and at the
inception of a fault, the protection scheme simulator
predicts the sequence of events which would ensue. The
diagnostic program hypothesises the causes of the
transition between the initial and final configurations
of the network. The Matcher compares the two configurations for similarity. The Blackboard is a database
for messages which may include raw of processed data
such as commands, hypothesis and help requests. The
programs are written in OPS5 and have been developed on
a VAX/780 with UNIX and tested on a VAX/750 with VMS.
The VMS versions which are compiled rather than
interpreted are considerably faster. Simulation and
diagnosis of a network of the size shown in Fig. 4 is
completed in a few seconds on VAX with VMS. Extension
of the work to more complex networks will require
increased capabilities without any significant increase
in processing time and will probably be achieved by
distributed processing and hierarchical logic to
identify the network regions on which attention should
be concentrated before deployment of the simulator and
diagnostic programs.

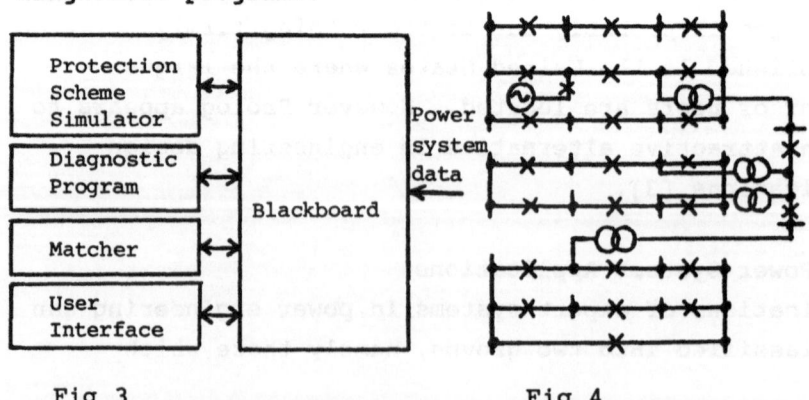

Fig 3   Fig 4

The on-line diagnosis of turbine generators using artificial intelligence has been described by Gonzalez et al [5]. An expert system shell using LISP and running on VAX 11/780 under VMS operating system is used as the basis for diagnosing the inputs of a number of plant sensors. The complexity of turbine-generator plant means that for economic reasons, not all sensors are connected as direct inputs to the diagnostic system interface. The diagnositic system therefore has the facility to generate instructions to the plant operator as to which additional sensor indications should be accessed or which test procedures should be initiated to provide additional evidence to support the diagnosis. Each diagnosis is accompanied by a confidence factor within the range -1 to +1 where -1 represents a definitely false conclusion and +1 a definite certainty. The turbine-generator has a number of modes of operation including run-up to synchronisation onto the grid system, generator synchronised, and turbine on turning gear during cooling following de-synchronisation. Different diagnostic rules may be applicable in the different modes and therefore each rule has an associated context which must be true for the rule to be used. The diagnosis system thus switches from one set of rules to another. The rule base for the generator is divided into five different sub-systems:
  (i)   stator winding and magnetic core
  (ii)   cooling hydrogen system
  (iii) oil system and seals
  (iv)   generator field excitation
  (v)   other miscellaneous systems.
The knowledge-base for each sub-system represents a different area of engineering expertise. The total system initially developed comprised approximately 900 rules which identified 185 different plant conditions, supported by 102 on-line sensors.

An expert system which assists power system control engineers in making decisions on action to be taken in

the control of reactive power and voltage is described
by Liu and Tomsovik [6]. The knowledge base contains
empirical rules which generate appropriate control
actions when voltage variations occur on the system.
Controls such as shunt capacitors, transformer tap
changing and generator voltages may be used. When
severe voltage problems ocur such that empirical judgements are identified by the knowledge base as being
unreliable, the expert system can assist in formulating
the problem in a way such that one of the analysis
software packages described in Secion 2 can be used.

An off-line application of expert systems is the
planning of generation expansion as proposed by
Farghal, Kandil and Abdel-Aziz [7]. The system is
implemented in Pascal on an IBM-AT microcomputer and is
based on the decision tree approach. The knowledge
required to perform the planning tasks is expressed in
terms of IF-THEN structures and the system described
has the capability to process and integrate the output
from planning models such as a simulation model, a
financial model and an environmental model. A variety
of decisions and uncertainties with measures of feasibility can be incorporated.

The design of cables is an area closely related to
electric power systems and an expert system for such an
application has been described by Hofstadler et al [8].
The system was developed in direct response to the loss
of expert designers due to retirement and the difficulties in recruiting replacement personnel. It has the
facility to produce layout designs and materials
choices for cables in a number of hostile environments
and is based on the accumulated knowledge of a retired
cable engineer and company design manuals.
Implementation was originally on VAX 780 using LISP but
was subsequently converted to operation on an IBM-PC.

An application which may be regarded as an
artificial intelligence-rather than an expert system-
approach is the allocation of loads in distribution
substations described by Wong and Cheung [9]. The

procedure is based on the application of set theory to
data sets consisting of loads which may be allocated to
a busbar section, and to other sections, and to one
section only. The extent of the human knowledge base
is in the specification of these initial conditions.
Subsequent generation of the solution or solutions
using Prolog on IBM PC-AT computer includes test
facilities to evaluate busbar capacity constraints and
supply security considerations. A typical substation
configuration is shown in Fig. 5.

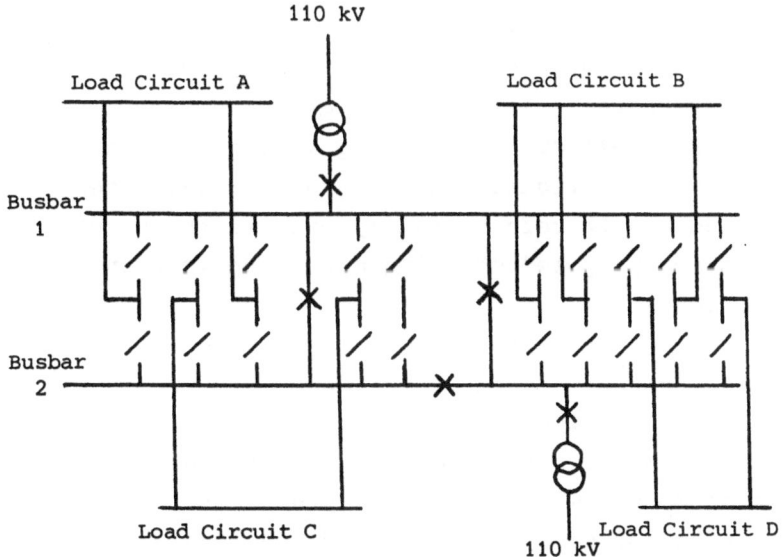

Fig 5

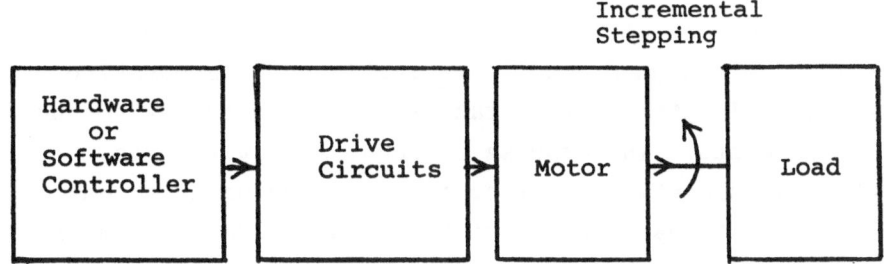

Fig 6

3.3 Electrical Machines Design

In the design of electrical machines, the incremental-change approach has been described in ection 2.2. When modifying an existing design or producing a new design, the engineer uses a number of sources of information together with accumulated experience. Much of this knowledge may be readily integrated into knowledge-based systems.

An IBM PC-based mini-expert system for the simplification of stepping motor system design has been described by Palmin [10]. The elements of a stepping motor system are shown in Fig. 6 and the expert system ensures optimum operation of the total system by considering the load, stepping motor, driver and system controller characteristics. The system designer can evaluate such effects as changing voltages, motors, motor winding configurations, or full or half step modes, and can therefore readily explore the practicalities of any proposed stepping motor application and control scheme.

In the design of induction motors, the use of the equivalent circuit model of the motor has been described in Section 2.2. An extremely practical expert system for the design of 3-phase cage rotor induction motors which effectively uses this approach is described by Landy, Kaplan, and Lun [11]. The system has been developed for both the tuition of trainee designers and for providing assistance in many of the routine design aspects for experienced designers. The design procedure is initiated by suggesting suitable length/diameter ratios depending on the torque requirements. In response to input supply voltage, winding connection (star or delta), winding type, and stator slot shape are suggested. The conductor dimensions are determined by the designer using available sizes. The expert system then provides advice on winding arrangements and insulation class. An important feature of induction motor design is the correct selection of stator/rotor slot combinations and the knowledge base suggests practicable combinations and also prohibited combinations in order to avoid cogging and crawling torques in the motor. Subsequent stages in the design process specify rotor configuration, airgap length, ventilation requirements and mechanical features such as bearing types and teeth supports. Overall evaluation of the design performance followed by possible modifications and refinements then result in an iterative process.

Similar iterative procedures for the design of dc motors by expert systems have been described by Freeman and Mukolera [12] and Shui, Blakley and Cross [13], using Prolog. A particular feature of variable-speed dc motor design is the requirement for satisfactory sparking performance at the brush-commutator interface over the entire speed range of the motor. Rigourous analysis of the commutation process is extremely complex and is therefore beyond the capabilities of expert systems or normal CAD techniques. Design of this aspect of the motor performance is usually based on the designer's experience and subsequently validated by commutation performance testing on the completed machine, using the well-known black band test as shown in Fig. 7. An expert system for the integration of this design-and-test approach to identify the cause of commutation malfunction from a graphical display of the black band curve, and to suggest necessary design modifications, is currently under development by the authors.

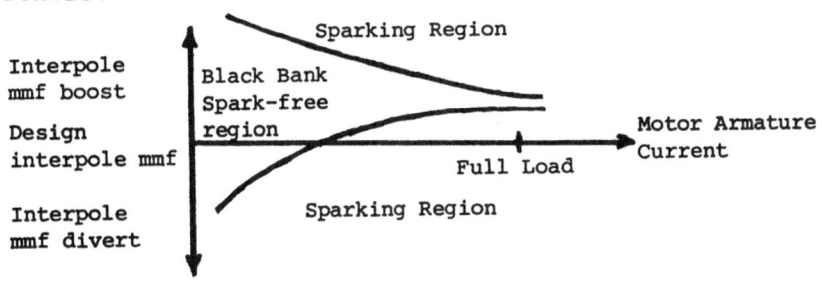

Fig 7

4. CONCLUSIONS

The electrical engineering manufacturing industry is a
highly competitive one. The potential problems arising
from erosion of the knowledge base due to the retirement of experts are widely acknowledged. The more
widespread use of knowledge based systems will assist
in archiving existing expertise and in releasing scarce
manpower resources to address non-routine problems.

The electrical power supply industry has also been
shown to have considerable potential for the application of artificial intelligence to system operation,
control and planning. There is currently considerable
worldwide interest in the application of expert systems
in these areas but such applications tend to be
developing on a somewhat piecemeal basis. To achieve
optimum power system operating economy and security
requires a more co-ordinated approach. Within a United
Kingdom context, it remains to be seen whether or not
such co-ordination is achievable in the proposed
complex structure for privatisation of the electricity
supply industry.

REFERENCES

1.   Kellock, B., 'Systems that turn Novices into
      Experts'. Engineering Computers, June-July 1987,
      pp 32-35.

2.   Freeman, E.M., 'CAE for Electrical Engineers'.,
      Keynote Address, Universities Power Engineering
      Conference, 1988.

3.   Mukolera, J.R., and Freeman, E.M., 'The Potential
      of Knowledge Based Systems in Engineering Design',
      E.E.D., Imperial College of Science and
      Technology.

4.   Talukdar, S.N., Cardozo, E., and Perry, T., 'The
      Operator's Assistant - An Intelligent, Expandable
      Program for Power System Trouble Analysis', IEEE
      Trans. Power Syst., Vol PWRS-1, No. 3, August
      1986. pp 182-187.

5.   Gonzalez, A.J., Osborne, R.L., Kemper, C.T. and
      Lowenfeld, S., 'On-line Diagnosis of Turbine
      Generators using Artificial Intelligence',
      IEEE Trans. Energy Conv., Vol. EC-1, No. 2,
      June 1986, pp 68-74.

6.   Liu, C.C., and Tomsovic, K., 'An Expert System
      assisting Decision-Making of Reactive
      Power/Voltage Control', IEEE Trans. Power Syst.,
      Vol PWRS-1, No. 3, August, 1986, pp 195-201.

7.   Farghal, S.A., Kandil, M.S., and Abdel-Aziz, M.R.,
      'Generation Expansion Planning: an Expert System
      Approach', IEE Proceedings, Vol 135, pt.C, No. 4,
      July 1988, pp 261-267.

8.   Hofstadler, P., Machin, E., Marder, B., and
      Palmer, D., 'Beyond Design Guides and Veteran
      Engineers - An Expert System for Cables and
      Connectors', Proceedings 35th Electronic Comp.
      Conf. (Cat. 8SCHZ184-0), 1985, pp 132-136.

9.   Wong, K.P., and Cheung, H.N., 'Artificial
      Intelligence Approach to Load Allocation in
      Distribution Substations', IEE Proceedings, Vol
      134, Pt.C, No. 5, Sept. 1987, pp 357-365.

10.   Palmin, S., 'Mini-Expert System Simplifies
      Stepping Motor System Design', Powerconverters
      and Intell. Motion (USA), Pt. 12, Vol. 12, 1986,
      pp 44-50.

108

11. Landy, C.F., Kaplan, R., and Lun, V., 'An Expert
     System for the Design of 3-Phase Squirrel Cage
     Induction Motors', IEE Electrical Machines
     Conference Proceedings, 1987, London, England,
     pp 127-131.

12. Freeman, E.M., and Mukolera, J.R., 'Knowledge
     Based Systems for the Design of DC Machines', IEE
     Electrical Machines Conference Proceedings,
     1987, London, England, pp 132-137.

13. Shui, F., Blakley, J.J., and Cross, G., 'Computer-
     Aided Design of Electrical Machines : The role of
     Expert Systems', Proceedings 24th UPEC, Sept,
     1989, pp 487-490.

Artificial Intelligence Applications in Geology: A Case Study with EXPLORER

M.D. Mulvenna, C. Woodham & J.B. Gregg

ABSTRACT

The modern exploration geologist deals with large amounts of data. This paper describes a system, EXPLORER, which was designed to help manage data collected in gold exploration in Northern Ireland The system uses geologists heuristics of exploration and may be expanded to search for base-metals and precious-metals in regions of glaciated terrain. EXPLORER generates reports on each square km of the licence. It utilises a forward-chaining inference strategy, where the rules are fired from licence data held in a Prolog database, to produce the reports. The data includes information on geology, geomorphology, geophysics; and empirical data is analysed for both target and pathfinder elements.

INTRODUCTION

Modern exploration techniques can present the geologist with a plethora of data which he or she may have difficulty representing and interpreting using traditional presentation methods. As computing technology is making inroads into geoscience, and generating large databases, data management will occupy an increasingly greater percentage of most geologists time. To quote an example, "over a five week period, a survey involving the collection of 500 samples per week with ten field observations at each site and an analytical suite of only 30 elements will provide 750,000 items of information" [1].

Increasingly, because of the large data sets that are generated, the field geologist will need to rely on "exploration assistants" [2]. These expert decision-support systems will ensure consistent treatment of data, and importantly, when fully developed, tutor the inexperienced geologist in under-developed nations.

EXPLORATION BACKGROUND

Review

Traditionally, gold prospecting in Ireland has been carried out using the tried and tested method of stream panning. This can be advanced to sampling the soil around the streams in search of the bedrock source of gold. Nowadays, these methods are still used, but are complemented with a wide range of other methods (Table 1). These empirical, or field methods, when combined into an integrated database, provide a comprehensive suite of data for analysis.

Another approach to gold exploration is to construct geological models which can aid the geologist enormously. For example, a model which uses information on the genesis of a geological structure which may contain gold (metallogenic models), could then be used in conjunction with what the geologist knows about a particular prospect. The model could perhaps reveal much about the prospect, and allow both qualitative and quantitative assessments to be made.

Prospector [3], was one of the first expert systems in geology, and it used a form of model-based reasoning, as outlined above, to evaluate a prospect. It "matched (volunteered) information with deposit models stored in a knowledge base".

Models developed for gold exploration in Ireland are too weakly focused to restrict the spatial extent for exploration sufficiently. However, they can target areas for recce evaluation (Figure 1).

Exploration Detail

The building of any expert system to aid gold exploration in Ireland must employ the geologists heuristics of exploration. The system should try to emulate the geologists interpretation of the field data.

Once designed, the system could then be generalised to explore for most minerals in glaciated regions.

Geological criteria from known gold mineralization can be built into the expert systems rule base to define favourable areas for further follow-up. The targeting criteria can assist at the ground selection and prospecting stages of exploration.

There are an abundance of empirical methods available to the exploration geologist, and results from these field methods, i.e. geological mapping, soil and drift geochemistry, outcrop geochemistry and geophysics, must be accessible to the expert system. Ranking of geochemical anomalies [4], where a score is accumulated for each area according to the results of the data, may be incorporated into future systems.

Generally the geochemical data includes readings (in parts per million/billion) for target elements and pathfinder elements. Pathfinder elements are so-called because they can lead to the target. They are frequently of higher concentration and more widely dispersed than the target element. In this version of the EXPLORER

Table 1. The fifteen data sets used for gold exploration in Northern Ireland

Data	Sampling Method	Analysis/Information
1. Drift	Minimum of 1 sample per square km	Information on drift type and thickness
2. Geology	Minimum of 1 sample per square km	Information on bedrock geology
3. Recce Soil	Carried out at 1 sample per square km (generally at roadsides)	Analysis for gold and arsenic
4. Recce Stream Sediment	Carried out at all road/stream intersections	Analysis for gold and arsenic
5. Stream Sediment HMC	Involves panning of stream bank sediment at all road/stream intersections in licence	Gold grains in the pan are counted and the concentrate is analysed for gold and arsenic
6. Till HMC	Involves panning of stream bank till sediment	Gold grains are counted and the concentrate is analysed for gold and arsenic
7. Grid Soils	Involves shallow soil sampling: generally on a 100m x 50m grid	Analysis for gold and arsenic
8. Grid Deep Sampling	Involves deep overburden sampling on a 100m x 50m grid	Analysis for gold and arsenic
9. Detailed Roadside Soils	Involves shallow soil sampling along roadsides at 0.1 mile intervals	Analysis for gold and arsenic
10. Detailed Roadside Deep Sampling	Involves deep overburden sampling along roadsides at 0.1 mile intervals	Analysis for gold and arsenic
11. Detailed Traverse Deep Sampling	Carried out as follow-up to grid deep sampling at 5 m intervals	Analysis for gold and arsenic
12. Outcrop Geochemistry	Involves geochemical analysis of outcrop samples	Analysis for gold and arsenic
13. Float Geochemistry	Analysis of boulders/cobbles, on surface or in stream sections	Analysis for gold and arsenic
14. Till Geochemistry	Analysis of the fine fraction of glacial overburden	Analysis for gold and arsenic
15. Geophysical Methods	Generally targetted to interesting areas	Induced Polarization/Resistivity collected

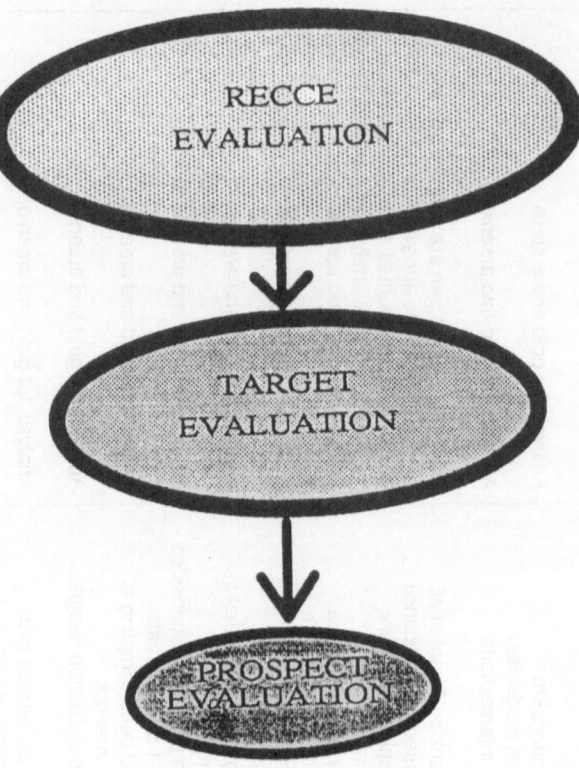

Fig. 1. How the exploration geologist
seeks to constrain the area of exploration.

system, gold (Au) is the target element and arsenic (As) is a pathfinder.

The current generation of geoscience database systems have many benefits, including complex statistical procedures, graphics, plotting, and windowing facilities. Windowing is the constraining of data sets to a given area. These facilities can allow complex spatial modelling to be carried out. However, to cover every square km of a licence area for every data set, with various statistical routines and different window sizes can prove extremely time consuming. Also, consistency is not guaranteed. In short, complex quantitative spatial reasoning is required [5]. This is a task at which most humans are not naturally adept, especially when the licence area can be 1000 square km in size. The procedure on how to carry out exploration varies from geologist to geologist and company to company, but the data set is finite to an extent; usually changing only by modification of an existing technique or old techniques becoming redundant and superceded through time.

The goal of exploration is to find mineral deposits by successfully restricting and testing the target area(s) (Figure 1), therefore methods which cover a larger area are tried first, and so on, until the area is either rejected or the target is well enough delineated to allow small scale exploration, and possibly, core drilling.

This brings up the question of cost. Exploration is very expensive, and when an area is rejected early, monies can be allocated instead to more promising prospects. Therefore, if a method existed to consistently examine the total licence area and assist in target definition and evaluation, resources would be optimised, and planning of future exploration could be more concisely pinpointed.

THE EXPLORER SYSTEM

A rule-based expert system to implement such a method was considered feasible (Figure 2). Benefits could include:
- rapid report generation and target definition for licence areas.
-consistent follow-up recommendations for target evaluation, optimising objectivity.
- a flexible exploration method able to be updated as new successful techniques are introduced or discovered and able to be used in different terrains for different minerals given new expert exploration criteria.

Reports are built up for each square km and kept on file for reference and comparison with updated reports generated when new data is added. As the exploration method is stored as rules with an understandable syntax, these can be modified easily by an exploration geologist using a text editor. Thus methods which represent the knowledge and expertise of a geologist (or company) can be captured.

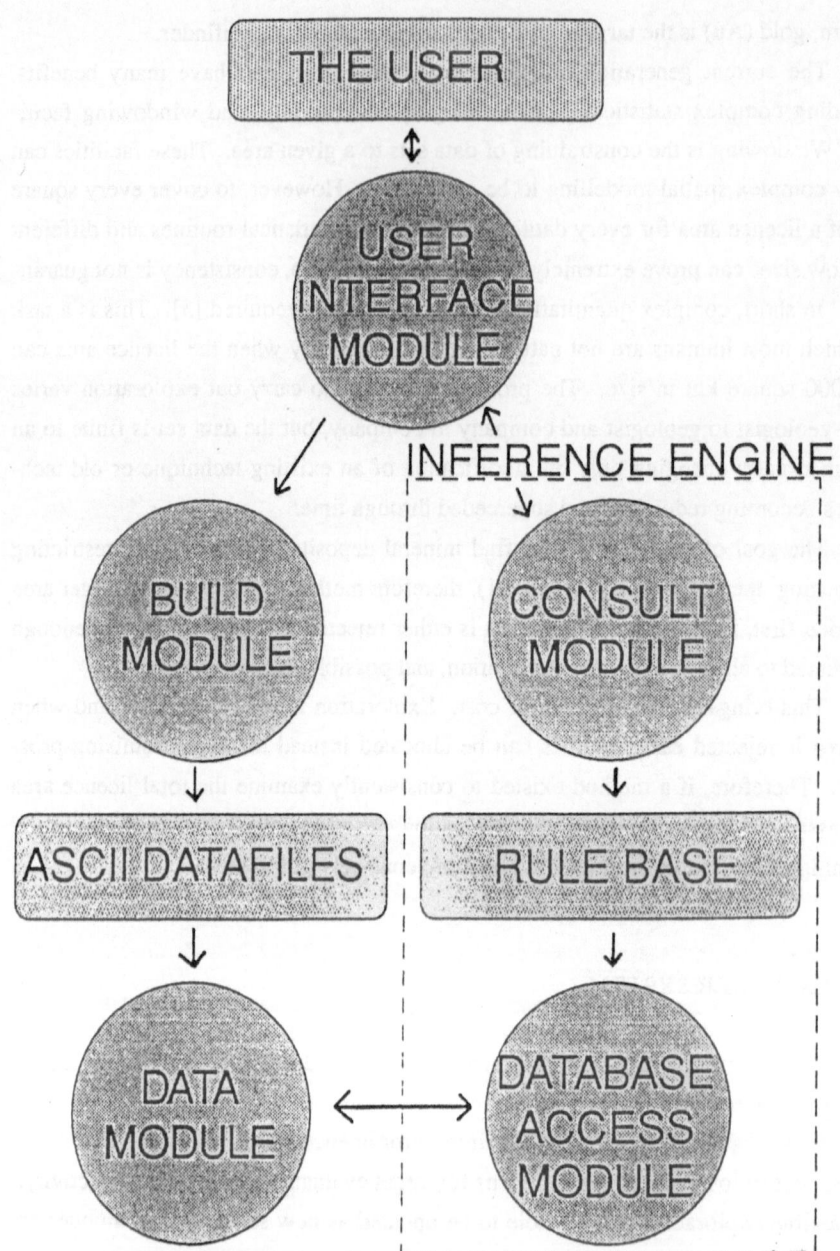

FIGURE 2. EXPLORER SYSTEM ARCHITECTURE

Initially the system was to link up to some commercially available database product, preferably in the geoscience field. Unfortunately the database which held the test licence data was unsuitable in its present form, due to an inflexible menu system, and no other geoscience products were available at the time. Rather than use a commercial database product such as dBase IV, where it was unrealistic to expect the geologist to maintain two separate database applications on one busy PC, the exploration databases were dumped to standard ASCII text datafiles. These datafiles could then be input to the BUILD module of the EXPLORER system.

The Build Module

When converting from ASCII files to Prolog clauses, the BUILD module (Figure 3) creates an imaginary 100 x 100 km grid over the National Grid and calculates the 1 km square in which the sample is located. It then adds this 'squares' information to the Prolog fact for that sample. This method considerably shortens access time, as the X and Y values of the square are instantiated when the system looks for a sample in the current square. Because the X co-ordinate is hashed, only those clauses with the same X square co-ordinate are considered by the ACCESS module.

The BUILD module compiles the data into 'heap' format, a machine representation which does not need to be syntax-checked or tokenised, and so loads quickly into memory. The DATA module created by BUILD is stored in virtual memory, which allows the size of the module to be much greater than the memory capacity of PC's running DOS, up to a limit of one megabyte. In practise, data sets for licence areas are less than this 1 megabyte limit. If more than 640 kilobyte RAM is present on the PC, the EXPLORER system can automatically use this to store the data as a virtual module on a virtual RAM disk, so optimising speed. This storage method bypasses the problems of large Prolog database storage in main memory, although ultimately the intention would be to revise the EXPLORER system to form a tightly coupled expert database system [6].

INFERENCING

The system forward chains through the rules initially from the data and then from inferred facts and data. The rule base has fifteen rule groupings each representing a data set. The inferencing method emulates the geologists thinking by first looking at general rules within the chosen 1 square km window. All the general rules are

116

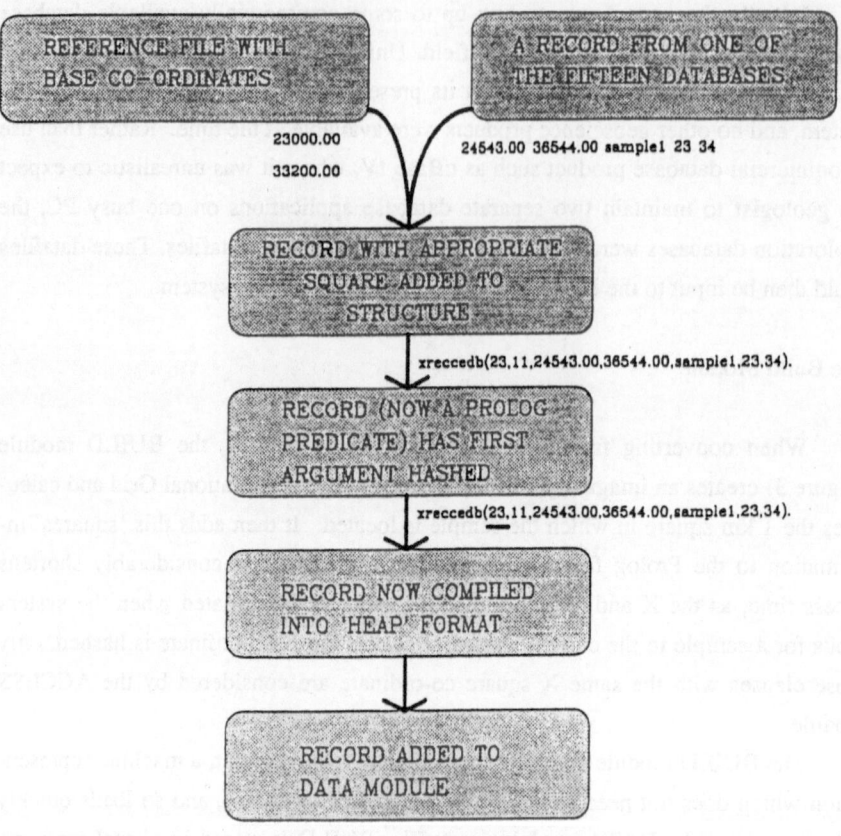

Figure 3. How the BUILD module works

gathered up and processed through conflict resolution. Conflict resolution is determined by refractoriness, recency and specificity, in that order, and the system picks
the first rule in the list that emerges from the process.

Meta-rules

Meta-rules, embedded within each of the fifteen rule groups flag two situations:
- the system should look at the eight squares adjacent to the current square, and
apply a selected rule grouping to those squares;
- the system should apply a different rule set to the current square.

The meta-rules proved necessary because the geologist often decides to look
around the area currently being investigated. An example of this would be to expand
the window, i.e. look at adjacent squares. If data from one rule group is positive, the
geologist must look around for support in adjacent squares. One example would be
to look for support as geochemical dispersion trains located in a 'down ice' direction
perhaps the result of glacial smearing from a bedrock source. Also, if one rule group
showed nothing for a square then the geologist looks at the data from another source
for that area.

When the system has processed all applicable rules and returns to the main
menu, all items flagged are dumped into two lists, the first for applying flagged rules
groups to the current square, and the second for applying flagged rule groups to
adjacent squares (windowing). These lists are displayed to the user as sequences of
two menu sets. Each menu gives the user the option to carry out the flagged option,
ignore it and look at the next flagged option, or see why that item has been flagged.

The main output from the system is a report generated from text and variables
tagged onto individual rules. Explanation facilities in the form of 'why' are implemented for flagged items. This gives the user an explanation generated from the
rules to aid decision-making, and guide the EXPLORER exploration process.

The ACCESS Module

A subset of the inference manager is the database ACCESS module. This looks
at the database for each square and tries to find the best value for that square. For
example, with the outcrop geochemistry database, it searches for a sample with gold
and arsenic above their respective thresholds. If this fails, it attempts to find a
sample with gold above threshold. In turn, if this fails it searches for arsenic above
threshold. The bottom line is that, if all the above fail, it attempts to find any

reading for the current square. If this does not succeed, then the ACCESS module fails to find a reading and the conditional part of the rule which called the access fails. This causes the rule to fail, and the inference manager tries the next rule in the queue, or cycles again.

USER INTERFACE

The EXPLORER system was designed to be user-friendly. This entailed a design with almost total menu control. The pull-down menus are mouse-driven. Context-sensitive help is provided in the form of a help screen displayed with associated help nodes shown as an additional menu on an adjacent screen. These items can, in turn, be selected allowing the user to peruse the help information.

While the user is navigating through the system, all menu items are dynamically checked to see if their selection is valid. This style of dynamic menu-checking traps many potential conflicts in the system and therefore reduces errors.

FUTURE DIRECTIONS

EXPLORER has been applied to a data set representing exploration primarily in a prospecting licence (DE5) covering 130 km^2 in Co. Tyrone, N. Ireland. This licence has been held by Celtic Gold plc since 1987 and has known bedrock gold mineralization thus enabling the development of the EXPLORER rule base [7]. The system is currently, as of May 1990, being evaluated by a consulting geologist attached to the company. Future work is planned to integrate a more general system for exploration, perhaps employing blackboard architecture, and using empirical results, with a relational database, and graphics system to provide an Intelligent Geological Information System (IGIS)[8], a step further than Geological Information Systems [9]. Initially, the scope of the IGIS will be to reason spatially about precious metal and base-metal deposits overlain by glaciated terrain.

It is recognised that geological reasoning systems will be in demand to help understand and filter the large data sets which cover many disciplines, and areas of the earth [1]. It may be that by developing IGIS's, we can explore and more fully integrate field data, satellite data and qualitative data to provide a richer working environment for the exploration geologist.

SYSTEM IMPLEMENTATION

The system was built on an IBM 80286 PS/2 DOS computer. The Prolog system used was Prolog-2 Professional Plus. The computer was chosen because it was compatible with that used by the geologist, while Prolog-2 was picked primarily for its virtual memory management, window facilities and modular architecture, which enabled large database construction, a good user interface development environment and incremental program development, respectively.

Other methods of knowledge representation, including frame- and object-based systems, were not used.

Prolog-2, PS/2 and dBase IV are trademarks of Chemical Design, IBM, and Ashton-Tate, respectively.

REFERENCES.

1.   Plant J. A. Hale M. and Ridgway J. Developments in regional geochemistry for mineral exploration. Trans. Instn. Min. Metall. 1988 Sect. B;97:116-140.

2.   Martin L. Expert systems and their use as exploration assistants. Unpublished paper presented at 'Exploration 87' the Third decennial international confer ence on geophysical and geochemical exploration for minerals and groundwa ter, Toronto, Canada.

3.   Gaschnig J. G. PROSPECTOR: An expert system for mineral exploration. In: Michie D (ed) Introductory readings in expert systems, Gordon and Breach science publishers New York, 1982, pp 47-64

4.   Bonnefoy D. Jebrak K. M. Rousset M. C. and Zeegers H. SERGE: an expert system to recognise geochemical anomalies. J. Geochem. Explor. 1989; 32:343-344.

5.   Yatabe S. M. and Fabbri A. G. Putting AI to work in geoscience. Epi sodes,1989;12,1:10-17.

6.   Torsun I. S. and Ng M. Y. Tightly-coupled expert database systems interface. In: Kelly B & Rector A (eds) Research and Development in Expert Systems V.Cambridge University Press,1988, pp 211-223

7.   Woodham C. Finlay S. and Holman R. Gold Exploration in the Dalradian of Northern Ireland. Trans. Instn. Min. Metall. 1989; Sect. B;98:63-65.

8.   Smith T.R. Artificial Intelligence and its Applicability to Geographical Prob lem Solving. The Professional Geographer. 1984;26:147-158

9.   Coupez Y. A micro-computer based geological information and processing system for exploration. In: Farrell L & Jones G Ll (eds) Irish Assoc. Econ. Geolog. Annual review. IAEG, 1989 p 48

An Expert Assistant For High Frequency Electronic Circuit Design

M Brennan, V F Fusco, J A C Stewart

ABSTRACT

An expert system to aid in the design of Microwave Integrated Circuits is described. The design process is discussed and the areas of need and the system objectives identified. The paper concludes with a view of the overall system architecture and a discussion of one of the knowledge bases to be included in the system.

1. INTRODUCTION

The application of an expert system to a design process is a challenging task. Simon [1] characterises design as an ill-structured problem ie one which is difficult to formalize and thus difficult to solve. However, some parts of the design process are formalizable and have been automated. Other areas have been explicitly described as technical knowledge in the form of textbooks etc. Begg [2] argues that there still remains unformalized knowledge which constitutes the engineers expertise. This unformalized knowledge represents a domain in which knowledge based systems may provide considerable help to the designer.

The design of Monolithic Microwave Integrated Circuits (MMICs) is a process characterized by high complexity, a mixture of heuristic, empirical and theoretical knowledge, and a scarcity of domain experts. MMICs are of ever increasing importance in commercial and consumer electronic applications such as home satellite receiving systems etc. Designing such devices, which provide a complete circuit function on a single chip, is computer intensive mainly because non-recurrent and design costs are high, and because the circuit may not be 'tuned' after fabrication. Therefore correct first time design is the paramount goal. This results in the extensive use of CAD to maximise the probability of 'first pass' designs meeting specification. In this paper we present a 'birds eye' view of an expert system currently under development in the Microwave Research Group at Queens University to assist in the

design of MMICs. This work has been undertaken in collaboration with Barnard
Microsystems Ltd (schematic capture and layout software), the Plessey Three-Five
Group (a commercial foundry design house), and under the auspices of SERC (Science
and Engineering Research Council).

Section 2 contains a short summary of the MMIC design process. (This will be
expanded upon as necessary in the remaining sections). Section 3 discusses the
objectives of the project and Section 4 describes the system architecture. Finally
section 5 describes in some detail GALEDA, a small knowledge base to be
incorporated into the full expert system.

2. MMIC DESIGN

At present, MMICs may be classed as small scale integrated (SSI) devices. It is
uncommon to find a MMIC with more than around twenty components. However the
trend is upwards and MMICs combining several circuit functions (eg oscillator, mixer,
amplifier etc) are appearing on the market. The design of such multifunction circuits
may be considered largely technology independent and yields to conventional
hierarchical problem decomposition techniques. We concern ourselves here only with
single function circuits at that point in the design cycle where fabrication effects must
be taken into account.

The MMIC design process may be represented as a four level subtask hierarchy:

(a)   Specification;
(b)   Trial design with ideal elements;
(c)   Design with 'real' elements;
(d)   Layout.

Tasks (b) and (c) define our domain of interest. Specification remains a unique
function because of the need for human interaction at this point [2]. Layout poses its
own problems in a different state space from (b) and (c) and remains, at present,
something of a black art.

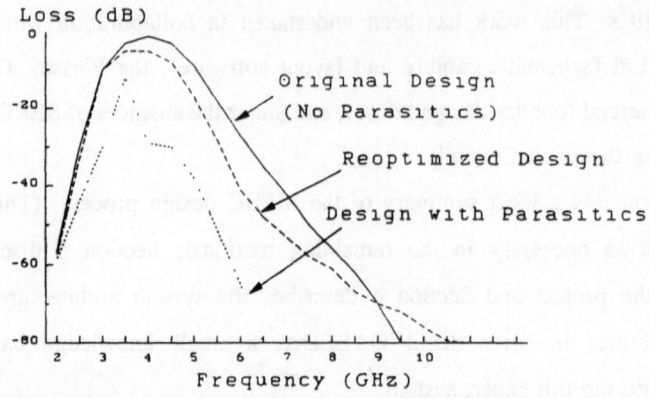

Fig.1 *Effect of parasitics on MMIC filter response.*

In stage (b) trial designs using ideal elements are simulated, optimised and finally analyzed to ascertain the feasibility of them forming the basis of the final circuit. Stage (c) may be considered a process of refinement in which unwanted parasitic effects due to the fabrication process must be accounted for to achieve a more realistic simulation of the final circuit performance. At microwave frequencies, greater than 1E9 Hz (used for satellite signal transmission), stray capacitance, inductance, and the finite resistance of connecting leads can affect circuit performance considerably; these strays are termed parasitic effects. Fig.1 gives an example of how the response of a MMIC filter is affected by parasitics [3]. The performance degradation caused by parasitic effects is not generally recoverable through optimization. This means that an engineer involved in stage (b) of the design process must attempt to predict in some way the loss of performance caused by stage (c). This is done mainly through a blend of heuristic rules and experience.

A circuit element, the inductor, may be realised in a MMIC in the form of a rectangular spiral metal track (Fig.2). The inductor is physically built up as a multilayer insulator-conductor sandwich. The inclusion of

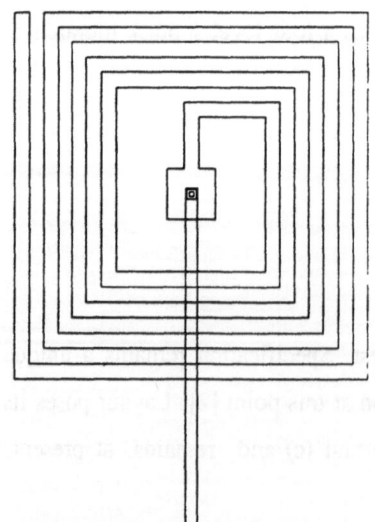

Fig.2 *MMIC Spiral Inductor.*

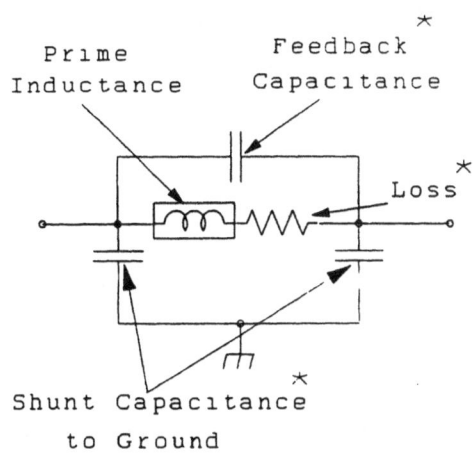

Prime
Inductance

Feedback
Capacitance $\times$

Loss $\times$

Shunt Capacitance $\times$
to Ground

Fig.3 *Spiral Inductor Electrical Model.*
(* = Parasitics)

parasitic effects leads to the electrical model for this element shown in Fig.3.

To allow this step (c) in the design process to be carried out, commercial foundries (integrated circuit manufacturers), supply detailed information to the designer in the form of a Foundry Guide, which contains the rules and equations for the design of viable structures for a given prime component value. This foundry guide provides the knowledge base for the system and can amount to several hundred rules.

During the design process the MMIC engineer will employ several stand-alone industrial standard CAD packages (eg for simulation, for layout) each of which will use a separate method of representation for the design (electrical behaviour for simulation, polygons for layout). Input to simulation programs invariably consists of writing circuit files describing electrical connectivity (netlists) in the specific command language of the simulator. Layout software packages, also, utilize their own command languages for the creation of complex geometrical structures. Automatic links between these varieties of CAD tools (with a few notable exceptions) do not at the moment exist; communication requires human intervention, a process both time consuming and wearying. Fig.4 gives an example of a simulator circuit file, describing in this case the spiral inductor electrical model (Fig.3).

3. SYSTEM OBJECTIVES

Before the design of any computer program can commence, a clear definition of the requirements is necessary (as with any design task). In this case the questions of paramount importance are:

3.1   Which areas in the design process will benefit from additional software support?

3.2   How much, and what kind of support can be provided?

To develop question 3.1 further we quote from Begg: "..an ideal CAD system should provide:

1   automatic generation of design data compatible between design tools;

2   high level design languages and tools;

3   a supportive and adaptive human-computer interface;

4   access to information in the system, combined with security against accidental damage or loss of files;

5   a dedicated, customized system which provides good relevant information quickly.

```
' SPIRAL INDUCTOR CIRCUIT FILE

DIM
   FREQ    GHZ
   CAP     PF
   RES     OH
   IND     NH

CKT
   IND 1 2 L=.7
   CAP 1 3 C= 017
   RES 2 3 R=4 97
   CAP 1 0 C= 061
   CAP 3 0 C= 041
   DEF2P 1 3 SPIND

OUT
   SPIND DB[ S21] GR1
   SPIND DB[ S11] GR2

FREQ
   SWEEP 1 20 1

GRID
   RANGE 1 20 1

OPT
```

Fig.4 Example of a Simulator Circuit File

We wish to pursue a two-level approach to the MMIC design task, corresponding to stages (a) and (b) described above. Stage (b) can be thought of as a technology independent level of abstraction but as mentioned earlier, low level constraints must also be considered. Fig.5 represents this step in the design process as a 'generate and test' cycle. During the 'generate' process the topology of the circuit is designed using the traditional techniques of network synthesis. To 'test' the circuit a microwave simulator is used to produce a graphical (or tabular) display of the circuit response. This response is then compared with the specifications and the circuit modified (and retested) if they do

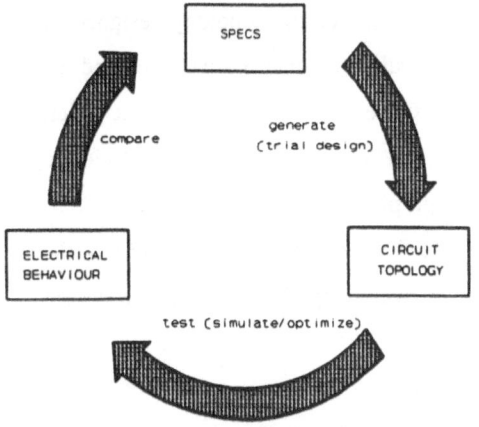

SPECS

generate
(trial design)

compare

ELECTRICAL
BEHAVIOUR

CIRCUIT
TOPOLOGY

test (simulate/optimize)

Fig.5 *First level 'generate and test' cycle.*

not match sufficiently. As already mentioned the engineer interfaces with the simulator by writing circuit files. This situation is far from ideal (consider points 1 and 3 above); engineers are much happier working with the schematic diagrams and symbols common to their profession, and the provision of a graphical interface here would carry significant benefits, not least in eliminating the human errors inherent in the manual creation of large netlists.

Many different types of circuit are today being implemented on MMICs. Each kind of circuit requires a different design method, and hence the partitioning of the 'generate' process above would require a different subtask hierarchy to be defined for each one. Therefore, at this point we do not contemplate creating a rule-based guide to the overall design strategy. The system accepts a previously determined circuit topology as its functional starting point, with one exception. The first operation to be performed in the design of almost all circuit functions is FET (Field Effect Transistor) selection. The transistor(s) is chosen from those available according to its device parameters and passive networks designed around it. This represents an application for a knowledge base to be incorporated as part of the main system.

We can now identify the following objectives for the first level of the system.

O1.1 Graphical (schematic entry) input of circuit description.

O1.2 Interactive (question and answer) interface to elicit additional relevant simulation/optimization information.

O1.3 Automatic generation of simulator/optimizer circuit files.

O1.4 Automatic translation between schematic diagram and optimized circuit files.

O1.5 Yield and performance predictions based on parasitic degradation and component sensitivity to technological variations.

O1.6 Expert advice on the selection of appropriate transistors.

The second level of the system takes as its input the circuit topology designed in stage (b). Here the geometrical structures of the circuit components are chosen and the parasitics calculated and added to the circuit file. Generally speaking, a circuit which fails to meet specification through parasitic degradation and/or undue sensitivity to process variations must be discarded and the design process begun again from scratch. This illustrates the importance of objective O1.5 above. Section 5 describes the addition of parasitics in more detail. We define the following objectives for the system second level as:

O2.1 Interactive addition of parasitic element models to the circuit file (with guidance, problem analysis, advice etc).

O2.2 Guidance in choice of geometrical structures for floorplan (standard library component cells, custom cell compilation).

O2.3 Full integration of Foundry Guide into the system.

4. SYSTEM ARCHITECTURE

Fig. 6 shows the architecture of the complete system. It consists of three main components; a controller, a generator, and a block of knowledge bases.

The controller has several functions: to keep track of the progress of a particular design; to call the knowledge bases pertaining to a specific design task into action when needed, and to control the operation of external CAD tools. The controller is a 'Meta' knowledge base. External (external CAD tools) and internal (lower level knowledge bases) control functions are seen as extended right-hand side parts of rules. Using an expert system as a controller in this way provides a means of separating conventional formalized algorithms from heuristics based on informal procedures.

The generator has the task of maintaining a communications link between the layout software and the simulator. It also performs the important function of formalizing the specifications elicited from the user at system start-up. This formalized specification then provides a reference for the entire system during a design. The generator can also however perform another useful task, albeit outside our original domain definition. That is, during the process of layout, it can continue to communicate between the

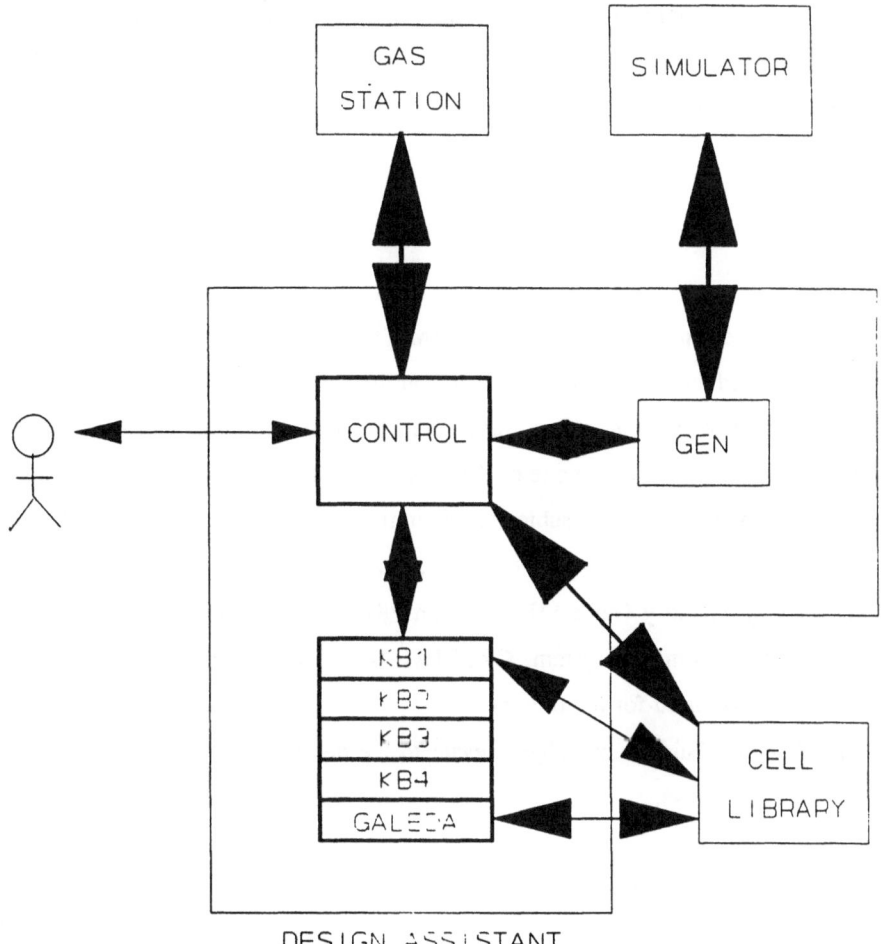

Fig.6 *Design Assistant System Architecture.*

external CAD tools thus providing a valuable guide to the further degradation caused by layout effects.

Each of the shown knowledge bases deals specifically with one of the objectives defined in the previous section. That is, KB1 deals with acquiring the circuit specifications, KB2 provides guidance to the selection of (satisficient) transistors, KB3 predicts yield and performance by heuristically estimating the effect of parasitic degradation. At the system second level, KB4 interacts with the designer to add parasitic elements to the circuit description file while KB5 (GALEDA) deals with the geometric construction of the circuit components. During a design, the engineer will interface directly with the knowledge base 'expert' at the design task he is performing.

By partitioning the domain of interest in this way, the tasks of knowledge acquisition and formalism may be suitably apportioned, easing the overall system design and narrowing the range of each rulebase.

5. GALEDA

Components used in the construction of MMICs include: resistors; metal-insulator-metal (MIM) and interdigital capacitors; ribbon, loop and spiral inductors; diodes; and metal semiconductor field effect transistors (MESFETS). The rectangular spiral inductor illustrated in Fig.2 is one of the most difficult of the passive components to design, and was chosen as the subject of the initial pilot scheme, called GALEDA for Gallium Arsenide Lumped Element Design Assistant. Design information was provided by the Plessey Foundry Guide [3]. Although PROLOG has been chosen as the host language for the complete system, CRYSTAL [4], a commercially available expert system shell, was used for the pilot system.

The Foundry Guide knowledge relevant to the design of spiral inductors can be categorised as follows;

Empirical and Theoretical.

Complex mathematical equations, tables and graphs.

Heuristic.

Informal recommendations on geometrical structure. An example is ".. the inside dimension of the inductor should be greater than eight times the track width, since winding the spiral any tighter results in little increase in spiral inductance, and that at the expense of increased parasitic capacitance and resistance".

Constraint.

Hard and fast limitations on geometrical dimensions.

Implicit.

Some geometries have less associated parasitic information available on them than others. This category sorts spiral structures according to the electrical information (or lack of it) available.

GALEDA uses forward chaining to achieve its goal. It first requests general electrical information (eg prime inductance value, current to be carried etc) which should be known at this stage in the design. This data is used to form an educated guess at a good geometry. A conventional iterative technique (seen as an extended consequent) refines and calculates the exact component dimensions, which are then compared with the Foundry constraints. Assuming the resultant structure is viable and is accepted by the engineer, it is saved in a format compatible with the layout software. If the engineer does not wish to accept the structure GALEDA will behave as an 'expert calculator' allowing a quick and easy investigation into other geometries.

The inherent inference method used in CRYSTAL is depth first search. The rulebase has been carefully structured so that geometries are considered in order, depending on how much parasitic information can be derived. The implicit information described above is thus implicit also in the rulebase. This is a disadvantage. Future versions of GALEDA will use a certainty function to allow this type of information to be expressed explicitly in the knowledge base, negating the need for procedural considerations. A further disadvantage associated with the CRYSTAL shell is the difficulty in separating facts from heuristic and mathematical procedures. In PROLOG, predicates are used to represent facts which can easily be kept separate from the rulebase, facilitating the future updating of the knowledge base as the fabrication process is improved.

The big advantage offered by GALEDA is speed. A design task which typically may take an experienced engineer several hours to perform is reduced to a matter of minutes. The use of "what if" analysis allows an easy investigation into the effect of other geometries on circuit performance. Also, any component designed using GALEDA will be entirely consistent with the foundry process. Thus the component may be used with confidence, and the final task of checking the circuit for design rule faults simplified.

6. ACKNOWLEDGEMENTS

Some of the work presented in this paper is based on ideas developed by Vivienne Begg in her book "Developing Expert CAD Systems". We would like to acknowledge our indebtedness.

7. REFERENCES

1. Simon, H.R. The structure of ill-structured problems.
   Artificial Intelligence 4, 145, 1973.
2. Begg, V. Developing Expert CAD Systems. London: Kogan Page, 1987.
3. Plessey Foundry Guide, Plessey Three-Five Group Ltd, Towcester.
4. CRYSTAL, Intelligent Environments Ltd, Richmond.

SWEEP: Social Welfare Entitlement Evaluation Program

H. Sorensen & O.M. Flynn

ABSTRACT

This paper describes SWEEP, an expert system developed at University College Cork for determining the entitlements and payment rates due to Social Welfare Applicants. Used by Social Welfare Officers, it is a rule-based system which applies the regulations to the case at hand, and includes extensive explanation facilities. The paper describes the architecture, rule representation and user interface of SWEEP.

1. INTRODUCTION

In this paper, we describe "SWEEP"[1], an expert system used to evaluate an applicant's entitlements under the Irish Social Welfare code. It has been designed and implemented for use by Social Welfare Officers, who would heretofore have evaluated any claims submitted using available regulation books. The package described is a rule-based system written in Prolog, and constitutes a working system. It has been tested and proved viable in practice, though certain enhancements would appear to be desirable. These enhancements are currently being implemented and are outlined in a later section of the paper.

In the next section, we describe the domain area of SWEEP, the Irish Social Welfare code. We outline the benefits and assistances available to individuals under this code, and the regulations which govern entitlement to same. We indicate, by means of a sample interview with an applicant, how a Social Welfare Officer applies these regulations in a particular case, thereby determining the benefits, if any, which are due to the applicant. Rather than describing the entire Social Welfare code, which encompasses a vast collection of regulations and legal requirements, we shall restrict our discussion to two representative sub-areas: *Unemployment Benefit* and *Deserted Wives Benefit*.

In the following section, we show how the regulations can be organised into a format and order which is more suited to integration into an efficient rule-based expert system. We also note that the Social Welfare code is not always *linear*, in that

---

1. Social Welfare Entitlement Evaluation Program.

qualitative decisions must sometimes be made by the officer concerned, based on knowledge of the case at hand - such decision-making techniques must be built into the resulting expert system. We outline the extensions required of the rule representation mechanism so that such judgemental assessments can be handled.

Following this description of the domain area, we outline the design and implementation of SWEEP. It will be shown how the Social Welfare regulations were encoded into a form suitable for integration as rules into a Prolog-based system, and how a search methodology was devised over these rules which corresponded to the method used by a Social Welfare Officer in determining satisfaction of the regulations pertinent to a given case. We shall also indicate how non-linearities (or uncertainties) were represented and implemented in the developed system. The user interface is also described here, demonstrating how a user (the SW Officer) converses with SWEEP, providing the information pertinent to the applicant's case, and how the system responds with a description of the benefits, if any, due to the applicant. We shall also indicate how the Officer (or applicant) might query SWEEP as to to the line of questioning which it is following and/or as to any conclusions at which it has arrived. A set of screen shots, charting the progress of a particular case, is used to highlight the user interface facilities.

Finally, we include a review section, whereby we describe the current use of SWEEP, and indicate its strengths and weaknesses. We describe enhancements which are currently being designed and implemented. These enhancements address the weaknesses perceived to exist in the current implementation of the system, and purport to make the system easier and more 'natural' to use by a Social Welfare applicant.

2. SOCIAL WELFARE SYSTEM

In recent years, a number of expert systems have been designed for use in legal / administrative domains [1, 2, 3]. The legal area has proved a fruitful application area primarily due to the fact that its regulations can intuitively be represented by rules, which can then be embedded in a rule-based expert system. Such rule-based systems can be implemented using available expert system tools, ranging from logic-based languages (i.e. Prolog [4]) to various expert system shells developed around the rule-based and logic-based paradigm (e.g. CRYSTAL, XI, etc). While the rule-based approach to legal reasoning has been justifiably called into question [5], it has nevertheless survived and proved successful within certain restricted domain areas. In this paper, we outline the design, implementation and use of one such expert system, SWEEP, which covers the interpretation and application of the regulations governing the Irish Social Welfare System (hereafter termed the Social Welfare *code*).

The Social Welfare code comprises a set of allowances and benefits (e.g. Unemployment Benefit, Disability Benefit, Deserted Wives Allowance, Free Medical Care, etc) which are granted to individuals and families who are deemed, through circumstances, to be in need of state assistance. An applicant (individual or family) may be deemed eligible for one/more benefits based on their satisfaction of a set of regulations and qualitative assessments. The entire system of allowances and benefits is administered at a national level by the Department of Social Welfare and at a local level by various district Health Boards. Details of the regulations and benefits are contained largely in two booklets published by these official bodies [6, 7]. Fig. 1 contains a synopsis of some of the regulations and benefits contained therein, formatted here for brevity and simplicity. From this, it should be noted that an individual may be entitled to several benefits based on their fulfilling the requirements for each benefit. In the ensuing discussion, rather than surveying the entire Social Welfare code, we shall limit our interest to two specific, but representative, areas: *Unemployment Benefit* and *Deserted Wives Benefit*.

A Social Welfare Officer decides on each application, determining the entitlements and rates of assistance due to the individual or family concerned. The decision process usually takes the form of an interview between the applicant and the Social Welfare Officer, with the latter gathering information from the former and using this to determine the entitlements due and the rates of assistance applicable. Fig. 2 summarises a typical such interview process.

3. SOCIAL WELFARE RULE REPRESENTATION

It is interesting to note that, in the dialogue of Fig. 2, the Officer's questions all depend on the previous reply received. This particular method of knowledge elicitation works very well, but is not sufficiently structured for direct incorporation into a computer-based package. Bearing this in mind, we would propose that a more structured method would be to follow a fixed line of questioning, the line to be taken depending on the previous answer received. This would mean that the Officer would take the initiative, asking the first question, with the applicant replying either "yes" or "no", or quantifying some information previously given (such as no. of children). Depending on the answer received, a subsequent line of questioning may be determined. This method is more tedious to follow in practice, but leads to equally accurate results. SWEEP is therefore designed on the premise that an alternative, but functionally equivalent, interview mechanism is possible, with a limited range of responses permitted on the part of the applicant. This alternative mechanism permits the efficient representation and application of the regulations of the Social Welfare code, the expert system developed emulating the rule search and application methods of the (SW) Officer in the above interview process.

IF YOU ARE:	YOU MAY BE ENTITLED TO:
Unemployed:	Unemployment Benefit and Pay-Related Benefit (if you are an insured person).
	Unemployment Assistance (if you are not entitled to Benefit).
Injured at Work:	Injury Benefit
	Pay-Related Benefit
	Disablement Benefit
	Medical Care
Suffering from a Prescribed Occupational Disease:	Injury Benefit
	Pay-Related Benefit
	Disablement Benefit
	Medical Care
Sick and Unable to Work:	Disability Benefit
	Pay-Related Benefit
	Invalidity Pension (permanently incapable of work)
	If so, you may also be entitled to:
	free travel
	free electricity
	free t.v. licence
	free telephone rental
	free bottled gas
	prescribed relative allowance
	fuel schemes
	living alone allowance
A Widow:	Widow's Pension

Fig. 1 - Benefits and Regulations

Based on this approach, we have adopted a particular knowledge representation mechanism for the Social Welfare regulations. Similar in style to the RUBICS system [8], we have adopted a two-level information representation mechanism - and, correspondingly, a two-level inferencing mechanism is employed in SWEEP. At an upper level (corresponding to the earlier questioning during an interview), we have organised the regulations into a binary tree structure. Each tree node represents a regulation to be tested, and the branches emanating from this node correspond to the applicant

> Applicant: I was made redundant yesterday, and am now unemployed. I have been working for the last four years, and I think my pay-related contributions are satisfactory. I have registered with the necessary job-finding agency, and want to know what benefits I am entitled to until I find another job. Also, how can I claim for these benefits?
> Officer: As you are insured you will probably be entitled to Unemployment Benefit. I presume you are under sixty-six years of age. How many contributions did you make in the last year?
> Applicant: Fifty.
> Officer: That means that you will qualify for the highest basic rate of Unemployment Benefit, which is £36.75 per week. Are you married?
> Applicant: Yes, with three children.
> Officer: You will then have an additional dependants allowance of £55.20 a week. Have you claimed Unemployment Benefit or Disability in the last thirteen weeks?
> Applicant: No.
> Officer: That means that payment will be made from the fourth day of unemployment. You will also be entitled to Pay-Related Benefit. You will be able to claim these benefits for as long as 390 days, after which you will have exhausted your benefit and will be entitled to Unemployment Assistance. Following is how you may claim your current benefit entitlements: (etc.)

Fig. 2 - Interview

satisfying or not satisfying the regulation (i.e. responding with a "yes" or "no" to the question which tests for regulation compliance). A path is thus traced through the regulation tree during an interview, based on the responses of the applicant to the questioning of the system. The leaf node of a particular path, if reached, indicates the entitlements of the applicant. The advantage of this binary tree representation mechanism is that of computational efficiency: first, a fixed line of questioning can be followed, leading to a determination of entitlements; second, a considerable degree of pruning is possible, as rules need be tested only along the current path being followed (this corresponds to the pruning applied by a experienced SW Officer, who would not, for example, consider the Unemployment Benefit regulations in the case of an applicant known to be working); and, third, the binary tree structure is extremely suitable for implementation within a logic programming language. Fig. 3 displays a section of the regulation representation mechanism relating to Unemployment Benefit entitlement, where a "yes" answer at each node causes the left-hand branch to be taken and a "no" answer results in following the right-hand branch.

On arrival at a leaf node of this binary tree, one/more entitlements are indicated as being payable to the applicant. The problem now is to compute the amount of assistance due for each entitlement. This is where the second level of knowledge representation is required. The amount due depends on such factors as spouse's income, other income, no. of children, etc. It is computed from formulae or from tables of values (e.g. for Unemployment Benefit, there is an increase in payment of £9.40 for

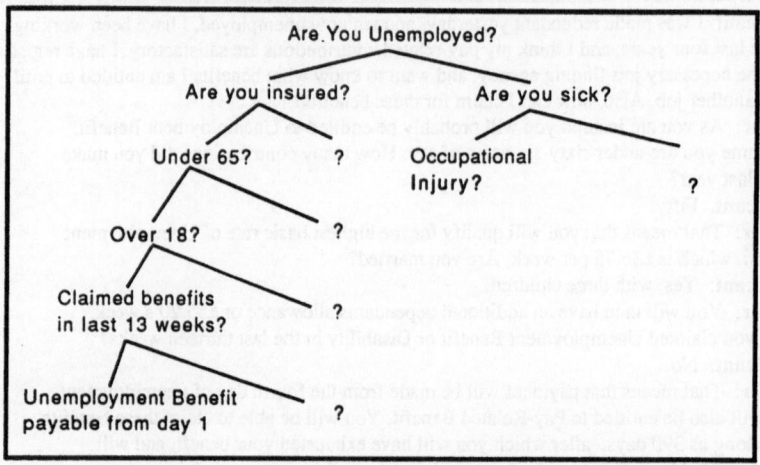

Fig. 3 - UB Regulations as a Binary Tree

the first child dependent, £10.50 for the second, £8.70 for the third to fifth, etc). Formulae can easily be computed. To determine which of a set of values (e.g. from a table) is pertinent in a particular case, we use a general tree structure, with root as entitlement (i.e. leaf node of binary tree) and branches for each condition relevant in computing payment due. By directing an applicant through this non-binary sub-tree, we may correctly compute the amount of each benefit. Fig. 4 includes an example of such a sub-tree used for computing Unemployment Benefit.

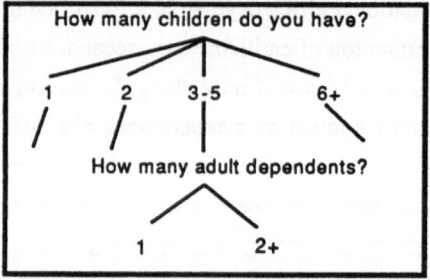

Fig. 4 - Computation of UB Payment (Non-Binary Sub-Tree)

In summary, therefore, we have arranged the regulations covering Social Welfare applications on two levels. Those governing *entitlements* are arranged into a binary tree structure, with satisfaction or non-satisfaction of a regulation determining the next regulation to be tested. Satisfaction of a set of regulations, corresponding to a path in the tree, yields the set of entitlements of the applicant. The second set of regulations covers *rates of payment*, and here we use a general tree structure, with branches for each option available to the applicant. Thus, in an overall context, an applicant will navigate through a tree structure, part binary and part general, through the responses which he/she provides. Arriving at a leaf node at the lower level results in computation of the amount of payment due, while failure to arrive at such a node at the upper level (binary) tree indicates non-entitlement to benefit.

There remains the problem of uncertainties. Not all regulations are deterministic in the sense that a yes/no response suffices - instead a qualitative assessment may be needed on the part of the (SW) Officer who is interviewing the applicant. This assessment may be made on the basis of information provided by the applicant, knowledge of the case at hand, information obtained through investigation of the case, etc. One such area where assessment is needed is provided by *Deserted Wives Benefit*, where a number of points of contention may arise. For example, two of the four criteria satisfying the 'desertion' claim are that there was an 'absence of consent to live apart' and that the husband had 'absence of just cause for leaving'. Similarly, once a wife is judged 'deserted', it must be decided that the husband is 'failing to provide regular and adequate maintenance' and that the wife has 'made adequate attempts to secure maintenance'. Clearly, all such conditions require some judgemental control on the part of the (SW) Officer, who must decide the case. It should also be noted that these conditions are not entirely independent, but accumulate to build up a picture of the applicant's situation.

Our (initial) solution to this was to build a probabilistic model of such decision-making processes. We can thus retain our binary tree decision-making strategy (such uncertainties affect entitlements rather than rates) but, rather than a yes/no answer, a probability is determined by the Officer as to whether or not a regulation has been satisfied. This probability in turn determines the next regulation to be tested, i.e. whether the left or right branch is taken. By compounding the probabilities at successive levels, a cumulative approach to decision-making can be effected. An advantage to this approach is that a uniform organisation of, and approach to, the regulations governing all entitlements is possible.

In the previous section, we have shown that the Irish Social Welfare code constitutes a system of regulations, benefits and payment rates. We have indicated that the application procedure is that of a (SW) Officer interviewing a claimant, applying the regulations in order to determine entitlement to benefit, and computing the amount of benefit based on payment rates. Here, we have suggested that a more structured approach to the interview session is possible through the ordering of the regulations and the payment rules as described. We have also suggested that the notion of qualitative assessment can be superimposed on the regulation structure through use of probabilistic methods. It should be noted that this entire approach might be taken irrespective of the computational process - the current administrative methods could be adapted accordingly. In the next section, we show that the real advantage of the new approach lies in the possibility for automation, and, in particular, to the integration of such a rule structure into an expert system.

Using the previously mentioned booklets [6, 7], and in consultation with Social Welfare Officers, we organised the regulations into a structure as described. It was agreed with the Officers that our new structure appeared to reflect the SW code as presently constituted, and that entitlements and rates as determined by our method coincided with general case studies. The only doubt on the part of the Officers related to the correctness of a probabilistic model for assessments. As it was agreed that this method provided a close approximation to the problem, we chose to proceed with this approach.

4. SWEEP

SWEEP is an expert system, written in Prolog, based on the rule structure outlined in the last section. We chose Prolog because its goal-directed, backward-chaining inferencing mechanism was well suited to the processing of the tree-structured rules. It also provided the flexibility and transportability which we found not to exist in the various shells available for expert system building. Essentially, the inferencing mechanism of Prolog allowed us to consider sets of benefits, at the leaf nodes of the binary tree segment, and to test for satisfaction of the conditions or regulations which would deem the applicant eligible for this set of benefits. We are, therefore, effecting a goal-directed search from the root of the tree towards the leaf node denoting benefit(s). For example, to test if an applicant is eligible for both Unemployment Benefit and Pay-Related Benefit, the rule of Fig. 5 would be applied. Each clause of the Prolog rule causes the condition to be tested through dialog with the user, who responds with a Yes/No answer. If each clause evaluates to true, then the applicant is deemed eligible for both of these benefits - a path has effectively been traced from the goal, through each condition, to the top of the tree. Note that, as Prolog clauses are evaluated from left-to-right (until some sub-clause proves false, in which case the evaluation can be terminated), the applicant will be asked if he/she is unemployed, insured, capable of work, etc. in that order.

Each clause of the form found in Fig. 5 causes a question to be asked of the applicant, whose response results in the satisfaction or otherwise of that clause. The response also causes a *fact* to be inserted into the Prolog database, denoting the applicant's position. By doing this, we can guarantee that the same question is never asked twice of the applicant, since the database of facts is consulted before issuing questions. Thus, for example, an applicant who is deemed ineligible for Unemployment Benefit and Pay-Related Benefit because of not satisfying the 'contribution conditions' may still be considered for Unemployment Benefit only, covered by the 'rule_is(5):-' Prolog rule. This latter rule, which is also predicated on the applicant being 'unemployed' and 'insured', need not query these conditions, as they are already asserted as facts into the database.

```
rule_is(4):-
        positive (unemployed),
        positive (insured),
        positive (capable_of_work),
        satisfies (contribution_conditions),
        not (disqualified (employment)),
        positive (under_65),
        positive (claimed_benefits_in_last_13_weeks).
```

Fig. 5 - Test for UB + P/RB

A similar rule-based strategy was adopted towards the non-binary segment of the database, representing the rates of payment. Also, by judicious ordering of the entitlements tested for, the efficiency of SWEEP has been improved - essentially, the more popular benefits (or groups of benefits) are tested for first, so that irrelevant questions are avoided to the greatest degree possible.

Uncertain reasoning methods have been implemented in expert systems using a number of different techniques, such as certainty factors [9], Bayesian probabilities [10] and fuzzy logic [11]. The first and second methods, being probabilistic in nature, would appear to be more applicable for legal / administrative assessments of the type being made here. As Prolog was the language of choice for implementation, however, the use of certainty factors appeared most appropriate. Using this technique, each rule to which there is not a definite Yes/No response is assigned a certainty factor between -1 and +1, with positive values suggesting satisfaction of the rule. The certainty factors for combinations of uncertain regulations can be compounded mathematically (see [12] for full details) to yield a 'likelihood' of a set of conditions being satisfied. A combined certainty factor of 0.2 or greater would suggest satisfaction of a set of conditions, while a value less than -0.2 would suggest certain rejection. Values in between would suggest a high degree of uncertainty as to whether or not the applicant had satisfied the regulations concerned. As presently constituted, SWEEP would compute the entitlement in this latter event, but would inform the applicant of the doubt associated with the outcome. The implementation of certainty factors to model qualitative assessments by a SW Officer merely involved the association of such a factor with each fact asserted into the database, and the combination of factors when dealing with sets of conditions. It integrated well into the general Prolog paradigm, and had little effect on efficiency.

From an architectural point of view, we may best describe SWEEP by the diagram of Fig. 6 - here we follow the methodology of [13] for describing such systems. We have effectively covered aspects of the system below the top level of the user interface. Knowledge, as denoted here, comprises *domain* or *procedural knowledge* [14], *control knowledge* and *declarative knowledge*. The domain or procedural knowledge relates to the goals (entitlements) and rules (regulations) of the Social Welfare code, and is stored as described above, and shown in Fig. 5. Control Knowledge generally determines the order in which goals and rules are tested - as we have utilised Prolog, which has an in-built ordering, control knowledge is also determined by the stored goals/rules (i.e. the domain knowledge acts as the control knowledge). Finally, the declarative knowledge corresponds to the facts associated with a particular applicant's case. These are asserted into the database as the applicant responds to queries, possibly with certainty factors attached. We have adopted a very simple structure for such facts - e.g. 'unemployed.', 'insured.', 'children(3).', etc.

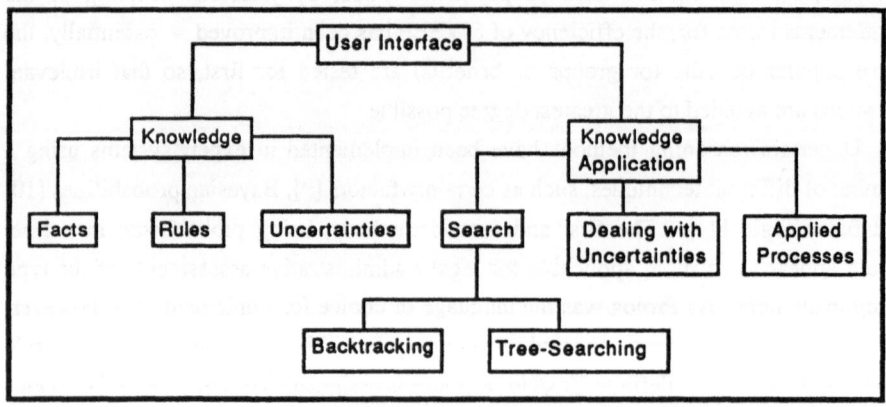

Fig. 6 - System Architecture

'Knowledge Application' in Fig. 6 refers to the use of the stored knowledge. The search is directed by the standard Prolog semantics, while we have explained above how we extended these semantics to deal with uncertainties. The 'Applied Processes' relates to the computational procedures which we have written to evaluate the payment due to an applicant, based on information provided. In this segment of the system, we simply used the standard algebraic operations inbuilt in Prolog.

We have adopted a straight-forward interactive interface, with optional windowing facilities. It essential operates in three different modes: *normal*, *why* and *how*. In *nor-*

mal mode, the user responds positively or negatively to queries posed by the system, which in turn stores facts in the database representing those responses. The *why mode* is entered by the user responding to a query with a '?', and denotes that the user wishes to know why the particular query was posed. For this reason, each question asked has an associated regulation or rule, with a fixed internal *rule-number*. This rule-number is, in turn, associated with a short textual description of the rule concerned. In the why mode, the textual description of the rule associated with the current question is displayed. Finally, the *how mode* allows a user to determine how the system arrived at the current question or conclusion. It effectively requires a trace of the facts asserted into the database which led to the question being asked or to the conclusion being drawn. To this end, each fact asserted in the database is assigned an incremental *trace-number*. When a user requires a trace (by entering a 't', or responding 'y' when asked at the end if a trace is required), a counter is used to sequence through the facts in trace-number order, and a very simple paraphrase of the fact is generated. We have implemented a simple procedure by which only relevant facts need be displayed.

We conclude this section with a number of screen shots of SWEEP in use for a particular case investigation. For clarity, we have suppressed all windowing - the entry of certainty values, shown below in bold face, is entered by the SW Officer though a callable windowing mechanism. We have provided both an online help facility (again using windows) and an offline guide for the determining of certainty factors.

```
┌──────────────────────────────────────────────────────────────────────┐
│                   Expert Aid on Welfare Benefits                       │
│                                                                        │
│ Are you physically separated?                                          │
│ ?                                                                      │
│ This can only be answered 'Yes' if you and your husband are not physically living │
│ together at the same abode. If you are, then a 'No' answer must be returned.      │
│                                                                        │
│ Are you physically separated?                                          │
│ y                                                                      │
│ Have you agreed to separate initially?                                 │
│ ?                                                                      │
│ If the couple agree to separate, then neither is in desertion. Although a wife may │
│ possess a certificate of separation, if they were already separated at the time of │
│ obtaining the certificate, this would not prevent her from qualifying for deserted │
│ wives benefit. If the initial separation was agreed by both parties, then answer  │
│ 'Yes' to this question; otherwise, answer 'No'.                        │
│                                                                        │
│ Have you agreed to separate initially?                                 │
│ n                                                                      │
└──────────────────────────────────────────────────────────────────────┘ 1
```

Expert Aid on Welfare Benefits

Were you left by your husband?
y
 Did you give him cause to leave?
?
We assume no major provocation from you.

Did you give him cause to leave?
n
    C = 0.41354

It is relatively uncertain as to the acceptability of the information provided. A
certainty value of 0.41354 has been determined; so we assume the facts to be
verified, although not very certain as to the truth of the hypothesis.

Are you receiving sufficient payment (regular or irregular)?
n

2

Expert Aid on Welfare Benefits

Have you made sufficient attempts to get maintenance?
y
Is your husband barred or wanted by the police?
?
If a man is forced to leave home by a barring order, then he did not 'leave home of
his own volition'. The same applies if the Police are looking for him. However, the
decision may be made that the husband effectively deserted the wife before she obtained the barring order.

Is your husband barred or wanted by the police?
n
Are you living with another man as husband-and-wife?
n

    C = 0.4069

3

Expert Aid on Welfare Benefits

It is relatively uncertain as to the accepatbility of the information provided. A certainty value of 0.4069 has been determined; so we assume the facts to be verified, although not very sure as to the truth of the hypothesis.

- - - - - - - - - - - - -
Final Conclusion
- - - - - - - - - - - - -

The reliability of our conclusion is relatively high. A certainty value of 0.65 out of 1 has been computed as to the likelihood of your qualifying for the benefit applied for

---

Are you married to an insured person?
y

4

Expert Aid on Welfare Benefits

Are you over 40?
y
You are a deserted wife, with insurance. You are also over 40, so you qualify for deserted wives benefit.
Deserted Wives Benefit cannot be claimed unless the desertion has lasted three months. Claims should therefore not be submitted until three months have elapsed. The necessary claim form is available from:

    Dept. of Social Welfare,
    Phibsboro Tower,
    Dublin 7.

Do you wish to compute the amount due?
y
How many children do you have?
t

5

```
┌─────────────────────────────────────────────────────────────────────┐
│                    Expert Aid on Welfare Benefits                     │
│                                                                       │
│  Trace:                                                               │
│                                                                       │
│  These are the facts which were verified during the interview:        │
│                                                                       │
│  You think you are a deserted wife.                                   │
│  You are physically separated                                         │
│  You have agreed to separate initially                               │
│  You were left by your husband                                        │
│  You gave him no cause to do so, with certainty 0.4135                │
│  You are not receiving sufficient payments                            │
│  You have made sufficient attempts to get maintenance                 │
│  Your husband is not barred or wanted by the Police                   │
│  You are not living with another man as man-and-wife, with certainty 0.4069 │
│  you are married to an insured person                                 │
│  You are over 40                                                      │
│                                                                       │
│  How many children do you have?                                       │
└─────────────────────────────────────────────────────────────────────┘
```
6

5. REVIEW

SWEEP has been constructed and implemented as described above on a MicroVax 3400, with 20Mb of memory using DEC10 standard Prolog. It has been tested on numerous cases by SW Officers, who generally agree that it conforms to their own decisions on these cases. There has been some (justifiable) debate on the efficacy of using certainty factors, or any probabilistic method, for handling qualitative assessments. The argument is not just about the outcome reached, but more so about the justification produced by SWEEP for that outcome. In fact, the outcome frequently corresponded to that of the Officer - particularly when we added a facility so that an Officer could manipulate certainty factors after each question as they were compounded mathematically. However, it it difficult to explain to an applicant that he/she has failed to qualify for some benefit for probabilistic reasons. Suppression of such explanations is not seen as a reasonable alternative. For this reason, the system has not yet been used by actual applicants for SW benefits.

One solution to this problem of uncertainties, which we are currently designing, is that advocated in [5], which argues that legal decisions are made on a *case* basis rather that on a purely logical basis. Under this scheme, if two applicant's cases are judged substantially similar, then the same outcome is likely in both. We shall thus record details of a number of representative case studies (called *prototype cases*), and we shall attempt to match an applicant's details with those of one such case. The outcome of the sample case shall then determine the outcome of the applicant's enquiry and a textual justification of the first will be parameterised to justify the second. We shall use a frame-based mechanism for representation of case details, with inheritance capabilities between frames. This approach, we feel, is more realistic when dealing with qualitative judgements. The knowledge base of SWEEP shall then comprise three components: *logical regulation* rules, *payment* rules, and *assessment* frames.

There is one further modification which we feel necessary. This relates to the mechanism for gathering information from the applicant, and the discrepancy between SWEEP and general interview techniques. During an interview with a SW Officer, an applicant generally (i) knows which benefits he/she is applying for, and (ii) can provide a substantial amount of information, without prompting, to justify their claim to benefit. We plan to accommodate this strategy within the system in two ways. To cater for these issues, we intend to extend SWEEP with a general-purpose volunteering facility, as advocated in [14]. This will require that the applicant begin with a form-filling option, whereby the form is used to volunteer information. The information volunteered will result in facts being asserted into the database, and these facts need not be further enquired into during the course of the questioning. Such a facility should, we feel, ease the frustration of users.

Finally, we have attempted to implement SWEEP on a Personal Computer, but have not succeeded due to memory constraints. We have managed, however, to modularise the rule base in such a way that all rules relating to a particular benefit, or set of benefits, is located in one module. A file is then created for each module and, provided an applicant knows which benefit(s) he/she is claiming (which is generally the case), the relevant rule module can be *consulted* on disk by the top-level Prolog code of SWEEP. In this way, by a mere addition of a menu-based front-end, through which a user specifies the benefits of interest, we have made the system available on PCs. By combining this technique with the form-based information volunteering facility mentioned above (and, using different forms for particular benefit(s)), we would hope that the system could become cheaply available and quite natural to use.

In conclusion, SWEEP has been constructed as outlined. It has been tested and, while the outcome has generally been favourable, certain problem areas have been identified. We are now attempting to address these problems. We might make one final point: SWEEP was initially conceived as a system which might be used by the applicants themselves, rather than by the SW Officers - this may be evident in the format of the questions asked. Because of the use of probabilistic reasoning methods (or any other method for handling uncertain data), however, we now feel that this objective is unattainable in its entirety from a legal point of view. SWEEP is, nevertheless, a useful tool for applicants who wish to determine their rights under the Social Welfare code and who subsequently wish to pursue these rights through the intermediary figure of a SW Officer.

REFERENCES

1. Kowalski, R.A. et al. The British Nationality Act as a Logic Program. Communications of the ACM, 29, 5, May 1986.

2. Capper, P. Latent Damage Law - The Expert System. Fifth International Expert Systems Conference, London, 6-8 June 1989, p.101-6.

3. Whitely, E., Doukidis, G. & Singh, A. An Expert System to Assist in Filing Tax Returns: The Case of Indian Income Tax. Fifth International Expert Systems Conference, London, 6-8 June 1989, p.115-29.

4. Clocksin, W.F. & Mellish, C.S. Programming in PROLOG. Springer-Verlag, 1984.

5. Leith, P. Fundamental Errors in Legal Logic Programming. The Computer Journal, 29, 6, 1986.

6. Summary of Social Insurance and Social Assistance Services. Published by the Irish Social Welfare Department.

7. Rates of Payments. Published by the Health Boards.

8. Moldona, D. & Wu, C.I.; A Hierarchical Knowledge Based System for Airplane Classification; IEEE Software Engineering, Dec. 1988.

9. Shortcliffe, G.H. Computer-Based Medical Consultation: MYCIN. American Elsevier, New York, 1976.

10. Shen, S. & de Mora, C. Expert System Implementation in the Bayes Theorem Approach and the Expert Edge. Library Software Review, 7, 4, Jul-Aug, 1988.

11. Zadeh, L. Commonsense Knowledge Representation Based on Fuzzy Logic. Computer, Oct. 1983.

12. Flynn, O.M. Expert Systems and their Application in the Area of Social Welfare. M.Sc. Thesis, Computer Science Dept., University College, Cork, Ireland, 1989.

13. Erman, L., Lack, J. & Hayes-Roth, F. ABE: An Environment for Engineering Intelligent Systems. IEEE Software Engineering, Dec. 1988.

14. Forsythe, R. Expert Systems: Principles and Case Studies. 1984.

ARCHITECTURAL ISSUES IN KNOWLEDGE-BASED SIGNAL PROCESSING

Reza Beic-Khorasani, Fabian C. Monds and Terry J. Anderson

ABSTRACT

In this paper the complex nature of Signal Processing (SP) is highlighted. Some basic architectural considerations for Knowledge-Based (KB) signal processing systems are discussed. A "multi-task" KB system called SPES (Signal Processing Expert System) is proposed to provide solutions for multiple SP task that employ common SP functions in their solution. Some features of the system to deal with multiplicity of tasks are discussed in detail.

1. INTRODUCTION

In the last few years, researchers in the field of Signal Processing (SP) have shown increasing interest in the concepts of Knowledge-Based (KB) systems (often known as expert systems). Signal processing KB systems have been used to deal with limitations of conventional SP functions such as the failure of some adaptive signal processing algorithms to achieve expected performance when model parameters vary with time [1]. For example, Knowledge-Based (KB) techniques have been successfully used for a variety of applications ranging from improving the performance of spectral analysis techniques [2,3] to biomedical image processing [4].

Although there appears to be a better understanding of the problems involved in merging SP with KB concepts, no coordinated effort has been made to provide architectural guidelines for the design of KB signal processing systems mainly due to the apparent lack of interaction between knowledge engineers and SP researchers. This is, of course, not surprising when one considers the distinct processes involved in the two disciplines. While SP involves the application of numeric processes, KB concepts deal mainly with symbol manipulations, knowledge representation, logic and inference.

Signal processing tasks usually involve the application of different processes in a noisy environment. The SP domain, therefore, tends to be complex and ill-structured. Although the complex nature of SP applications calls for more sophisticated inference and knowledge representation techniques, the overall organisation of the system is extremely important. The success of a KB system is not only dependent upon the quality of its knowledge, but also on how this knowledge is organised and used. Although the complexity of the application task and the type of knowledge available determine the most appropriate overall architecture, there is sufficient commonality in most SP applications to establish a set of architectural requirements which may be helpful in the design of more effective and efficient KB systems for SP applications.

In this paper, the complex nature of the SP domain is briefly highlighted, and some architectural considerations are discussed. The benefits of using the blackboard architecture for SP applications are also outlined. An overview of a KB system, which is currently under development, called SPES (Signal Processing Expert System) is given, and the implementation of a prototype system is also discussed.

2. CHARACTERISTICS OF THE SIGNAL PROCESSING DOMAIN

In the following sections, important characteristics of the SP domain are briefly discussed in order to discern some basic architectural requirements for SP applications.

2.1 Types of Error

There are essentially two types of error in this area. Firstly, an important and common characteristic of this domain is error due to noise and distortion in signal data which could cause uncertainty in the results. Variability in signal data due to time varying parameters could introduce errors in the results too. Also there may be missing data arising from faulty components for example, which could cause errors. As a result, there can not be a fixed set of SP algorithms for all problems particularly in adaptive SP systems. For example, in the case of adaptive filters the response of the filter is adaptively adjusted according to the input signal. Secondly, error can result when the available algorithm is bounded by factors beyond its control such as gaps in mathematical theories particularly when the signal is unknown. The complexity of the SP domain is partly due to noise, variability in signal data, and incomplete knowledge or algorithms.

2.2 Hierarchy of Processes

Signal processing operations can usually be divided into three main stages; namely preprocessing, signal processing and postprocessing [5]. In the preprocessing stage, signal data is usually transformed into a form suitable for the application of the main SP function. Postprocessing usually deals with estimation of parameters such as estimating the spectrum of the signal for example.

SP operations can be viewed as a hierarchy of sub-processes, partitioning the problem into various levels of abstraction. Raw data represents the lowest level in the hierarchy while subsequent preprocessing and SP processes represent intermediate stages of operation. This sequence of processes could produce the overall solution in the highest level of the hierarchy.

2.3 Numerical Algorithmic Solution

SP algorithms are based on mathematical models, and the conventional SP solution can be considered as a *low level* algorithmic procedure. For every stage of the processing, there may be a number of available solutions, and the expert is faced with the difficult task of selecting the most suitable solution. Typically the processes involved in the overall solution deal with one aspect of the problem, and are usually independent of each other.

3. ARCHITECTURAL REQUIREMENTS OF A KNOWLEDGE-BASED SYSTEM FOR SIGNAL PROCESSING APPLICATIONS

The formulation of the architectural requirements of an application and their relative importance is critical for comparing and selecting design alternatives. Although the specific requirements are primarily dependent on the problem and could involve many issues, it is felt that the following set of basic requirements, based on SP domain characteristics, is particularly essential in the design of KB system architecture for SP applications.

3.1 Multiple Lines Of Reasoning

As highlighted earlier, the SP domain is characterised by error due to noise, variability in data and incomplete mathematical theories. Multiple lines of reasoning [6], could be used to reduce the effect of uncertainty by selecting the most appropriate course of action. This means that multiple arguments would have to be developed to support the selection of potential SP functions. At each stage of the processing all plausible conclusions leading to the selection of an optimum SP function are considered for the solution. Uncertainty could then be reduced by evaluating the potential of every alternative towards the incremental development of the solution. This evaluation is dependent on the degree of confidence that each conclusion receives from current available data and the knowledge used, and also on the mutually supporting conclusions.

3.2 Multiple Sources Of Knowledge

As explained, major SP applications involve the use of a number of fairly independent processes at various levels of analysis, each dealing with a particular aspect of the overall problem. This implies that multiple knowledge representation and inference techniques may have to be adopted in complex situations. Because it is quite inefficient and impractical simply to place all knowledge in a single large knowledge base, knowledge about different aspects of the problem and control strategy can be partitioned between separate and independent knowledge sources (KS). These KSs are responsible for generation and evaluation of conclusions. The KSs would have to be organised and invoked hierarchically under the control of a higher level control KS - a "meta" control KS - similar to the way in which SP functions are operated conventionally. The structure of KSs in this way can clearly increase the scope of the system as each KS could be equipped with its own problem solving method. Cooperation among various KSs (i.e cooperation among various problem solving methods) can be used to solve complex SP problems that may be too difficult for a single KS to handle. Also this modular structure permits knowledge to be represented using different knowledge representation techniques such as production rules and frame-based methods [7]. A communication medium would have to be provided for the KSs to inform each other of their conclusions so that an overall solution may be developed.

In addition, real-time capability is another important requirement of many signal processing systems [8]. This means that KSs should be organised hierarchically and be capable of performing inference concurrently.

3.3 Transformation of Data

As we have seen, SP systems are essentially numeric systems. The transformation of numeric SP data into its symbolic form is central to the successful operation of KB systems for SP [9]. This is because SP functions are mathematical programs which carry out *low level* numeric manipulations, and are devoid of any SP problem solving expertise. KB systems, on the other hand, perform *high level* symbolic manipulations based on logic and on heuristic knowledge embedded in the system. As the overall knowledge is hybrid, information flow should be *bi-directional* such that control information can be passed from symbolic processes to numeric processes. In this way, the main benefits of coupling numeric and symbolic computing techniques are [10] :

1. Improvements to the quality of numerically computed solutions.
2. Savings in the computational resources required by numeric processes.

More effective KB signal processing can, therefore, be achieved if numeric data is transformed into symbolic information before being passed to the KB system.

4. ARCHITECTURAL APPROACHES FOR SIGNAL PROCESSING TASKS

The architectural requirements as specified above provide basic guidelines for the design of KB systems for SP applications. Currently, two broad classes of architecture are used for building KB signal processing systems.

Firstly, in the simplest architecture, knowledge inferred by the reasoning system is added to a database. Examples of this type of system include; a system for the removal of ocular artifacts from Electroencephalogram signals [11], a system for improving the performance of adaptive SP algorithms [12], and a system for automatic patient realignment [4]. This structure does not accommodate the architectural requirements as outlined in the previous section since knowledge is encoded in a single Knowledge Base. Also, no provision seems to have been made for signal to symbol transformation as signal data is provided by a preprocessing system or through user interaction. While single knowledge base architectures may be useful in some small applications, the size and complexity of the some SP tasks requires a more sophisticated approach.

Secondly, a more robust approach has been to use the blackboard method [13]. The blackboard methodology provides the basic framework for the development of KB systems for signal processing because :

1. Large amounts of diverse expertise can be organised hierarchically in independent KSs each dedicated to a particular aspect of the problem, where these KSs communicate through a global database called the blackboard under the control of a control KS to achieve cooperative problem solving behaviour.

2. It supports incremental formation of solutions and opportunistic problem solving behaviour which are combined to resolve uncertainty and control combinatorics.

This type of architecture has in fact been used for knowledge-based detection, segmentation and classification of signals [14], for improving spectral estimation [2], and in the design of a computer vision system [15]. In addition, as reviewed by [16], it can be used to carry out signal to symbol transformation since individual symbolic and numeric processes can be incorporated as separate multilevel KSs. Also it provides the basic architecture for the development of real-time KB systems for signal processing and interpretation tasks [8].

5. AN OVERVIEW OF A KB SYSTEM FOR SIGNAL PROCESSING : SPES

It is clear from the above that current KB systems for signal processing applications are designed to deal with a single application area. Although the architecture used in these systems may be useful in other SP application areas, the systems are primarily dedicated and are designed for a particular signal processing task.

As some SP functions are commonly used in various signal processing tasks, it is believed that the potential of KB systems could be extended to develop a system capable of tackling a number of similar SP tasks within the same application area such as image processing. While traditional blackboard-based systems have been designed to deal with a single problem area [17], the system to be presented here builds on some of the foundational concepts of the blackboard architecture in order to provide a multi-task KB system for signal processing. "Multi-task" is used here to describe a system that aims to provide problem solving capability for a limited number of similar SP tasks. As will be explained later, a multi-task system is effectively a dedicated system when in operation, but provides the flexibility to be used for different tasks within the same application area.

5.1 Basic Architecture of SPES

Broadly speaking, there are three main types of knowledge in this design. Firstly, *task-specific knowledge* which is knowledge about the problem solving procedure that a particular task employs to provide focus of control; secondly, *signal processing control knowledge* which is control knowledge about a particular aspect of SP problem, and finally, *signal processing knowledge* which is used to solve a SP task.

The basic structure of the system and the way in which KSs are organised in SPES is illustrated in figure 1. For simplicity, the diagram shows the situation where 'task1' has been selected for SP operations.

As can be seen the system is composed of four main components as follows :

1. Task Knowledge Sources
2. Signal Processing Control Knowledge Sources
3. Signal Processing Knowledge Sources
4. The Blackboard

In the following sections, the roles of these KSs and their relation with the blackboard is outlined.

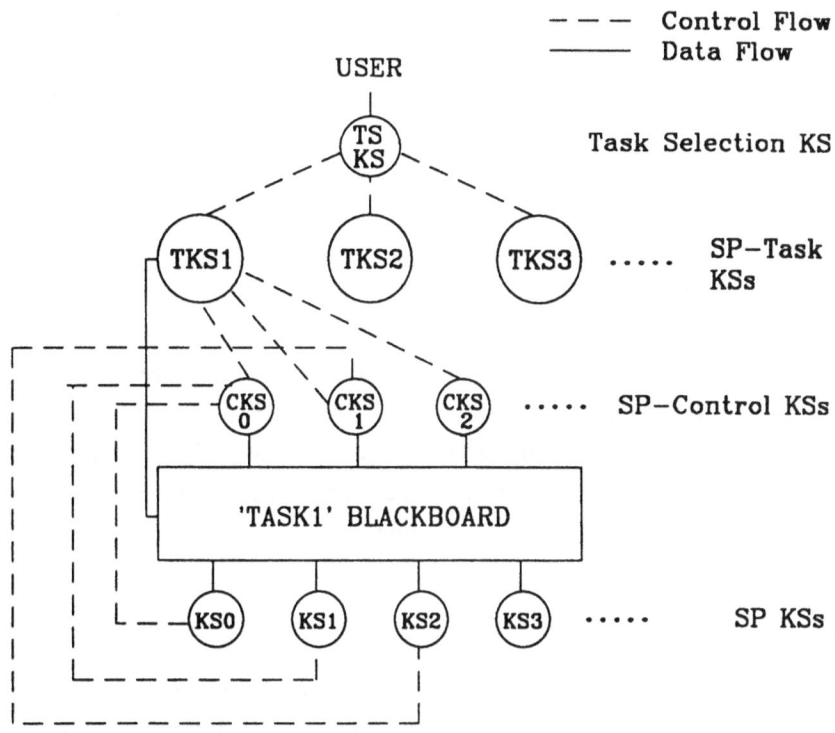

Figure 1. The basic structure of SPES (with 'task1' activated).

5.1.1 Control Strategy

It has long been recognised that the success of a KB expert system depends both on the quality of KSs and on the appropriate use of these KSs, and SPES in no exception. Unlike traditional blackboard based systems, however, SPES is faced with the difficult task of coping with distinct and multiple task-specific problem solving procedures. The control strategy is therefore very important in this application.

Although each SP task adopts a distinct solution procedure and, accordingly, has its own distinct focus of control mechanism, some signal processing control knowledge is commonly shared between the tasks as similar SP functions may be employed by the tasks in their solution.

Traditional blackboard-based systems such as the HASP/SIAP system [9], use a *single* strategy KS for scheduling tasks. This approach is clearly inappropriate for a multi-task system since it aims to provide solutions for a variety of tasks. It is believed that *multiple* scheduling KSs would have to be employed in SPES in order to deal with multiplicity of SP tasks. Thus task-specific control knowledge is to be encoded separately to provide focus of control for the task. On the other hand, we believe that effective problem solving could be achieved in SPES by adopting multiple problem solving methods and *multi-level control strategy* in hierarchically organised control KSs as will be explained later. In this way, the architecture of SPES provides a convenient framework for solving multiple SP problems. The complex problem of providing multiple control strategies and the way in which various SP tasks share common control and signal processing knowledge is described below.

5.1.1.1 Task Selection Knowledge Source (TSKS)

As SPES is to provide solutions for a number of SP tasks, it needs to establish first of all which particular task is required. TSKS is a simple *user-interactive* KS that queries the user about the selection of a SP task. Upon the selection of a task, TSKS performs a simple control operation by passing system control to the selected task KS (see figure 1), and the solution procedure commences.

5.1.1.2 Task Knowledge Sources (TKS)

As each SP task adopts a distinct and separate approach in its problem solving, it is distinct and independent of all others. Each TKS is dedicated to a particular task, and contains *task-specific* control and problem solving strategy which enables it to proceed with solving the problem. A TSK is under the control of TSKS (see figure 1), until it is activated by the user whereupon it takes over system control by providing problem solving policy for the selected SP task. It is a top level control meta-KS which provides scheduling capability for the activated task. In brief, an activated TKS performs the following functions :

1. Initialise activated 'task' blackboard to record precalculated numeric and symbolic data on the blackboard.
2. Decide on an appropriate course of action to provide focus of control for the SP task by selecting the most suitable action, based on current available information on the blackboard.
3. Invoke a suitable lower level control KS or a signal processing KS.

With the selection of a SP task by the user, overall system control is passed to the activated TKS. In this way, SPES effectively becomes a dedicated KB system to provide solution for the selected task. It is, however, fundamentally different from traditional KB systems as it represents a framework to develop solutions for other similar tasks, albeit at separate times (see figure 1).

5.1.1.3 Signal Processing Control Knowledge Sources (CKS)

The activated TKS is in control of a group of lower level SP control KSs (CKS). The CKSs control the operation of signal processing KSs according to the current policy of the TKS. Each CKS is itself a meta-KS which embodies control knowledge about a particular aspect of the signal processing problem. It is this knowledge which enables a CKS to invoke an appropriate signal processing KS. CKSs are *multi-purpose* control KSs providing control knowledge for all signal processing KSs. It follows that they could potentially be of use to more than one SP task.

5.1.2 Signal Processing Knowledge Sources (KSs)

Signal processing knowledge is to be embodied in a number of KSs, which are under the control of one or more CKS (see figure 1). Signal processing KSs contain expertise and heuristic knowledge about a particular aspect of the SP problem. Signal processing KSs usually have no direct knowledge of each other and communicate through the blackboard, under the guidance of a control KS. These KSs are divided into two categories; firstly, *multi-purpose* signal processing KSs are *task-independent* and could have potential application in more than one SP area. On the other hand, *task-exclusive* signal processing KSs are applicable to a single SP task only.

In order to cope with multiplicity of knowledge and data, SPES adopts multiple knowledge representation and inferencing techniques for more effective performance. Although there are no criteria for the selection of a particular technique, the choice depend, to a large extent, on the nature of KSs and on the power of the analysis model available. In SPES, the selection of an appropriate problem solving strategy is mainly dependent on Signal to Noise ratio (S/N). For example, problems which have inherently low S/N ratio are better suited to solutions by model-driven programs [9], where solutions are inferred from knowledge about the domain, and primary data are used solely to verify the results. On the other hand, a data-driven approach is used in situations with high S/N ratio where the solution is inferred from input data.

As the task is essentially an 'algorithm selection' task, knowledge will be represented using sets of production rules. The modularity of the system, however, allows knowledge to be represented using other representation techniques such as frame-based representations [7]. This should not cause any problem in the overall operation of the system because knowledge is embedded in separate KSs, and this allows a good degree of flexibility.

5.1.3 Hierarchy of Knowledge Sources

With the activation of a TKS, the structure of KSs could be considered hierarchical as all subsequent 'task' related KSs are organised hierarchically (see figure 1). On the lowest level of the hierarchy is a group of signal processing KSs whose main task is to place their inferences on the blackboard. At the next level are CKSs which have control knowledge about a particular aspect of the SP task. At the highest level is the activated task KS, or the strategy KS, which decides the next course of action. In the SPES architecture, therefore, TSKS is simply a meta-KS for TKSs which activates an appropriate TKS based on user's response. It is this multi-level control structure which enables the system to activate a particular task, and hence to provide solutions for multiple SP tasks.

6. STRUCTURE OF THE BLACKBOARD

As was explained, we are essentially faced with two types of data in this problem; numeric and symbolic. The transformation of numeric signal data into symbolic information is quite important in KB systems for signal processing. The blackboard can be conceptually divided into symbolic and numeric segments as shown in figure 2. The symbolic segment is used for representation of partial static and dynamic solutions, and for representation of pending activities, and it is the framework through which signal processing KSs inform each other of their findings

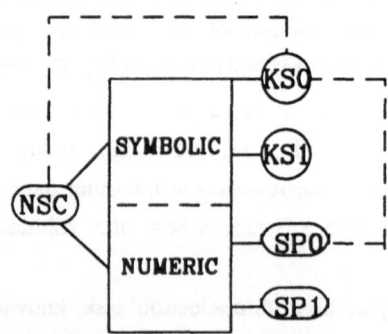

Figure 2. Integration of symbolic and numeric information in SPES.

Depending on the complexity of the task, the symbolic segment may have to be divided into a number of independent information and analysis levels, each representing intermediate stages of the solution. The numeric segment, on the other hand, is a global numerical data storage medium where numeric data may be stored in the form of data files for example.

6.1 Signal to Symbol Transformation

In SPES, cooperative numeric and symbolic processing is combined in order to achieve more effective KB signal processing problem solving. Systems that couple numeric and symbolic processes could be considered as coupled systems [16]. Accordingly SPES is a coupled system which aims to utilise numeric processes for improving the quality of symbolic solutions. As shown in figure 2, KSs have knowledge about SP functions that they may use in order to select an optimum algorithm from several candidate SP algorithms. The KSs, therefore, perform high level processing on global symbolic data in the blackboard in order to extract information that can not otherwise be obtained. Unlike traditional systems, however, SPES is to provide solutions for multiple SP tasks. This means that all numeric functions would have to be encoded separately so that they may be used by any activated SP task if required, similar to the way in which signal processing KSs are organised. Signal Processors (SP) are a series of programs that perform low level signal processing providing numeric information under the *control* of a KS over a bi-directional communication medium. Alternatively a signal processing package could be used to provide the necessary SP algorithms. Numeric to Symbolic Converter (NSC) is a program module that is composed of a number of programs which attempt to convert numeric data into its symbolic form. The integration of numeric and symbolic data in this way should enhance the performance of the system.

One of the main benefits of the SPES architecture is that multi-purpose signal processing KSs and numeric processes could be commonly used by a variety of tasks. Computing costs could be reduced if basic SP software is readily available as only task-exclusive functions, control and domain knowledge are needed to build the system.

7. IMPLEMENTATION OF A PROTOTYPE SYSTEM

At present, two SP tasks are being used for the implementation of a prototype system of SPES. These are 'interpulse period estimation' using Electrocardiograph (ECG) signals, and 'range estimation' using ultrasonic signals. As shown in figure 3, ECGSIG and USCSIG are the two task KSs. PREP, SIGP and POSP are preprocessing, signal processing and postprocessing control KSs respectively. RECN and MFLT are signal processing KSs which carry out preprocessing and signal detection tasks. TRSH and ESTR carry out thresholding and range estimation.

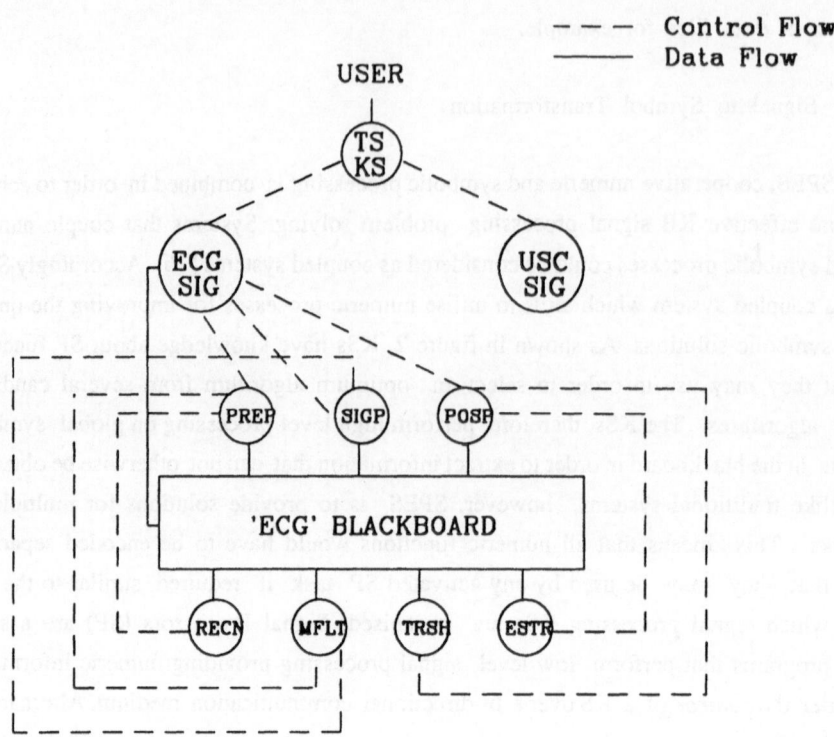

Figure 3. The structure of the prototype system.

In the implementation of the prototype, the Edinburgh blackboard shell [18], was initially considered, but it was observed to be very slow for SP purposes. The prototype system is being implemented in a traditional programming language - (the C language) - as it provides us with flexibility in building KSs and in the construction of the blackboard. The C language provides run-time efficiency and is a very effective language for implementing numerical algorithms. It is a suitable tool from a signal processing perspective because SP operations will be carried out faster. On the other hand, system maintenance and explanation generation for the system's reasoning is quite difficult in C particularly since inference processes could span several levels of analysis.

8. CONCLUSION

In this paper, the complex nature of signal processing involving multiple levels of processing is discussed. In particular the way in which the solution procedure is divided into distinct problem stages is outlined. Basic architectural requirements of a KB system for various SP application areas such as image processing and radar are considered.

These are :

1. Multiple lines of reasoning to develop multiple arguments to support potential conclusions.

2. Multiple sources of knowledge to encode information about different aspects of the overall SP problem, and to adopt multiple knowledge representation and inference techniques.

3. Transformation of numeric signal data into symbolic information for more effective performance.

The blackboard architecture is considered to be a suitable basis for the development of signal processing KB systems because it addresses the basic architectural requirements as outlined above, and provides a convenient framework to carry out signal to symbol transformation.

An overview of a multi-task KB system for signal processing has been presented. The system represents a framework which aims to provide solutions for multiple SP tasks that use similar processes in their problem solving. The SPES system is effectively a dedicated system when in operation, but provides the flexibility to be used for different SP tasks. The proposed architecture could have potential application in various SP problem areas such as image and speech processing. Also, SPES could prove valuable in other domains such as statistics, where solution processes are predominantly algorithmic.

REFERENCES

1. Hu Y. H. & Abdullah A. H. Knowledge-Based Adaptive Signal Processing. Proc. IEEE 1987; 1875-1878.

2. Daku B. L. F., Grant P. M. & Cowan C. F. N. et al. Intelligent Techniques for Spectral Estimation. J. of IERE 1988; 12:275-283

3. Li X., Morizet-Mahoudeax P. & Trogano P. et al. A Spectral Analysis Expert-System. IASTED Int. Symp 1985; in Sig Proc & Filter; 5-8

4. Young H. S., Dalton B. & Saeed N. et al. Expert System for Patient Realignment in MRI. Proc. IEE Colloquium on "The Application of Artificial Intelligence Techniques to Signal Processing" 1988; 8/1-8/3

5. Oppenheim A. Applications of Digital Signal Processing. Prentice-Hall Inc 1978.

6. Hayes-Roth F., Waterman D. A., & Lenat D. B. Building Expert Systems. Addison-Wesley Publishing Company 1983

7. Walters J. R. & Nielsen N. R. Crafting Knowledge-Based Systems. John Wiley & Sons 1989.

8.   Dai H., Anderson T. J. & Monds F. C. A Parallel Expert System for Real Time Applications. Cambridge Univ. Press, "Research and Development in Expert System VI", Edited by Shadbolt, N. 1988; 220-234

9.   Nii H. P., Feigenbaum E. A. & Anton J. J. et al. Signal-to-Symbol Transformation: HASP/SIAP Case Study. The AI Magazine, Spring 1982; 23-35

10.   Sharman D. B. & Durrani T. S. An Overview of AI Applied To Signal Processing : A Perspective on Coupled Systems. Proc. IEE Colloquium on "The Application of Artificial Intelligence Techniques to Signal Processing" 1989; 1/1-1/4

11.   Ifeachor E. C., Hellyar M. T. & Mapps D. J. et al. Intelligent Enhancement Of EEG Signals. Proc. IEE Colloquium on "The Application of Artificial Intelligence Techniques to Signal Processing" 1989; 4/1-4/9

12.   Sharman K. C., Chambers C. & Durrani S. Rule Driven Adaptive Signal Processing. Proc. IEE Colloquium on "The Application of Artificial Intelligence Techniques To Sensor Systems" 1987; 8/1-8/5

13.   Erman L. D., Hayes-Roth F. & Lesser V. R. The Hearsay-II Speech Understanding System: Integrating Knowledge to Resolve Uncertainty. Computing Surveys 1980; 12:213-253

14.   McDonnell E., Dripps J. & Grant P. The Knowledge-Based Detection, Segmentation, and Classification of Foetal Heart Sounds. Proc. IEE Colloquium on "The Application of Artificial Intelligence Techniques to Signal Processing" 1989; 6/1-6/4

15.   Martin D. L. & Shaheen S. I. A Modular Computer Vision System for Picture Segmentation and Interpretation. IEEE Trans. Pattern Analysis & Machine Intelligence, 1981; 3:540-556

16.   Kitzmiller C. T. & Kowalik J. S. (1987) Coupling Symbolic and Numeric Computing in Knowledge-Based Systems. The AI Magazine 1987; 85-90

17.   Nii H. P. Blackboard Systems, Blackboard Application Systems, Blackboard Systems from a Knowledge Engineering Perspective. The AI Magazine 1986; 82:106

18.   Jones J. & Millington M. An Edinburgh Prolog Blackboard Shell. Dept. Artificial Intelligence, Univ. of Edinburgh 1986.

Section 4:

Speech and Vision

NEURAL NETWORKS FOR SPEECH RECOGNITION

E Ambikairajah and S Lennon.

Abstract

This paper describes a multi-layer neural network model, based on Kohonen's algorithm, for which a physiologically-based cochlear model acts as a front-end processor. Sixty-dimensional spectral vectors, produced by the cochlear model, act as inputs to the first layer. Simulations on a 9*9 neural array for the first layer were carried out for the digits 0 to 9. Results show that the network produces similar trajectories for different utterances of the same word, while producing different trajectories for different words. The network also exhibits time warp invariance, a property which is desirable in speech recognition systems. Finally, a second neural array which accepts inputs from the first array is described.

1. Introduction

Neural networks are mathematical models based on the structure of the brain. They consist of many simple interconnected processing elements, operating in parallel as in biological nervous systems. In recent years, they have attracted a resurgence of interest, principally due to the failure of conventional von Neumann techniques in solving such diverse problems as speech and image processing, robotic control and pattern recognition. Such networks are particularly suited to carrying out speech and image recognition tasks where high computation rates are necessary. The more well-known neural networks [1] include the Multi-Layer Perceptron, the Hopfield net and Kohonen's self organising net [2]; the use of this latter network in speech recognition is investigated in this paper.

Recent research suggests that the brain contains various topologically ordered maps where different neural cells respond optimally to different signal quantities. The self-organising maps discovered by Kohonen [2] are based upon the behaviour of biological neurons. These networks map signals from a high-order dimensional space to a topologically-ordered low-order dimensional space (with dimensionality usually equal to two).

In this way, the method by which the brain accommodates high-dimensional incoming signals can be modelled.

Kohonen's network consists of a rectangular array, which, after sufficient training, forms a topologically-ordered map of the input data . This means that input data vectors which are close together in their N-dimensional space will be mapped to neurons which are physically close together in two dimensional space. This property may be exploited as follows: if a sequence of slowly varying input vectors is presented to the trained network, neurons with adjacent coordinates in the two-dimensional array become successively excited. A trajectory [3] is generated connecting the sequence of excited or "fired" neurons. This trajectory is then characteristic of the sequence of incoming high-dimensional pattern vectors. In this way, data from a multi-dimensional space may be mapped to a sequence of two dimensional coordinates, which when joined, produces a characteristic trajectory.

The front-end processor used is a physiologically-based cochlear model and is thus ideally suited as a preprocessor for a model of the brain [4]. The cochlear model carries out a transformation from the time domain to the frequency domain. When stimulated, the output of the model is a sequence of high-dimensional vectors, each vector element indicating the relative magnitudes of the stimulus frequency components at a particular time instant.

It is difficult to organise large quantities of high-dimensional data using conventional techniques. This is where Kohonen's self-organising net is so useful. The Kohonen net quantises the input vectors into classes according to their relative frequency of occurrence. Thus, for a commonly occurring input, a class representing that input will be found in the net. Due to the fact that the map is nonlinear, much of the original high-dimensional information is preserved, while a lower dimensional representation is obtained. Most front-end processors in common use produce pattern vectors of twenty-two dimensions or less; however, the physiologically-based front-end processor in use in this investigation represents the frequency spectral data in sixty dimensions.

In this paper, a description of Kohonen's algorithm is presented, together with a brief outline of the cochlear model used as a front-end processor. Also included are plots of trajectories produced by the network for different speech sounds. Examples to illustrate the property of time warp invariance are given and an account of how a second layer may accept inputs from the first is described.

2. Description of the Neural Array

The network is arranged as a two-dimensional array, the position of each neuron being defined by i and j coordinates. It contains M*M neurons, each neuron having N weights associated with it (Fig. 1). These weights, which represent the synaptic connections of the biological neurons, correspond directly to the N dimensions of the input vector. It is generally accepted that the process of learning is associated with the modification of these weights. There are no physical lateral connections between the neurons. The training of the array takes place in the following manner: an N-dimensional input vector, X, chosen at random from a large representative set of input pattern vectors, is compared with the weights of each neuron in the array (the weights are initially set to small random values). The neuron whose weights most closely match the input vector is chosen as the 'winner'. This is the neuron with the minimum Euclidian distance between its weights and the input vector. A set region, encompassing a number of neurons, is defined around the winning neuron, and is called its neighbourhood. The process of topological ordering requires that this neighbourhood initially include all neurons in the array and subsequently decrease in size. As more and more input vectors are applied to the network, different parts of the array become selectively sensitised to different classes of inputs.

2.1 Kohonen's Algorithm

The steps involved in implementing Kohonen's self organising network algorithm [5], [6] are as follows:

Step 1.   Initialise the weights
The weights of each of the M*M neurons are initialised to random values and the initial neighbourhood is set to cover all neurons.

Step 2.   Present the input vector
The current input vector $(X_1, X_2 .... X_N)$ is randomly chosen from a large representative set of input pattern vectors, and applied without supervision to the array.

Step 3.   Compute the distance between the input and all   neurons
The distance, d_{ij}, between the input vector and the weights of each neuron is computed as follows:

$$d_{ij} = \sum_{k=1}^{N} \{ X_k(t) - W_{ijk}(t) \}^2$$

where $X_k(t)$ is the current input at time t; $W_{ijk}(t)$ is the weight at time t;
$i = 1,2,...,M; j = 1,2,...,M; k = 1,2,...,N.$

Step 4.   Select the best matching neuron

The neuron with coordinates i,j such that d_{ij} is a minimum is selected.

Step 5.   Update weights to neuron i,j and neighbours

The weights of the winning neuron i,j and of all neurons within the current
neighbourhood are updated. The new weights are calculated from:

$$W_{ijk}(t+1) = W_{ijk}(t) + a(t) \cdot \{ X_k(t) - W_{ijk}(t) \}$$

i and j range through all values within the current neighbourhood. a(t) is a gain term
which decreases in time (0<a(t)<1); the size of the neighbourhood also decreases in time
(section 2.2). One can see from the above equation that the weight vector will tend to
take on values which match the values of commonly occurring input vectors.

Step 6.   Repeat Steps 2 to 5

until sufficient iterations have been carried out to tune the network.

2.2 Training parameters

During this investigation, the following values were found to provide good topological
ordering of the network. For improved ordering, lower values of gain and lower
shrinkage rates of the neighbourhood may be chosen, but only at the expense of more
iterations.

Training is carried out in two phases with varying gain parameter a(t). The gain
parameter is defined as:

$a(t)$ =   c1 (1- t/T1)   during the first training phase (0<t<T1);

=   c2 (1- t/T2)   during the second training phase (T1<t<T2);

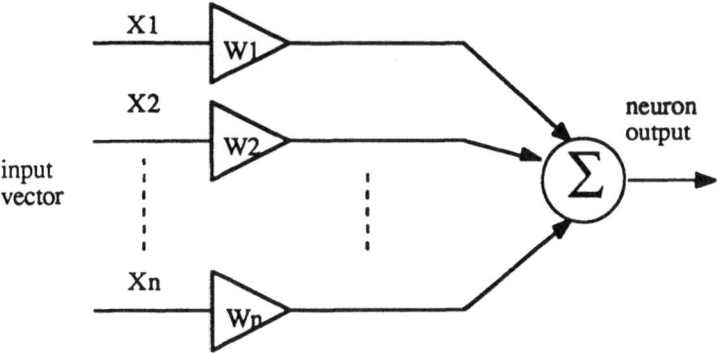

Fig 1. Model of a Single Neuron.

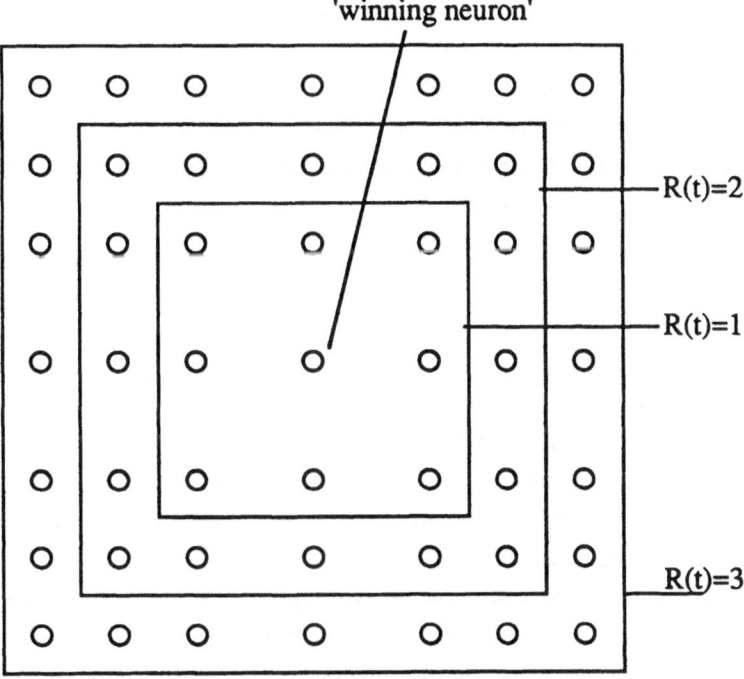

Fig.2 Neighbourhood around winning neuron at three different time instants during training. The same winning neuron is assumed for each size of neighbourhood.

c1=0.1, c2=0.008, T1=10000, T2=90000.

where T1 and T2 are the number of iterations of the first and second training phases respectively.

The size of the neighbourhood surrounding the winner, within which the weights of the neurons are updated, varies as follows:

$$R(t) = R + (1-R)t/T1 \quad \text{during first training phase;}$$
$$= 1 \quad \text{during second training phase;}$$

where R(t) is the current size of the neighbourhood (for example a 6*6 array will have R=6).

A final value of R(t)=1 means only the winner and the eight neurons which are closest to it are updated in phase 2 (see Fig. 2). There must be a sufficient number of statistically distributed inputs in order to obtain the topologically ordered map, especially when increasing the dimensionality of the input vector. A ratio of ten inputs to one neuron is recommended, and ordering improves if more inputs are added. According to Kohonen, between 500 and 5000 iterations are required per neuron, to completely tune a network.

3. Preprocessing of Speech Signals

There are a variety of front-end processors in use at present, for the purpose of speech analysis, including linear predictive coding, mel frequency cepstral coefficients, fast Fourier transforms and filter banks. In this investigation a physiologically-based passive cochlear model of the inner ear was chosen as being particularly suitable for interfacing to a model of the brain.

Studies of the cochlea have shown that the primary structure, the basilar membrane, transforms incoming time domain signals to their constituent frequency components. Each frequency within the auditory spectrum causes a maximum displacement to occur at a particular point on the basilar membrane. After sound has being converted to basilar membrane displacement, it then undergoes a further transduction process, which converts this displacement into electrical energy in the inner hair cells. The basilar membrane in this case is represented by a cascade of 128 digital filters. Out of these 128 filters, only 60 fall within the speech bandwidth of 250 Hz to 4 kHz. The outputs of the inner hair cells of these 60 filters provide the input pattern vectors for the network.

Speech samples were obtained at 8 kHz. Ten utterances of each of the single digit sounds (zero to nine) were recorded from one male speaker, thus creating a database of one hundred utterances. The speech sounds were manually endpointed and interpolated to 48 kHz as the cochlear model operates at this frequency. The samples were then band-pass filtered in the range 250 Hz to 4 kHz before being applied to the cochlear model.

Pattern vectors emerging from the cochlear model were filtered using a smoothing algorithm so that the formant content of the speech signal could be extracted. On average 29 sixty dimensional vectors were obtained from each utterance. These vectors were scaled and converted to integers before being used as inputs to the neural network.

4. Results

4.1 Tuning of a 2-D Network to 2-D Input Vectors

An 8*8 neural array was trained with a two-dimensional random elliptical input vector. This data is represented by the dots in Fig. 3. Also shown in Figs. 3(a) to (d) are the synaptic weight values of the array for four different time instants during the training period. Lines are drawn to connect the synaptic weight values of elements which are adjacent in the array. Fig. 3(a) shows the array after 20 iterations. Folding is evident in this diagram. The array is shown after 100 iterations in Fig. 3(b). The net is now beginning to spread out and cover the input data. Fig. 3(c) shows the situation after 500 iterations. After 1000 iterations (Fig. 3(d)), the synaptic weights have taken on positions which match the input data space in a topologically-ordered manner, in other words, the grid is not folded. Note that the input data was presented to the network in a completely random order, without supervision.

4.2 Tuning of a 2-D Array to 60-D Input Vectors

For accurate mapping from a higher-dimensional to a lower-dimensional space the inherent dimensionality of the data must be of the same order as the dimension to which it is mapped [3]. For example the inherent dimensionality of vowel sounds in speech is only two. Thus, these sounds may be mapped from a higher dimensional representation to a two dimensional array, without folding occurring between dimensions.

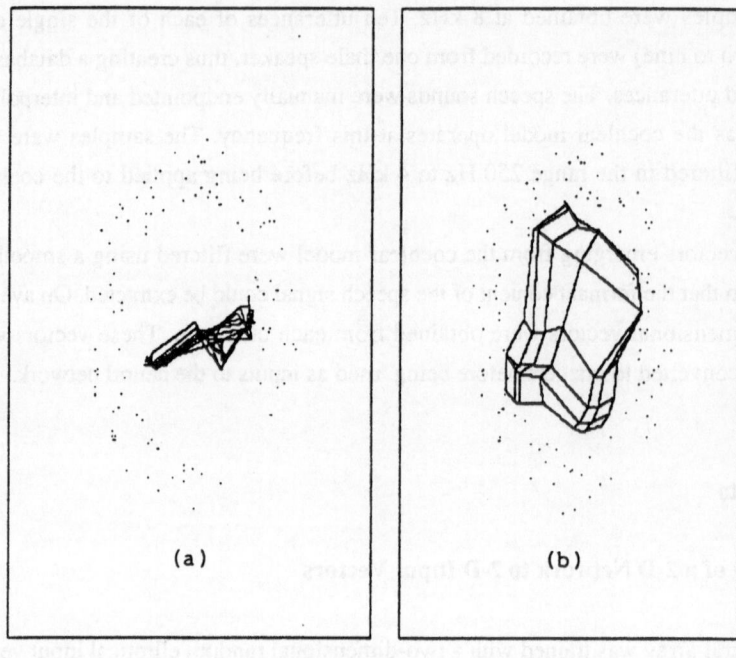

(a)

(b)

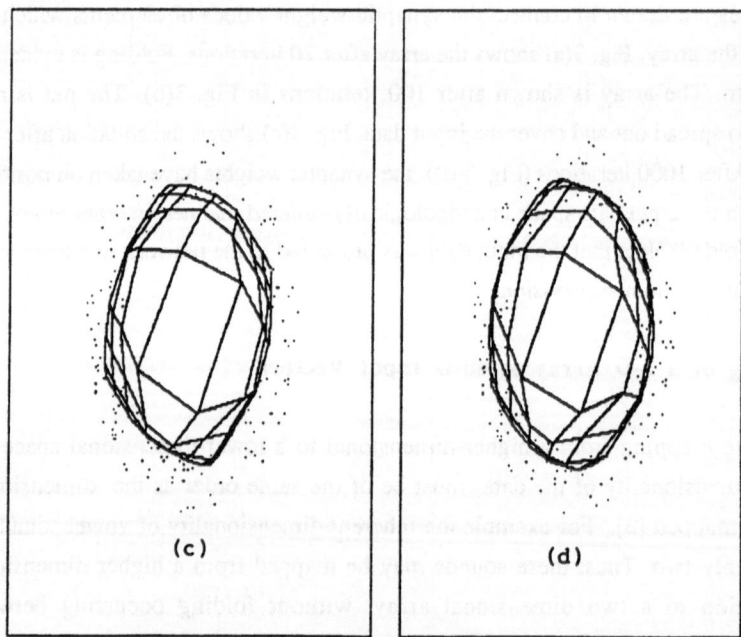

(c)

(d)

Fig. 3 Organisation of 2-D networks to 2-D input data.
The dots indicate the input data.

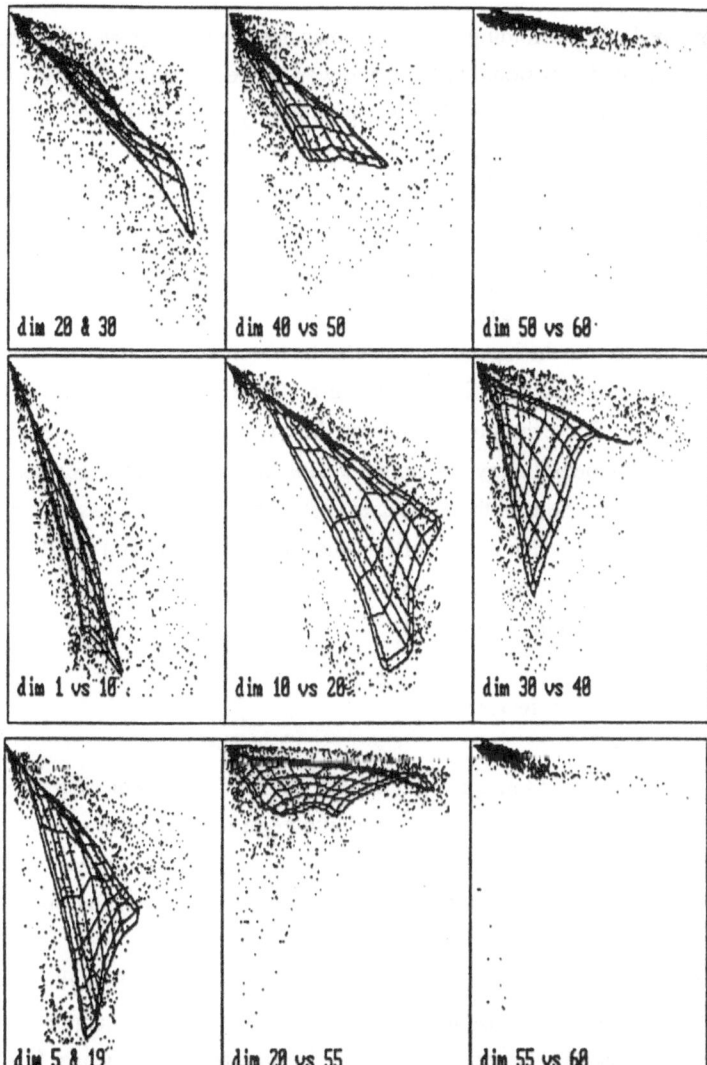

Fig. 4 Ordering of neural array across 60-dimensional
input data. The net gradually adjusts itself both to map
the input data and to remain topologically ordered.
Dots indicate the input data.

If data is mapped without folding between dimensions, it can be stated that the data has the same inherent dimensionality as the lower space to which it is mapped. The phenomenon of folding may be observed graphically during the training phase, by connecting neurons to their four nearest neighbours. For neurons located at the edges of the array, three connections will suffice. By plotting the weights of dimension p as an x coordinate and the weights of dimension q as a y coordinate, the ordering between the two dimensions may be observed. For a 60-dimensional map, p and q are any integers between 1 and 60. If the same dimensions p and q of the input data are also plotted, the tuning of the network to the input data may also be observed.

A plot of ordering between dimensions for a 10*10 array is shown in Fig 4., after 40,000 iterations. For example, in the top left corner of Fig. 4, the weights of dimension 20 of the neural array are plotted against the weights of dimension 30 of the array. The same dimensions of the input data vectors are plotted as dots. The absence of folding is evident, and is due to the small values of gain (0.004 and 0.001) being used to train the network. In this case, the input data vectors consisted of one hundred utterances, as mentioned in section 3.

4.3 Characteristic Trajectories

A 9*9 neural array was trained using ten digit sounds, 0 to 9. For each digit, the ten utterances collected (section 3) were divided into eight training utterances and two test utterances. Hence, the network was trained using eighty utterances, while twenty utterances were available for testing purposes. However, the results presented here apply only to tests carried out with digits 0 to 2.

A test utterance consists of a variable number of frames (each of length 16 ms), with each frame represented by a 60-dimensional vector. When this sequence of frames is presented to the trained network, neurons will fire in a certain sequence. It is possible that one neuron will fire for more than one frame. As a result, the number of distinct neurons which fire will be less than the number of frames in the utterance. By connecting the positions of the neurons which fire (i.e. the neurons most similar to the inputted sequence of frames) a characteristic trajectory may be plotted.

Trajectories for the digit sounds 0 and 2 are presented in Figs. 5 and 6. The test utterance for digit 0 consisted of 31 frame and for digit 2, 22 frames. The arrows indicate the direction in which the trajectory moves. These digits can be distinguished by the shapes of their trajectories.

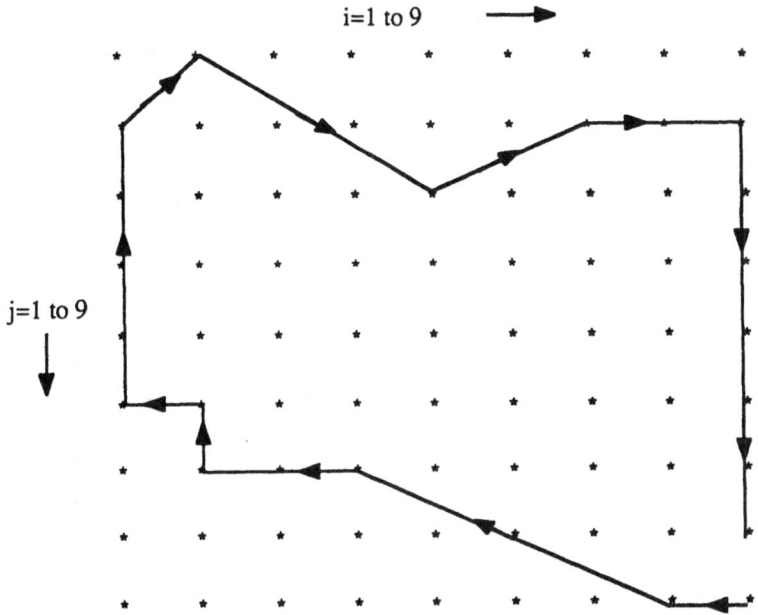

Fig 5 Trajectory of neuron firing for digit zero..

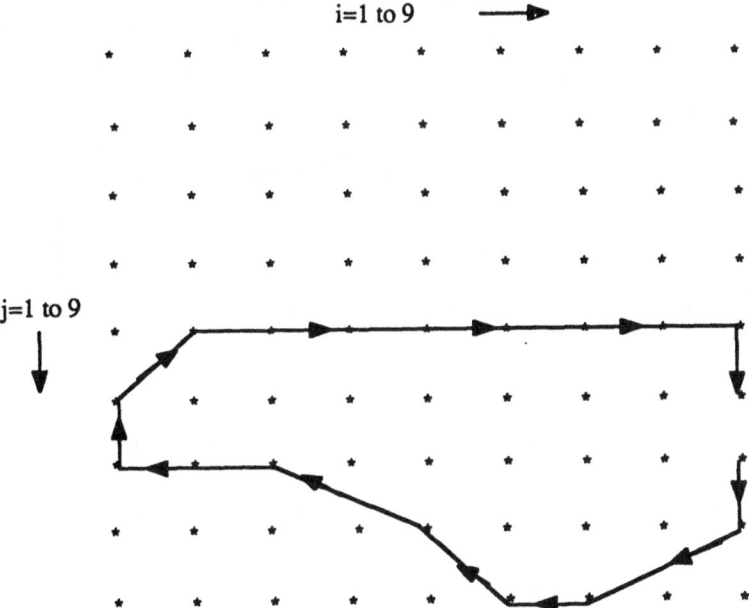

Fig. 6. Trajectory of neuron firing for digit two.

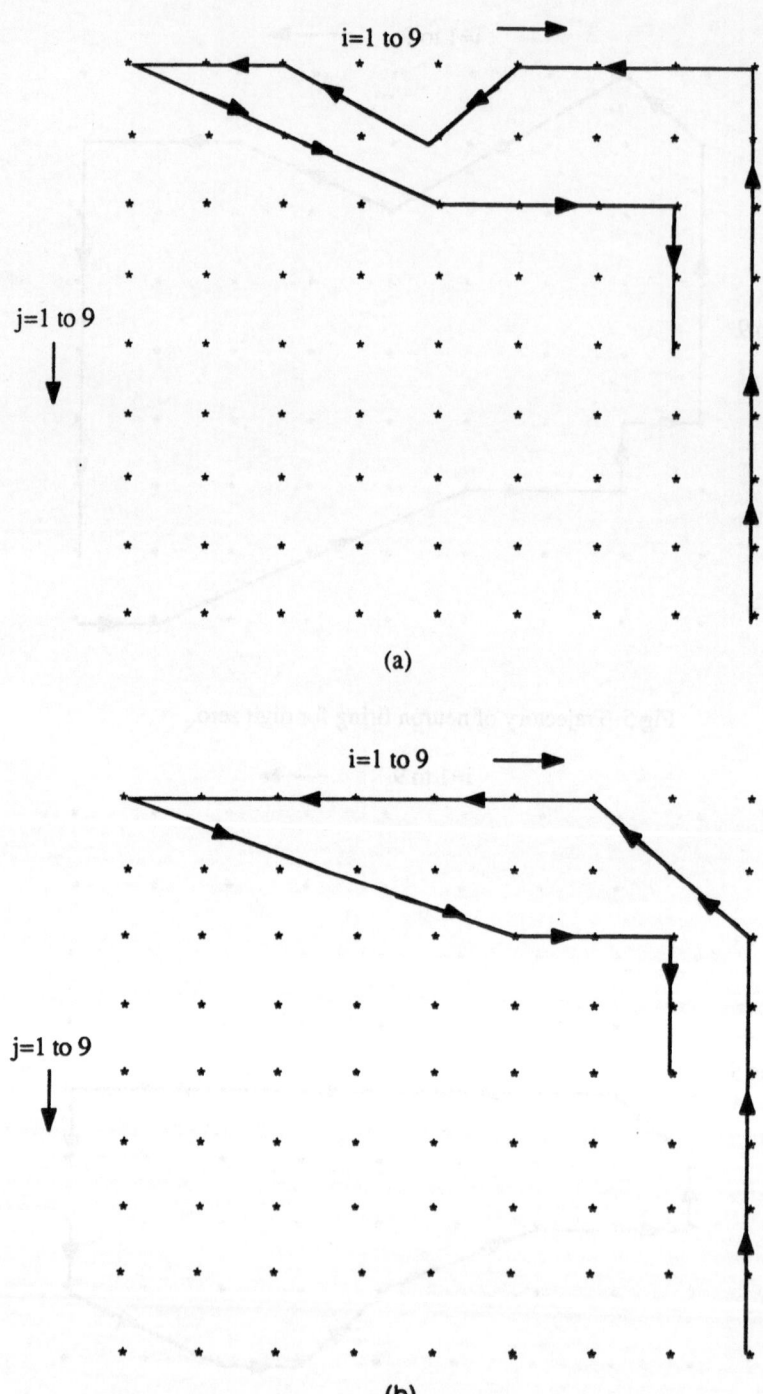

Fig. 7 Illustration of time warp invariance using the digit "one".

4.4 Time Warp Invariance.

When the same word is spoken more than once, even by the same speaker, the speaking rate varies. This causes problems in speech recognition systems, and is normally eliminated by using linear/dynamic time warping techniques. The feature of time warp invariance may be observed by comparing the trajectory of Fig. 7(a) with that of Fig. 7(b). The trajectories were produced by two different utterances of the digit "one". The first utterance was 22 frames long, while the second was 14 frames long. However, the network produces similar trajectories for the two utterances, by eliminating the differences in speaking rate (note that one utterance is almost twice as long as the other). This property of time warp invariance would prove useful in speech recognition systems using Kohonen's neural network.

4.5 The Second Kohonen Neural Layer

A second neural array may be used to accept inputs from the first array (See Fig.8). The inputs to the second layer are created by concatenating the sequence of "fired" neurons on the first layer. This then produces a P-dimensional column vector. In order that these inputs each have the same dimensionality it is necessary to pad the unused remaining entries of the vector with zero's. This ensures that the column vector inputs to the next layer are of a similar size. For each utterance one column vector is generated which then acts as an input to the second layer.

To fully exploit the inherent time warp invariance property of the trajectories on the first layer it was decided not to repeat consecutively firing coordinates in the concatenation process. In this way, the shape of the trajectory and thus its time warp invariance property is preserved while the sequence in which the neurons fire is also maintained.

After the second layer has been trained, using the same procedure as before, different parts of the array become sensitive to different words. Thus one neuron becomes excited when a particular utterance is inputted into the system. By labelling a neuron to the word to which it becomes most excited, different words may be mapped to different parts of the array. Neurons, located next to each other, which are excited by similar words, may be grouped to form distinct clusters (See Fig. 8).

Two clusters exist for some sounds, which indicates that global ordering has not occurred on the second layer. Global ordering will only occur if the inherent dimensionality of the input data is the same as the dimensionality to which it is mapped. However, global ordering is not necessary on the second layer for recognition purposes.

TWO LAYER KOHONEN NET

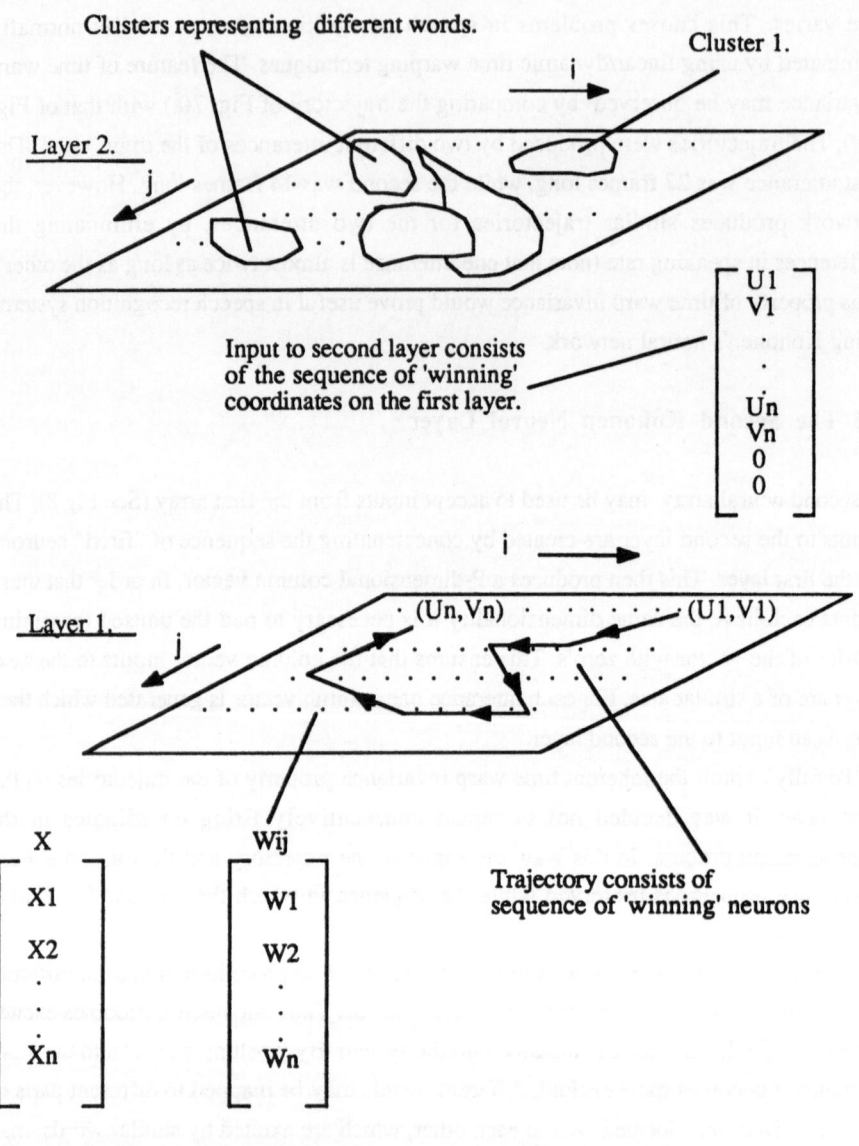

Clusters representing different words.

Cluster 1.

Layer 2.

i

j

Input to second layer consists
of the sequence of 'winning'
coordinates on the first layer.

$$\begin{bmatrix} U1 \\ V1 \\ \cdot \\ \cdot \\ Un \\ Vn \\ 0 \\ 0 \end{bmatrix}$$

i

Layer 1.

j

(Un,Vn)   (U1,V1)

Trajectory consists of
sequence of 'winning' neurons

X

$$\begin{bmatrix} X1 \\ X2 \\ \cdot \\ \cdot \\ Xn \end{bmatrix}$$

Wij

$$\begin{bmatrix} W1 \\ W2 \\ \cdot \\ \cdot \\ Wn \end{bmatrix}$$

Input Vector X   Weight Vector W

Fig 8. The Two Layer Kohonen Neural Network.

5. Conclusion.

This paper has presented an implementation of a two-layer Kohonen neural network. Results have shown that the first layer produces distinct trajectories for different sounds, yet similar trajectories for different examples of the same sound. Experiments with a larger database should further improve the accuracy and robustness of the network. The pattern of neural firings produced by the first layer acts as the input to a second layer. More layers may be added so that eventually only one neuron will be stimulated by a word. Current research will be directed towards the development of a multi-layer Kohonen net based on the two-layer model presented in this paper.

Acknowledgements

This research was funded by a grant from EOLAS, under the Scientific Research Programme 1989-1991.

References

1. Lippmann, R. P. (1987). "An introduction to Computing with Neural Nets", IEEE ASSP Magazine, April 1987, 4-22.
2. Kohonen, T. (1982). "Clustering, Taxonomy, and Topological Maps of Patterns", IEEE Proc. of the International Conference on Pattern Recognition, Oct. 1982, 114-128.
3. Tattersall, G. (1988). "Neural arrays for speech recognition", British Telecom Technology Journal, Vol. 6, No. 2, April 1988, 140-163.
4. Ambikairajah, E., Black, N. D. & Linggard, R. (1989). "Digital filter simulation of the basilar membrane", Computer Speech and Language 3, 105-118.
5. Kohonen, T. (1989). "Speech recognition based on topology-preserving neural maps", Neural Computing Architectures (ed. I. Aleksander), North Oxford Academic Publishers, 26-40.
6. Tattersall, G. (1989). "Neural Map Applications", Neural Computing Architectures (ed. I. Aleksander), North Oxford Academic Publishers, 41-73.

Using observer-controlled movement and expectations of regularity to recover tridimensional structure.

Roddy Cowie, Dearbhaile Bradley & Mark Livingstone

ABSTRACT

Movement has long been recognised as an important factor in the recovery of 3-d structure. Recent computational research has become interested in the special contribution that self-initiated movement can make. This paper reports experiments which illustrate an underexplored way of exploiting the ability to control movement, that is to manoeuvre for a good view. Results reinforce the widespread finding that expectations of geometric regularity play a pervasive role in human vision: regularity facilitated perception despite the rich information available from movement. Strategies which people use are identified and geometric relationships which may underlie them are considered.

INTRODUCTION

This paper is about discovering objects' shapes and movements from a changing image. That problem will be called the interpretation of visual motion.

The simplest approach to interpreting visual motion is to rely purely on a rigidity postulate, and that approach has been explored extensively (e.g. [1], [2]). However it seems increasingly unlikely that it is practical to rely on this single elegant constraint: even when objects are truly rigid, it leads to procedures which are computationally expensive and which are unacceptably vulnerable to errors in low level processes [3]. As a result computationalists have been increasingly interested in ways of drawing other types of knowledge into the interpretation process. Two main options stand out. The first is to interpret motion in accordance with expectations about object regularity. The second is to exploit the fact that observers have a good deal of control over the positions from which they observe objects.

Recently computational research has shown interest in both options. Observer-controlled motion is a high profile topic (e.g.[4],[5]). Although it is less high profile, one expectation about objects - that points will lie on a plane or at least a smooth surface - has been used in several schemes for interpreting motion ([3], [6], [7], [8]).

The general theme of this paper is that these analyses only scratch the surface of the potential for exploiting expectations of regularity and controlled motion. Two broad areas remain almost unexplored computationally, although observations of biological vision suggest that they are potentially rich. One of the key arguments in this paper is that progress in these areas can be facilitated by considering them together.

The first of the areas is the use of structural regularities other than planarity. Obvious examples are rectangularity, symmetry, and parallelism. Computational research has tended to avoid forms of analysis which are tied to objects with strong structural regularities, except at a relatively late stage when these regularities are embodied in models. However psychological evidence indicates that regularities of this order have a pervasive influence in human vision [9],[10].

The simplest term for the second area is exploration. In the kind of observer-controlled motion that has attracted computational interest, the observer exploits relatively precise interoceptive information about his or her path through space. Parallax from head movement is the archetypal example. It is not clear how important this way of using observer-controlled movement is to humans, despite its a priori attractiveness. Certainly its ability to yield quantitative information seems to be limited (e.g. [11]). However that is secondary to the present argument. The primary point is that there is a range of observer-controlled behaviours which aim at 'getting a good view', i.e. finding a viewpoint or a sequence of viewpoints from which key questions can be answered with a high degree of confidence. At first sight at least, this seems to represent a way of using the observer's control over his or her movements which is quite different from motion parallax, and it has received relatively little attention in the computational vision literature (though see [12]).

This paper reports experiments which indicate how computational research might approach the topic of exploratory vision. In the first instance, the idea that structural regularity needed to be considered in tandem with exploration emerged as an empirical finding. However, once the connection was noticed it was clear that it made sense. Exploration needs to be guided, and hypotheses about the structural regularities in a configuration are an obvious source of guidance, at several levels.

EXPERIMENT 1

The focus of this experiment is whether the ability to control viewpoint helps observers to recover the structure of a tridimensional object. Regularity of the structure was also examined, with particular interest in the question of whether active exploration would cease to help when the structure was regular. The kind of exploration which subjects used was observed informally.

Method

Subjects. Twelve subjects participated. All were students at Queen s University. None of them had specialised knowledge which was relevant to the task.

Apparatus. Stimuli were based on an H shape made of thin balsa wood sticks painted matt black. The left hand side bar was always 5cm long and the crossbar was 3cm long. The right hand side bar was varied in two ways:

1. Bar length varied from 3cm to 7cm in 1cm steps.
2. The bar was pivoted through 15°, 30°, 45°, 60°, 75°, or 90°, in a plane perpendicular to the crossbar. This construction means that the two side bars were at angles of 15°, 30°, 45°, 60°, 75°, or 90° to each other.

These features were varied factorially, giving 30 stimulus objects.

Objects were presented on a supporting plate which pivoted on a ball joint. The 5cm side bar was always attached to the plate. The objects were viewed monocularly through a half silvered mirror by two subjects at a time. The 'active' subject, whose line of sight passed through the mirror, could reach underneath the mirror and manipulate the supporting plate. The 'passive' subject observed the process reflected in the mirror. The mirror was angled at 45° to the active subject's line of sight, so that the active and passive subjects always had optically congruent views.

The whole apparatus was enclosed in a box. Its internal lighting and the matt black paint on the objects were calculated to ensure that information from brightness difference or shadows was minimal.

Procedure. On each trial, two subjects viewed an object for twenty seconds. One actively manipulated it, and the other watched passively. Then both drew the angle between the two side bars and estimated the length of the variable bar. Initially subjects were told the length of the constant bar and given five practice trials. Thereafter trials were grouped into blocks of 30. Each block contained a single presentation of every stimulus object. Order was randomised within each block. Each subject carried out five blocks as active subject then five as passive subject, or vice versa.

Results

Results were coded in terms of absolute errors in judging (a) the angle between the side bars and (b) the length of the variable bar.

Three factors significantly affected the accuracy of bar length judgements. Active viewers were more accurate ($F_{1,10}=19.6$, $p=0.002$). The bar's actual length had a significant effect (F 4,40=5.3, p=0.002): estimates were best when the constant and variable bars were equal. The angle between the bars also had a significant effect (F 5,50=2.7, p=0.03). This was due to reduced errors when the bars were at angles which one might imagine served as anchor points in subjects' representation of this type of structure. These were 90°; 15° (the smallest angle used); and 45°. There were no significant interactions.

Estimates of angle gave a more complex picture. Again active viewers were more accurate (F 1,10=36.8, p=0.0001). The angle between the bars also affected errors (F 5,50=22.0, p<0.0001): performance was best when the angle was 90°, intermediate when it was small (15° or 30°), and worst in the mid range (45°, 60°, and 75°). An interaction showed that this effect was more marked in passive viewing (F 5,50 =3.9, p=0.005). This has the useful implication that in passive viewing at least, the effect was perceptual rather than due to difficulty reproducing certain angles. The main effect of bar length was not significant (F 4,40=2.0, p=0.11) , but it interacted with the angle between the bars (F 20,200=2.8, p=0.0001). With most bar lengths errors tended to peak when angle was intermediate. However with the short bar, angle had relatively little effect on accuracy: it gave rise to better judgements than the other bars when the angle between the bars was intermediate, and worse judgements when the angle was either large or small.

Finally two effects involving order of presentation fell just short of significance. Accuracy was less dependent on structural regularity among subjects who experienced the passive condition first. This applied both to regularity of length (F 4,40=2.5, p=0.061) and to regularity of angle (F 5,50 =2.0, p=0.090). Both show the same kind of trend: the most marked dependence on object regularity occurred in subjects who were viewing passively after experiencing the active condition first. The obvious hypothesis is that attempting to exploit their control of motion led subjects into considering possible regularities of structure, and that tendency persisted even when they switched to the passive condition: whereas subjects who began in the passive condition adopted strategies which were less dependent on regularity.

Observation suggested a pattern in the way subjects exploited their ability to manipulate the object. They rarely moved the object smoothly and continuously, which would presumably have facilitated motion based analyses like those of traditional computational vision. Instead they manoeuvred it into a desired position, held it there, and observed it for a while before repositioning it. The positions to which they moved it generally created simple relationships between properties of the image and properties of the object - for instance, the cross bar was aligned roughly along the line of sight, so that the angle in the image reproduced the angle between the bars: or one of the side bars was aligned roughly at right angles to the line of sight, so that the length of its image was simply related to its distance from the viewer.

Discussion.

This experiment makes two clear points. First, it shows that active control over movement contributes to visual perception in a situation which is quite different from those considered by current computational work on active vision. Second, it shows that judgements are related to structural regularity even when observers have information not just from movement, but from self-generated movement. It is important that angular regularity affected judgements about bar length and (less straightforwardly) vice versa: this shows that the effects of regularity did not simply reflect response strategies biased towards regular outputs.

The most straightforward reading of these findings is as follows. Active subjects used their control to manoeuvre objects into 'good' positions, most obviously positions where key elements were aligned with the line of sight or at at right angles to it. Regular objects were more accurately described because subjects used multiple criteria to judge when these alignments had been reached: so they might, for instance, be reluctant to accept an alignment which was actually what they wanted because it implied that the object's arms were very unequal in length. Active control was an advantage because subjects' manoeuvres were designed to answer questions. Passive subjects had to infer the question that a manoeuvre was designed to answer, and if they had not done that successfully they might not realise what a potentially revealing view was telling them.

This reading is open to debate at several points. From the standpoint of contemporary computational theory, it is particularly tempting to suppose that proprioception was responsible for the advantage that active viewers enjoyed. This cannot be ruled out altogether, but it seems unlikely for two reasons. First, the proprioceptive information that subjects had at their disposal was probably not precise enough to make very much difference. Their only contact was with the rim of the supporting plate, which was connected to the test object by pivots which were neither simple nor smooth. Second, the approach that subjects used does not offer an obvious role for proprioception. They tended to manoeuvre the object into a 'good' position, and leave it static while they made their judgement.

A large part of this picture hinges on informal observation of the way subjects manoeuvred. It is clear that more formal data on that point is necessary, and so a second experiment was carried out to provide it.

EXPERIMENT 2

This experiment closely resembled the previous one except that the objects were generated on the screen of a Macintosh computer. Exploration was controlled by a mouse. This arrangement made it possible to keep a complete record of the views which subjects chose, so that their strategies could be studied systematically.

Method

Stimuli. The basic configuration can be conceptualised as follows. Imagine that the object was mounted at the centre of a sphere, with its crossbar vertical (i.e. running North/South). Both side bars were horizontal. One always ran East/West, and its length was always equal to the length of the crossbar. The other side bar will be called the variable bar. Its length was one of the object variables, and the angle between it and the East/West bar was the other. The variable bar was identified by being drawn with a thicker line than the others.

The screen showed a projection of the object taken from a point on the surface of the (imaginary) sphere around it. The point of projection was controlled by the mouse, as if the mouse controlled a video camera which was mounted on the surface of the sphere, perpendicular to the surface (hence it always pointed at the centre of the object). We used a very simple form of linkage between mouse position and 'camera': camera latitude was proportional to the vertical displacement of the mouse pointer from its (notional) origin, and camera longitude was proportional to the horizontal displacement of the mouse pointer from its origin. Pilot studies suggested that people were more comfortable with that than they were with geometrically more sophisticated forms of control. The one refinement was that subjects could move the pointer without moving the camera by dragging the pointer across the screen: each drag simply added a vector to the notional coordinates of the pointer, so that it could in effect move across an infinite plane rather than being confined within the boundaries of the screen. The equivalent of a drag was performed before an object was first displayed, so that starting position was random.

'Camera' position was recorded at every frame (i.e about 10 times per second).

Stimulus parameters were as follows. There were six angles between the two side bars - 15°, 30°, 45°, 60°, 75° and 90°. Four lengths were used for the variable bar - 3, 4, 6 and 7 units (taking the fixed length bars to be 5 units long for comparison with experiment 1). There was an appreciable amount of perspective in the projection: focal distance was 6 times the length of the fixed bars, which corresponds to viewing the objects in experiment 1 from a distance of 30cm.

Subjects. Subjects were 12 undergraduates at Queen's University. They had no prior knowledge about the task.

Procedure. The task was explained to each subject beforehand using a physical model. Subjects were allowed to practice manipulating the display with a randomly selected stimulus for as long as they liked before starting the trials proper, but given no feedback. Each subject then examined all of the objects, using a different random order for each. As in experiment 1, lengths were estimated numerically, and angles were reported by drawing.

Results.

To study 'camera' movement, the imaginary sphere round the object was divided into regions of equal area by lines of latitude and longitude, and the proportion of each trial spent in each region was calculated. Although fuller displays were examined, eventually latitude and longitude were considered independently. This makes analysis much simpler, and seems to lose relatively little information.

Experiment 1 generated a strong expectation about latitude: we expected that subjects would often move the camera to the 'poles', so that their line of sight was aligned with the object's cross bar and at right angles to the side bars. This is functionally reasonable since from that viewpoint, the projected angle between the cross bars is equal to the angle between them in three dimensional space, and the relationship between the projected lengths of the bars is almost the same as the relationship between their actual lengths. Hence it is an obvious exploratory strategy to seek out views where these simplifying relationships exist.

This expectation was confirmed. The sphere was divided into ten zones of equal area, each comprising identically positioned bands in the Northern and Southern hemispheres. Zone 1 comprised the two poles, and zone 10 straddled the equator. The percentage of his or her time that a subject spent in each zone was calculated, and these measures were subjected to analysis of variance. Data from zones 5-9 was omitted from the main analysis to produce a manageable data set: this is no loss, since previous analyses had found no interesting differences between the non-polar zones.

As expected, there was a strong effect of zone (F 4,44=33.5, p<0.0001). This occurred because as anticipated, subjects spent a relatively large amount of their time in the polar zone - 26% as against around 10% in any of the others. Tukey post hoc tests confirm that time spent in the polar zone was significantly higher than time spent in any other zone (p <0.01), and the other zones did not differ significantly among themselves.

In line with the previous results, this basic pattern is linked to object shape. A strong interaction (F 20,220=2.03, p<0.01) shows that the distribution of time across zones depended on the angle in the object. Simple effects analysis shows that angle has a significant effect on time spent in the polar zones (F 5,55=2.662, p<0.05), but not on on the way time was distributed between the other zones. The means indicate that the proportion of time spent at the poles is high for the acute angled objects (31% when the angle is 15°, 37% when it is 30°) and low when the object angle is right or nearly so (23% when the angle is 75°, 18% when the angle is 90°). As a result the simple effect of zone is not significant for the object where the arms are at 90° (F 4,44=1.688, p=0.17). Arm length may have some bearing on this effect, though it is less clear. The patterns of exploration associated with small and large angles seem be more sharply distinct when the variable arm is longer than the constant arm. However the interaction which shows that is not particularly strong (F 20, 220 = 1.647, p< 0.05).

To study longitudinal effects the sphere was partitioned into twenty 18° sectors. Data from the polar regions was ignored at this stage because positions close to the pole have a special significance irrespective of longitude, as was shown above. Diametrically opposite sectors were then combined except in the case of the two sectors flanking the Greenwich meridian and the two sectors opposite them. Because of a programming error three of these were combined and the last was ignored. However the error makes no difference to the pattern of results.

Again a very marked pattern emerged. Over 40% of subjects' time was spent in the two zones flanking 90° and 270° latitude - i.e. close to the plane containing the crossbar of the object and its fixed arm. Separate analyses of variance were carried out for the objects whose variable arms were longer than the constant arms and those whose variable arms were shorter. Both showed that the distribution of time across longitudes was highly uneven (F 8,88=28.8, p<0.0001 and F8,88=13.5, p<0.0001 respectively). In both cases Newman-Keuls analyses showed that the subjects spent significantly more time in each of the zones flanking 90° and 270° latitude than in any other (p<0.01). The only complicating effect involves which of these two zones predominates: an interaction involving longitude, angle, and length (F 40,440=1.5, p<0.05) indicates that one object (the object where the variable arm is at right angles to the constant arm, and is maximum length) differs from the others in this respect. That seems unlikely to be theoretically important.

The strategy of choosing viewpoints in or near a key plane was not foreseen, but once recognised it is easy to see that it makes sense. Its advantages will be considered in the discussion.

Attention can now be turned to the errors that subjects made. These are important for two reasons: first, because it should be established that the strategies considered above have some bearing on performance; and second, to indicate how closely this experiment relates to the more natural situation studied in experiment one.

The effects of strategy were examined by assigning subjects to groups. Decisions were based on cases where more than half of a trial was spent in a significant zone - either the poles or the zones flanking 90° and 270° longitude. Subjects were rated highly polar if a quarter of their trials or more were spent mostly in the polar regions, low polar otherwise. Correspondingly subjects were rated highly east/west if a third or more of their trials were spent mostly in the zones flanking 90° and 270° longitude, low east/west otherwise. Fortuitously, these divisions split subjects into four equal groups (high polar and high east-west, high polar and low east-west, etc.). Analyses of variance were carried out using these divisions. To make broad trends clearer, angle was described by two variables, broad group (15° or 30° vs 45° or 60° vs 75° or 90°) and subgroup (15°, 45°, or 75° vs 30°, 60°, or 90°).

Errors judging length may be considered first. The overall average error was 7.2% of the length of the constant bars. This was remarkably close to the level in the active condition of experiment 1, where the figure was 7.5%. As in experiment 1 actual bar length affected errors ($F3,24=4.7$, $p<0.01$), but this time errors were highest when the variable bar was most similar to the constant bars. Strategy had an intriguing effect. A significant interaction ($F1,8 = 5.6$, $p<0.05$) showed that errors were low in those subjects who favoured a single clear strategy - i.e. those who were classed high polar or high east/west, but not both (at 5% and 5.6% respectively); and high in those who used both strategies or neither (at 9.4% and 8.9% respectively). A significant four-way interaction ($F6,48=2.8$, $p<0.05$) seemed to indicate that the high east-west group had special difficulty with individual objects, notably the one with the longest bar and the most acute angle. It will be seen that this object caused problems for the high polar group in angle judgement.

The overall mean of errors in judging angles was 10.4°, considerably higher than the mean for angle errors in the active condition of experiment 1 (which was 4.1°) but close to the mean for the passive condition (9.0°). The angle in the object had no overall effect, but it interacted with east/west strategy ($F 2,16=4.4$, $p<0.05$). Simple effects analysis showed that the high east/west subjects were significantly affected by the actual angle in the object ($F 2,16=5.2$, $p>0.05$): their errors were well above the overall average (at 15.4°) when object angle was intermediate (i.e. 45° or 60°, and well below it (at 6.1°) when object angle was large (i.e. 75° or 90°). The polar grouping featured in a three-way interaction with length and angle which was close to significance and sufficiently orderly to deserve attention ($F 6,48=2.1$, $p=0.065$). Comparing high polar subjects with the rest, the high polar group fared better than the others at every combination of broad angle category and arm length but two. Their advantage was consistently high in the largest angle category (i.e. 75° and 90°). In the smallest angle category (i.e. 15° and 30°), the high polar group again showed strong advantages when arm length was intermediate. But when arm length was either at its shortest or at its longest, their average errors on the small angle stimuli were well above those of the low polar subjects.

Discussion.

The results of this experiment identify two patterns of movement which subjects tend to adopt, and the assumption that these patterns have a perceptual function is borne out by the finding that the patterns are related to subjects' judgements. The main aim of this discussion is to show that the patterns of movement, and the effects which they produce, make sense in terms of the geometry of vision.

Consider first the strategy of moving to the pole. The advantage of this strategy is obvious: it means that the picture contains the lengths and angles that subjects have been set to find. (In the case of length this is slightly oversimplified: the picture shows the ratio of fixed to variable arm length subject to some distortion by perspective. This qualification will be taken as read.)

The problem that must be solved to use this strategy is to find the polar view. Four ways of doing this are worth considering.

Logically the simplest strategy is to maximise the projected length of both side arms. This is a correct strategy if projection is parallel: perspective creates complications, but they can be ignored for the present. To find the pole according to this criterion, a subject could first maximise the projection of one arm, then move at right angles to that projection until the projection of the second arm reaches its maximum. Intuitively this strategy feels unnatural, and it seems unlikely that it was used by subjects who sought out polar views: if it had been, their judgements should not have been affected by object characteristics. It is interesting to speculate why this control strategy feels unnatural. Intuitively it seems unsatisfying to give an arbitrary line the status of an axis of rotation, which suggests that people expect the control of exploratory movement to mesh with object-based frames of reference.

A second strategy which works in this experiment is to find the view where the cross bar is end-on to the viewer (i.e. its image is a single point). Again, the effect of object characteristics suggests that it was not given overwhelming weight. This is logically sound, since the strategy only works because the cross-bar happens to have been perpendicular to both side bars in this experiment. This is to say that the strategy depends on a special kind of object regularity, and the weight attached to it should only be as great as subjects' confidence that the regularity in question is actually present. Conversely it is right to be influenced by other regularities if the evidence that they are present is on a par with the evidence that the cross-bar is at right angles to the side bars.

The third and fourth strategies involve assuming that other regularities may be present. It is convenient to describe them by considering the triangle which would be formed if half of each side arm were slid down the crossbar until the two halves met in the middle. It may be called the implicit triangle. The essence of these strategies is to manoeuvre the camera until the observed image has properties reasonably like those which the implicit triangle is expected to have. When that happens, the viewer can check whether he or she is close to the pole. The simplest test is that if the pole is close, then the relevant properties should remain stable when viewpoint changes by a moderate amount. In effect the observer is trying to obtain a good view of a regular structure: his or her goal is only achieved if the input qualifies on both counts.

The third strategy involves assuming that the implicit triangle may be roughly isosceles, with a narrow apical angle. There is a geometric reason for considering these properties together. If we consider all possible projections of an isosceles triangle with a narrow apical angle (ignoring perspective), then a high proportion of the projections have relatively narrow apical angles and sides of relatively similar lengths [13]. Isosceles triangles with large apical angles do not give a corresponding relationship. Hence it is a reasonable bet that a projected triangle with a relatively narrow apical angle and sides of relatively similar lengths may represent an isosceles triangle with a narrow apical angle. Strongly polar subjects may try to locate the pole by manoeuvring until the image shows this kind of triangle. They make their largest errors in angle estimation when the arms are at a small angle, but quite far from equal; and that is what would happen if they treated equality of length as a significant indicator when apical angle was narrow. The length-angle interaction in experiment 1 shows a related pattern, though it seems to be stronger when the variable arm is too short than when it is too long.

The fourth strategy involves assuming that the implicit triangle may be right angled. That assumption indicates that if the camera is at the pole, then the images of the side arms should be at right angles. Favouring positions which have that property, or which come close to it, would be expected to promote accurate judgements when the object angles were actually 90°, and to induce error otherwise. That fits the observed behaviour of subjects in the highly polar group.

No extra explanation is needed for the fact that subjects who consistently adopt a strategy of seeking the poles tend to estimate length well. At first sight it looks puzzling that length judgements do not show the complex of effects that angle judgements do, but in fact length judgements are simply less sensitive to exact position than angle judgements. There are only two easily identified points (the poles) where the image of an angle is equal in size to the angle itself. But (if perspective can be ignored) then the projected length of an arm equals its true length for a whole family of positions, those where the arm is perpendicular to the line of sight (these form a great circle on the sphere of viewpoints). These positions are easily identified because they are where projected length is greatest.

The strategy of moving to 90° longitude (or 270°) can now be considered. Unlike the strategy of moving to a pole, it was not recognised as a possibility before the experiment. However it too makes a great deal of sense theoretically.

One property of the strategy is obvious and strongly linked to the data. Suppose that the viewpoint is restricted to the plane which contains the crossbar and the constant side arm (call this the reference plane). This gives a simple way of testing whether the variable arm is at right angles to the constant arm: if it is, and only if it is, then the image of the variable arm will be at right angles to the image of the crossbar wherever the observer moves in the reference plane. The availability of this test explains one of the outstanding features of the data for highly east/west subjects, the fact that their angle estimates are particularly good when the object angle is 90° or close to it.

The second distinctive feature of these subjects' data is that they misjudge the length of the variable arm when it is longest and most acutely angled to the constant arm. This is presumably a simple consequence of the viewing position that the subjects favour. Adopting that viewpoint means that there is a large depth change from the near to the far end of that bar, since it is long and almost perpendicular to the image plane. Consequently perspective has a substantial effect on its projected length, and it is not surprising that it causes difficulty.

The data also show that length judgements benefit when subjects consistently adopt this strategy. This is observation is considerably more interesting than it looks, because it is not immediately obvious how positions in the reference plane can be exploited for judging length. However some interesting possibilities can be identified.

A key issue is finding the angle between the image plane and the crossbar - call it IC. Assuming that projection is effectively parallel, there is one straightforward way of finding IC. First, move through the reference plane until the image of the crossbar reaches a maximum length and begins to shorten. The maximum that it reaches is the true length of the crossbar. That knowledge can now be used to find IC: at any given time its cosine is the projected length of the crossbar upon its true length. This corresponds to a strategy which intuitively humans seem to use to orient themselves, using the foreshortening of a vertical to estimate their elevation.

A different possibility arises if the observer knows the rate of change of IC - i.e. the angular velocity of the camera. Then it is easy to show that if l is the projected length of the crossbar,

$$\tan IC = (dl/dIC) / l$$

However it is found, IC can be used directly to recover the length of the variable arm. Suppose that the projection of the variable arm makes an angle A with the projection of the crossbar, and its length is lp. Then the length la of the variable arm is given by

$$la^2 = lp^2 * (1 + cos^2 A / tan^2 IC).$$

This expression is neat enough geometrically, but it suffers a drawback. Using IC gives a similarly simple expression for the angle between constant and variable arms, and so it gives no clue as to why moving in the reference plane might benefit judgements of length, but not of angle. That remains an interesting question.

GENERAL DISCUSSION

This paper has shown that when human observers are given the opportunity to control their own viewpoint, they make good use of it. At least some of the strategies which they use can be identified, and make sense geometrically. Contrary to the default assumption that computational theory tends to make, observers' strategies were not independent of object shape. The simplest reading of the data is that observers operated by seeking out well behaved views of well behaved structures.

Three major questions remain to be asked about the demonstrations. They are how the tasks used here relate to exploration in the real world; how the exploratory strategies considered here relate to the more familiar strategies for interpreting visual motion which are relatively independent of viewpoint; and how the ideas relate to computational vision.

On a very broad level, it is obvious that the kind of behaviour which was described here has counterparts in real perceptual activities. When a referee at bowls has to judge which ball is nearest the target, he or she moves up to the target and looks directly down on the bowls; a surveyor stands in the plane of a gable when he or she is deciding whether it bellies out; a tourist moves round the tower of Pisa to see how far it actually leans; a dentist manoeuvres his mirror to get a clear view of a tooth; and so on.

At a finer level, it would be wrong to claim too close a correspondence between the strategies described here and real life exploratory vision. Most straightforwardly, it has been noted that angle judgement was better with the physical stimuli of experiment 1 than it was with the computer displays of experiment 2. The difference probably reflects factors which are important for real life exploration, but which are not available in the simulated displays. The most obvious example is lighting, which is probably useful for judging when a surface (or a strut) is frontoparallel. It is also likely that a good deal of manoeuvring is concerned with finding views whose 'goodness' is not related to the information that lines convey about object geometry, but with the distribution of lighting or the presence of features which are important for recognition - a topic which has received some research, though it has not been located in the general context of exploratory vision [14]. All this is a bonus rather than a problem, since the aim of this paper is to indicate that exploration is a large topic.

A more awkward question is whether the findings reported here bear any relation to the exploration of objects which contain larger numbers of points. Classical structure from motion algorithms require a certain number of points to operate, and it is unsurprising if expectations other than rigidity affect the interpretation of stimuli which do not meet that requirement. Our H shaped stimuli were designed with this in mind, because the six points which they contain are sufficient to let Ullman's structure from motion algorithm operate. Informal studies have also shown that the need to explore, or to invoke assumptions about structure, do not disappear when subjects are presented with parallelepipeds rather than the H shapes used here. These offer eight points, which is sufficient for any of the well known structure from motion schemes to operate. However, it remains possible that presenting substantially larger numbers of points would lead to performance more reminiscent of classical structure from motion schemes.

This leads on to the question of how the present ideas relate to standard approaches to the interpretation of visual motion. The simplest assessment is that the strategies consided here are adjuncts to the standard viewpoint-independent methods, used when the observer needs precise information about specific relationships. This may be correct, but it is worth considering a more integrated view. A classical problem with the interpretation of motion is that small errors in recovering the input can lead to massively wrong conclusions about tridimensional structure (e.g. [3]). This can be linked to the present discussion by supposing that observers continually estimate the potential for error in their interpretation of the image, and can manoeuvre to reduce that estimate whenever necessary. This presents exploration as an integral part of a general approach which is sensitive to the possibility of error rather than a separate alternative. The idea clearly invites mathematical development.

The strategies which have been sketched here are clearly limited, but they point to a very large area for development: building machine vision systems which understand how to see better by manoeuvring for a good vantage point. Since this is something that humans do very well, it is an area which invites a true cognitive science approach where empirical research and computation support each other.

References

1. Ullman, S. The interpretation of visual motion. Cambridge, Mass., M.I.T. Press, 1979

2. Longuet-Higgins, HC and Pradzny, K. The interpretation of a moving retinal image. Proceedings of the Royal Society of London B 1981; 223: 165-175

3. Scott, G. Local and global processing of moving images. London, Pitman, 1988

4. Ballard, DH. Reference Frames for Animate Vision. Proc 11th IJCAI, Detroit 1989. pp. 1635-1641.

5. Brown, CM. Predictive Gaze Control. Proc. 5th Alvey Vision Conf. Reading 1989, pp. 103-108.

6. Longuet-Higgins, HC. The visual ambiguity of a moving plane. Proceedings of the Royal Society of London B 1984; 208: 358-397.

7. Subbarao, M. Interpretation of visual motion: a computational study. London, Pitman, 1988

8. Murray, D. Algebraic polyhedral constraints and 3-d structure from motion. Proc. 5th Alvey Vision Conference, Reading 1989. pp. 215-220.

9. Cowie, R and Clements, D. The logical basis of visual perception: Computation and empirical evidence. Irish J. Psychol. 1989; 10: 232-246

10. Isiguchi, A. Perception of structure from two-, four-, or eight-dot oscillatory motion: restoration of rigidity with the aid of structural information. Perception 1990; 19: 197-205.

11. Reinhardt-Rutland, A. Perception of slant under reduced viewing conditions. Perceptual & Motor Skills 1981; 53: 146.

12. Cowie, R. The Viewer's place in theories of vision Proc. 8th IJCAI , Karlsruhe 1983. pp. 952-958.

13. Cowie, R. Modelling people's interpretation of line drawings. D.Phil. thesis, University of Sussex, 1982

14. Langdon, P , Mayhew, J E W and Frisby, J P S. In search of 'Characteristic View': 3-D object representation in human vision using ratings of perceived difference between views. In J E W Mayhew and J P S Frisby (ed) 3-D model recognition from stereoscopic cues. Cambridge, Mass., M.I.T. Press, in press.

Measuring the 'Rubber Rhomboid' effect.

David Clements and Roddy Cowie

ABSTRACT

The 'Rubber Rhomboid' effect is an illusion where a rigid skeleton parallelepiped is rotated and appears to undergo rubbery deformation [1]. This paper reports the first attempt to map the effect systematically. Three subjects assessed the level of deformation of computer generated rotating parallelepipeds. The main finding was that objects containing equal angles at a vertex were more stable across all orientations, especially when the equal angles were 90°. The findings are discussed in the context of computational accounts of human vision.

INTRODUCTION

The 'Blocks World'

In 1965 Roberts [2] published a report that was to pave the way for future 'real world' vision research with its attempt to analyse pictures of three dimensional solids. His program used a description in terms of lines, junctions and polygons to select plausible correspondents from *a priori* models of solids stored in memory. The program then demonstrated interpretation by generating a different view of the scene than the one presented as input. His method was limited in that the program could only operate on scenes which contained the polyhedral, plane faced, straight edged objects which were fully characterized in its memory. This explicit and full description of the contents of the scene to be analysed became known as the 'blocks world' approach, because of the similarity of the *a priori* models to the block toys of children's play. However, despite its relative success, many theorists believed that progress in machine vision could only be made with the development of programs that began with more general procedures instead of trying to fit models of specific objects from the outset as Roberts did.

For example, Guzman [3] discovered that useful interpretations of an image could be performed without recourse to explicit models stored-in memory. His procedure instead segmented a picture by classifying the vertices into a few basic types.

This ran concurrent to work advocating a linguistic approach to scene analysis. Following the lead of Chomsky, with his emphasis on the use of anomaly, Huffman [4] and Clowes [5] explored pictures of impossible objects in order to extract the rules with which they could assign structure and meaning to a scene.

The classical 'Generalists'

It became clear in the middle to late seventies that the prevailing approach to computer vision was getting bogged down. It had been felt that if one could deal with the essential aspects of the scene analysis of simple worlds then it would be straight forward to extend the analysis to include, say, the fuzzy world of teddy bears [6]. Instead, this approach was leading to a point where programming was simply in the business of developing patches to deal with each unexpected ambiguity.

In 1978 Barrow &Tenebaum [7] concluded that if the performance of previous domain specific programs was any thing to go by, the development of a successful, general purpose vision system might well be impossible. However, they did point to the phenomenology of human vision to provide clues as to how to approach vision in computational terms. In this they were echoing the feelings of Horn [8, 9, 10] and of Marr [11, 12, 13, 14]. These workers argued that the limitations of 'blocks world' programs stemmed from their unreliability in extracting features - an unreliability that occurred because these procedures had no basis for evaluating which intensity differences corresponded to scene events significant at the level of objects. Marr went further stating that to understand vision fully it was necessary to have a clear idea which of the possible representations available was the most appropriate. Only then was it possible to tackle the task of analysing the computational problems that could arise, in obtaining and manipulating each representation. He too believed that the efficiency of the human visual system provided important clues to the representations that would prove the most appropriate and likely to yield successful solution. Further, Marr's approach made explicit the move away from the use of specific knowledge as typified by the 'blocks world' approach. The crux of Marr's work was the desire to exploit only the most general of knowledge about the scene to be analysed. This was to be achieved by isolating constraints that were generally true of the world yet powerful enough to allow a process to be defined [15].

One example of this kind of approach came from Ullman [16]. He exploited the general assumption of rigidity to overcome the problem that three dimensional structure is under determined by the projected two dimensional transformation of a system of points. His treatment of rigid motion, under orthographic projection, involved finite displacement considerations closely resembling those relating to the analysis of stereo depth. Ullman and Fremlin (cited in [16]) found that given three distinct views of at least four non-coplanar identifiable points, it can be determined unambiguously whether or not they represent a single rigid body. If they do, then their three dimensional structure can be recovered uniquely, up to a possible reflection about the image plane. Longuet-Higgins [17, 18], in a general procedure derived from the eight sets of image co-ordinates, showed that if a scene contains as many as eight points in polar projection, where the points can be located in each image, the relative orientation of the two projections can be computed.

Problems with noise and a reaction to general constraints

During the 1980's it became clear that procedures based on Marr's approach were not without their practical difficulties. The most acute problem was their inability to handle noisy input data. The Longuet-Higgins algorithm, for instance, accepts any eight points and returns a solution. These eight points can be the result of real image points or they can be the result of noise. Further, methods that rely on differentiated information tend to suffer from a crippling magnification of any noise in their input data.

These noise considerations have begun to move workers away from the strict Marr doctrine and its reliance on only general constraints. This 'post-modernism' so to speak, is concentrating on the development of robust vision systems, systems that could potentially handle real, noisy data. To accomplish this, the exploitation of specific information is once again attracting attention as a means of improving performance.

The exploitation of specific information such as the expectation that a viewed object contains rectangular corners [19] forms part of this 'post-modernism'. So does the more mathematical approach which stabilizes interpretation by exploiting a number of assumptions together. An early example of this is the Longuet-Higgins [20] planar algorithm. In this case he not only assumed rigidity but he also assumed that the object was planar. Other researchers concerned with motion have moved further from the strict 'generalist' approach. Scott [21] stated that noise resistant systems must incorporate specific expectations such as symmetry and compactness and Subbaro [22] had as a main constraint, smoothness of image, the assumption that the object will move at a constant rate. Object rigidity was only then applied as an additional constraint. Poggio,Torre and Koch [23] saw early vision as being a set of 'ill-posed problems' corrupted by noise, which could be approached by applying regularization methods. These methods included a restriction of the classes of admissible solutions by the introduction of *a priori* knowledge. This knowledge could be in the form of statistical properties of the solution space or the exploitation of variational principles that imposed constraints on possible solutions.

The evidence suggesting the use of more situation specific postulates

It is of great interest as a psychologist to note the extent to which the mathematical constraints introduced, by the 'post modernists', to solve purely mathematical problems echo the results and suggestions from a body of psychophysical research. Much has been made, by Marr and others, of the general assumption of rigidity in considering how to extract meaning from, say, motion information. It has not, however, been established that humans employ the kinds of general, pervasive constraints 'generalist' often suggest we do.

Ames, in his classic 1951 experiment [24], found that when a trapezium is rotated about an axis in line with its parallel edges, apparent reversals of its direction occur. That is, the object seems to rotate so far, then reverses direction and swings back the way it came. He explained his findings by stating that from previous experience with rectangular structures the observer assumes that the transformations he sees are being made by a rectangle even when they are in fact being made by a rotating trapezia.

A variant on the Ames' trapezium involves the addition of a bar inserted through a solid trapezia, which effectively makes it three dimensional [25]. Computational theory suggests that this addition should reduce reversals, and to a limited extent it did. However, when the object did reverse, the trapezium appeared to go transparent. It seems our visual system can ignore what seem to be sensible general considerations: rigidity, occlusion, lighting and an objects optical qualities. Instead, it seems that we are relying on expectations which involve some form of geometric regularity and some sort of evenness of motion.

The 'Rubber Rhomboid' effect

One particularly striking illusion telling against general assumptions is the 'Rubber Rhomboid' effect. Cowie [1] presented a single wire-framed parallelepiped to twenty-two naive subjects. The angles at an arbitrary vertex were 65°, 75° and 125°, and the object was rotated about a vertical axis at either 1.5 or 12 rpm. The object was rigid, but it appeared to deform in a rubbery way as it moved. Cowie argued that reversals were not a major feature of the illusion and suggested instead that the rubberyness perceived by the subjects was due to the observer attempting to impose postulates based on the angles between edges; namely rectangularity. Effectively, as the image of the parallelepiped changed, the only way for the viewer to continue with the expectation of rectangularity was to see the object as being non-rigid.

The 'Rubber Rhomboid' effect is of interest for a number of reasons. It is common practice to test machine vision systems by comparing their performance to psychophysical data. It is, therefore, vital for this psychophysical data to be collected systematically. One illusion which has been well documented is the Ames trapezium. However, this illusion has a number of associated problems. Firstly, the information that the display contains is often insufficient for computational analyses to be applied successfully. Secondly, by the nature of the illusion, the main data which is collected involves frequency of reversal. This does not translate in any obvious way into the kinds of decision that computational vision schemes seek to make. The 'Rubber Rhomboid' effect potentially solves these problems. The three dimensional nature of the illusion should provide sufficient information for vision systems to be applied successfully. Further, by recording the level of deformation, the behavior of a shape can be measured continuously. Finally, because of the large number of shapes that can be employed, it is possible to develop a map of the effect which is rich enough to constitute a powerful test of programs' psychological appropriateness. This paper reports the first attempt to map the effect in this way.

Measuring the 'Rubber Rhomboid' effect

Three undergraduates from Queens university Belfast served as unpaid volunteers. All had normal or corrected to normal vision. The stimuli employed were computer based (ATARI 1040), mathematically generated parallelepipeds, projected orthographically onto a monochrome monitor (black figures on a grey ground). The decision to present the objects via a computer was a practical one. Attempting to manipulate such a large number of 'real' objects with all their associated initial orientations have been hopelessly cumbersome.

Each object was a non-rhombic parallelepiped, with a horizontal edge of length two units, a height across the front face of one unit and a depth of 0.5 units. The shape of each object was determined by the 3 angles at a vertex.

Starting at 15,15,15 the angles were varied in 25° increments, one angle at a time. This yielded fourteen basic shapes. The shapes were then variously orientated prior to rotation, to test for influence on subsequent reported deformation rates. The first orientation of interest was the effect of initial orientation in the picture plane. If you imagine the plane of the VDU with the left hand side being the Y-axis and the bottom edge being the X-axis, then the Z-axis points directly into the screen. So rotation about the Z-axis is effectively a rotation in the picture plane, and was varied in clockwise steps of 30°, from 0° to 90°.

Related to this was the possible effect of initial rotation about the x-axis. This kind of rotation takes the top of the object away from the observer and brings the bottom closer. To study this the shape was turned about the X-axis again by 0°, 30°, 60° or 90°

into the screen. Finally, Smyth, 1978 (cited in [1]) suggested that relatively rigid interpretations occur when non-rectangular objects are rotated about their long diagonal. This led to the introduction of another condition where each object was upended so that each of its diagonals in turn was vertical. The object was then rotated about this vertical diagonal. Overall, the combination of object shapes and initial orientations gave 280 trials.

Each trial began with an object in the position produced by one of the above initial manipulations. The subject observed the object rotating clockwise about the Y-axis (the vertical). The illusion of movement was achieved by generating an array of the co-ordinates of each of the corners for each of the 180 frames that constituted one rotation. The program then used this information to draw the shapes and present them in sequence and each sequence of 180 frames lasted 25 secs, giving a rotation rate of 2.4 rpm.

The task for the subject was to observe the object on the screen and decide, during its rotation, whether or not the shape was undergoing deformation, and if it was, how large that deformation appeared to be. Input was via keys on the calculator pad of the Atari keyboard. The subject was instructed to press no key if the shape did not appear to be deforming at that instant; to press key 1 if there was slight deformation; key 2 if moderate deformation was seen and key 3 if a large deformation was observed. The testing was divided into sessions of about 1 1/2 hours duration, and overall testing required about 10 hours per subject. The shapes were presented in a random sequence and this sequence was different for each subject.

Information from the keyboard was continuously read into an array at a rate of 90 items per rotation, that is, every other frame. The values could be overwritten until the subject was confident as to the accuracy of his/her responses. The press of the 'Help' button halted further over-writing and moved the program on to the next trial. The results array for each subject was subsequently saved to file and consisted of 90 numerical values, in the range 0 to 3, for each object in each orientation condition.

Results

A correlation between subjects' responses was carried out. It gave a coefficient of concordance of 0.815, indicating a good level of agreement between subjects' ratings of the observed phenomenon.

Initial inspection of the data showed that the Z=0 orientation had very high deformation levels for all objects and all orientations. It was felt that such a mass of high scores would tend to swamp the other values, so this block was considered a special case. It is not included in subsequent descriptions unless stated otherwise.

The results from the three subjects were averaged and arranged so that shapes with at least two angles the same were grouped together: for instance, the shapes with angles of 90°, 90°, 40° ; 90°, 90°, 65° and   90°, 90°, 90° were considered as a sequence whose members differed only in the size of the last angle. This arrangement allows the data to be displayed in the form of a response surface (Fig. 1). One horizontal axis is the size of the equal angle pair, the second is the size of the last angle, and the vertical axis is the average reported elasticity.

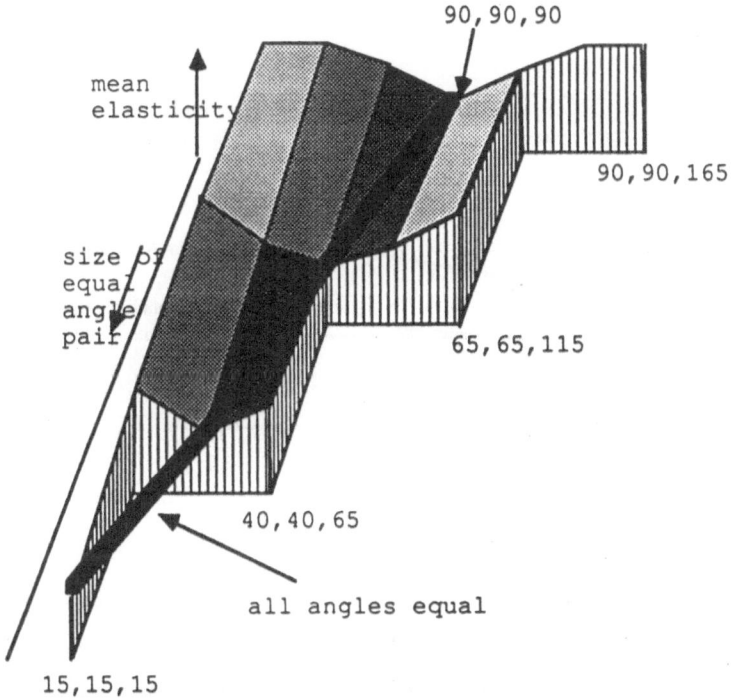

Fig. 1. Response surface showing the effect of object shape. Each point is the average of observations from 16 orientations.

The response surface result is challenging because it clearly shows the effect of equal angles at a vertex. It was expected that the cubic shape would evoke the lowest levels of perceived deformation, which indeed it did. But it was of great interest to discover that the equal angle condition had the lowest deformation levels for each of the object groupings. That is, for each object type, say 40°, the most stable shape, across all orientations,was the one which contained only equal angles at a vertex e.g. 40°, 40°,

40°

The next task was to assess the influence of the particular initial orientations on perceived deformation levels. Here the Z=0 condition was re-included. The object deformation values were averaged which yielded two graphs, one for X-orientation and one for Z-orientation.

Inspecting Fig (2) it is apparent that the X rotation has little effect on deformation levels, but a gentle trend toward instability as X approached 90° was evident.

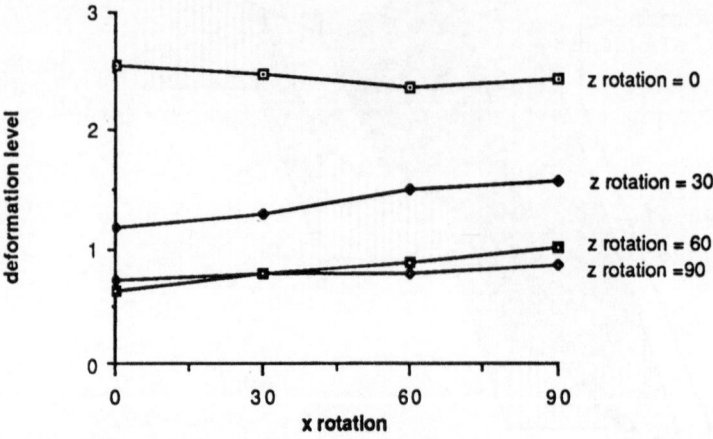

Fig. 2. Effect of x rotation on deformation rates

The exception was the Z=90° condition were the trend was reversed, indicating that perhaps some form of interaction was occurring. Importantly, the block in which Z orientation resulted in the object on the screen containing horizontal lines ie. Z=0, had the highest deformation rates across all X orientations. The other Z orientations were more closely grouped, but tended to be more stable as Z approached 90°, the condition that contained vertical lines. Fig(3) shows the effect of Z rotation more directly.

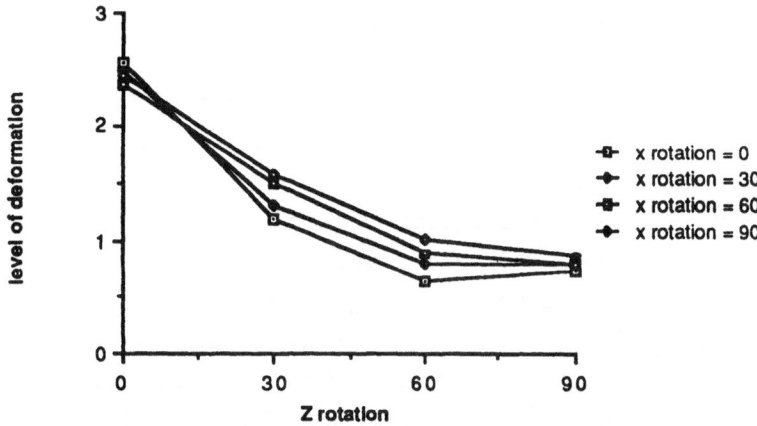

Fig. 3. Effect of z rotation on deformation rates

The traces are in the form of curves from a unstable high at Z=0° to the most stable condition at Z=90°. From the two graphs the important findings are:

1/ the Z=0° orientation yield by far the highest level of deformation

2/ and there is evidence to suggest that X-orientation has little effect on the level of perceived deformation.

The block in which the objects had been rotated about their diagonals was then considered. For each of the objects the lengths of the four diagonals were measured, and ordered from longest to shortest. The deformation values associated with these diagonals were then averaged across the objects to yield an average deformation level for each diagonal length. The resulting graph is shown in fig (4).

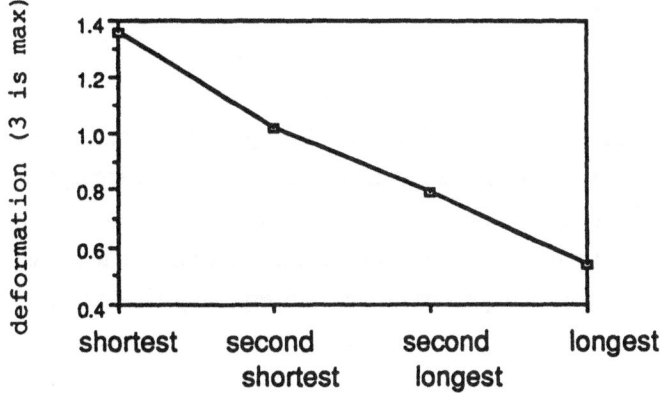

Diagonal about which the object was rotated

Fig .4. Effect of diagonal length on deformation rates

The effect of decreasing diagonal length is quite clearly to increase the level of perceived deformation. It must also be noted that deformation levels in this block were well below those of the other deformation conditions.

The study also examined another approach to predicting perceived non-rigidity. Computational accounts of human vision tend to assume that there is a sharp separation between representations of the two-dimensional image and representations of the three-dimensional things which it represents. From this perspective it seemed likely that the most successful geometric predictors of the behaviour of rotating parallelepipeds would be related to the three dimensional interpretation that the display subjectively appears to produce. However, psychologists have sometimes argued that perception tends to be tied to immediately accessible variables of the stimulus.

In order to assess how successful predictors in only two dimensions could be, various two dimensional characteristics of the objects were measured in each of the 180 frames. It seemed obvious that it would be the change in the calculated variable that would predict subjects' response, so the difference values of the predictors, from frame to frame, were correlated with the subject data. The surprising result was that there were strong correlations between perceived deformation and two of the measures. One was the perimeter of the shape, which is proportional to the total length of the lines which are involved in depicting it. The second was its horizontal width. Change in these measures correlated strongly with perceived deformation , with coefficients of 0.82 and 0.63 respectively. This clearly suggests that even interpretations which are subjectively quite tridimensional may be tied surprisingly closely to the two dimensional variables which give rise to them.

Discussion

Overall this study is important for a number of reasons. The high correlation between perceived deformation and the two dimensional geometric characteristic of changing size of perimeter, suggests that computational workers may have to revise their thinking. The expectation that interpretation can only follow from the information contained in a three dimensional representation, is called into question by the discovery that the human visual system may rely quite heavily on two dimensional information when interpretating a scene. It may well be that our visual system will tend to exploit only information at the lowest level of representation that makes a solution tenable.

The study also showed that we may look for interpretations where the angles at the corners of a body are relatively similar. It has often been argued that people favour interpretations which involve right angles, and a preference for equal angles is geometrically related to that: but this seems to be the first evidence which suggests that an equal angle constraint is used in perception.

It was also found that shapes which contain horizontal lines, with respect to the screen, that is the Z=0° condition, are perceived as exhibiting the highest levels of deformation of any condition. This result is reminiscent of a finding by Wallach & O'Connell [26] who, when partially rotating a T shape piece of wire, found that the horizontal shadow gave the appearance of a line in the plane of the screen expanding and contracting - that is, it did not suggest motion in depth. However, the effect is harder to explain in this context than it is in Wallach and O'Connell's study. A rigidity postulate alone is not enough to interpret the Wallach & O'Connell stimulus. This is due to the impoverished nature of the information which the display provides. However, this explanation cannot be applied to the tri-dimensional, parallelepiped display contained in this study. Hence it seems that there is something more general, and more interesting to the effect than has generally been assumed.

In total these findings call into question the ability of strict 'generalist' schemes to account for human visual behavior. It seems that not only do we not exploit general constrains when processing scene information, but also we can exploit simpler information to arrive at the stage of interpretation, than most computational systems would try to use.

For these reasons, it vital for computational workers to attend to psychophysical studies, like this one, in order to both test the performance of their systems against human data and to uncover the constraints that make the human visual system as robust and universal as it seems to be. For their part psychologists must supply high quality data relating to these constraints, in a form that can be exploited by computational workers. The 'Rubber Rhomboid' illusion is an important example this type of effect. It is a reliable illusion, capable of being systematically investigated, and it is both capable of generating surprises and supplying the quality of data necessary to test computational models against. This kind of test is important if computational theory is serious about modelling, and learning from, human vision.

References

1. Cowie R. Rubber Rhomboids: Nonrigid interpretation of a rigid structure moving. Perception and Psychophysics 1987; 42: 407-408

2. Roberts RG. Machine perception of 3-Dimensional solids. In: Tippett JT, Berkowitz DA, Clapp LC, Koester C J & Vandenburg A (eds) Optical and Electro-optical

Information Processing. MIT Press Cambridge, Mass., 1965

3. Guzman A. Decomposition of a visual scene into three-dimensional bodies. In IAFIPS Conference Proceedings 33. Thompson, Washington, DC 1968, pp 291-304

4. Huffman D. Impossible objects as nonsense sentences. In: Meltzer B & Michie D (eds) Machine Intelligence (Vol. 6). Edinburgh University Press Edinburgh, 1971

5. Clowes M. On seeing things. Artificial Intelligence 1971; 2: 79-116

6. Michie DM. On not seeing things. In On Machine Intelligence. John Wiley New York, 1974

7. Barrow HG & Tenebaum JM. Recovering intrinsic scene characteristics from images. In : Hanson AR & Riseman E M (eds) Computer vision systems. Academic Press New York, 1978

8. Horn BKP. Determining lightness from an image. Computer Graphics and Image Processing 1974; 3: 277-299

9. Horn BKP. Obtaining shape from shading information. In: Winston PH (ed) The psychology of Computer vision. McGraw-Hill London, 1975

10. Horn BKP. Image intensity understanding. Artificial Intelligence 1977; 8: 201-231

11. Marr D. Early processing of visual information. Philosophical Transactions Royal Society London 1976; B 275: 483-524

12. Marr D. Analysis of occluding contour. Proceedings of the Royal Society London 1977; B 197: 441-475

13. Marr D. Artificial intelligence-a personal view. Artificial Intelligence 1977; 9 : 37-48

14. Marr D. Representing visual information- a computational approach. In: Hanson AR & Riseman EM (eds) Computer vision systems. Academic press New York, 1978

15. Marr D. Vision. WH Freeman San Francisco, 1982

16. Ullman S. The Interpretation of Visual Motion. MIT Press Cambridge, Mass., 1979

17. Longuet-Higgins HC & Prazdny K. The interpretation of a moving retinal image. Proceedings of the Royal Society London 1980; B 208: 385-397

18. Longuet-Higgins HC. A computer algorithm for reconstructing a scene from two projections. Nature 1981; 293: 433-435

19. Cowie R. The alternatives allowed by a rectangular postulate, and a pragmatic approach to interpreting motion. In: Hallam J & Mellish C (eds) Advances in Artificial Intelligence.Wiley Chichester, 1987

20. Longuet-Higgins HC. The visual ambiguity of a moving plane. Proceedings of the Royal Society London 1984; B 223: 165-175

21. Scott GL. Local and Global Processing of Moving Images. Pitman London, 1988

22. Subbarao M. Interpretation of Visual Motion: a computational study. Pitman London, 1988

23. Poggio T, Torre V & Koch C. Computational Vision and regularization theory. Nature 1985; 317: 314-319

24. Ames A. Visual Perception and the Rotating Trapezoid Window. Psychological Monographs 1951; 65: whole number 324

25. Cowie RID. Rotating trapezia which appear transparent and luminous during apparent reversals. Perception 1989; 18: 173-180

26. Wallach H & O Connell DN. The kinetic depth effect. Journal of Experimental Psychology 1953; 45: 205-217

Section 5:

Cognitive Modelling

Knowledge, Cognition and Acting in an Environment
Computational Neuroepistemology – an Alternative Invitation to Cognitive Science

Markus F. Peschl[1]

Abstract

This paper presents a project in cognitive science having the following aims:

- the *consequent* integration of methods, concepts, knowledge, etc. from computer science, epistemology/philosophy as well as from neuroscience,

- the investigation of *basic questions* concerning both natural and artificial cognition;

- the construction of a *model of cognition* based on epistemological and neuroscientific issues rather than on computer science constraints;

- the investigation of a *constructivist* concept of cognition; the implications for the understanding of knowledge, knowledge representation, language, etc..

1 Computational Neuroepistemology – the Theoretical and Methodological Background

Cognitive Science is a young emergent field of research developing dynamically which is, very generally speaking, interested in the investigation of cognitive processes. On the one hand it is well known that these processes represent one of the most *complex* and difficult phenomena science knows; on the other hand we are confronted with these phenomena every second, whenever we are thinking, acting, talking, etc. – everything we do is based on these processes. Hence, it is not surprising that cognitive science is far from being a well defined discipline having a well worked out

---

[1]Dept. for Epistemology and Cognitive Science (University of Vienna), Sensengasse 8/9, A–1090 WIEN, AUSTRIA, EUROPE, Tel. –43 222 42760141; Fax: –43 222 488838.

methodology. This is only one reason why a short introduction to an alternative approach to cognitive science will be given in this section; it is called *computational neuroepistemology (CNE)* and represents the methodological foundation of the following considerations. Some other reasons will be discussed in sections 2, 3 and 5.

As such complex phenomena have to be investigated in an *interdisciplinary* manner the basic idea of this approach was to *consequently integrate* philosophical/epistemological, neuroscience as well as computer science issues. This restriction and choice of disciplines will cause, of course, some questions, because "traditional" cognitive science contains more disciplines, such as (cognitive) psychology, linguistics, Artificial Intelligence, etc.:

- In working *interdisciplinarily* it has turned out that this form of scientific cooperation is almost impossible because of the following reasons:

  - The *dominance* of one discipline: there is always one discipline which dominates the discourse, the questions, etc.; this implies that the methods and standards of this discipline are compulsory for each of the other participating disciplines. In the long run this will cause the breakdown of the cooperation.

  - In many cases an interdisciplinary discourse is impossible, because the *levels of discussion* are completely different (and the participating discussion partners do not even notice that they are talking at cross purposes). Their fields of investigation are in many (most) cases completely *incompatible* for real interdisciplinary cooperation.

  - Interdisciplinarity often is (mis)understood as a "method" for the *justification* of the results being produced in one's own discipline[2]. Thus, interdisciplinarity is reduced to comparing results from different disciplines – such a strategy can not be called interdisciplinarity (it could rather be called multidisciplinarity), because no progress in the *common* development of knowledge can be found.

What does this mean for traditional cognitive science as well as for our alternative approach of computational neuroepistemology?

- If we are looking at the results, models, simulations, etc. of traditional cognitive science we can see that they are not only influenced, but rather *dominated* by the methods of *computer science*. Think, for instance, of the models of cognitive psychology (e.g. *Anderson* [2], *Mandl* et al [25], etc.); they are – in many cases – adopted from computer science, Artificial Intelligence or traditional cognitive science. Another example for the dominance of computer science in the investigation of cognitive processes is *Newell & Simon's Physical Symbol Systems Hypothesis* [33, 34, 35] which says (in short) that cognitive processes can be reduced to the manipulation of syntactic structures (i.e. symbols). Their hypotheses have caused a whole *paradigm* (in the sense of *T.S.Kuhn*

---

[2] Artificial Intelligence and cognitive psychology make use of the other's results for justification.

[24]) in computer science, AI and cognitive science – most of the models are (implicitly) based on these assumptions which are in no way justified.

Having a closer look at the assumption of cognitive processes being manipulation of symbols it turns out that it is neither based on empirical evidence nor has it any theoretical (natural scientific or epistemological) justification. It is rather very comfortable and convenient to program cognitive processes as symbol manipulation processes on a computer – the *v.Neumann* architecture (and the software and programming languages) very much influence, dominate and reinforce this assumption. Hence, computer science concepts *constrain* the process of modeling, interpreting, simulating and understanding cognition – what is the justification for such an action?

- *Impossibility of a successful interdisciplinary discourse.* As has been mentioned above, in traditional cognitive science many different disciplines arguing on different logical levels are involved. It is clear that, for instance, AI people (being interested in automatically playing chess or in an "intelligent" reasoning mechanism for expert systems) and researchers from neuroscience (investigating very small circuits of neurons) cannot find a common level of (scientific) discussion as a wide gap between their methods, models, strategies, etc. can be found; this gap cannot really be filled, because their aims are different. This example illustrates in a very exaggerated manner the difficulties and problems arising in an interdisciplinary discourse. That is why we have to look out for an alternative concept of interdisciplinary cooperation in cognitive science; a concept complying with the following points:

  - *common level of discourse*;

  - *consequent integration* of disciplinary knowledge in the other disciplines' work;

  - approaching a phenomenon from the problem itself, not from the background of different disciplines and methods ("*problem oriented approach*").

- The reduction to three disciplines (epistemology, neuroscience and computer science) and their interaction being suggested by the computational neuroepistemology approach seems to meet these claims for the following reasons:

  - the common level of discourse is given by the *neuroscientific foundation* of this alternative approach to cognitive science: i.e. neural activities are assumed to be the basis for all our cognitive processes, knowledge, communication, understanding, etc. as well as for a scientific discourse. Hence, the level of discussion is based on this assumption – "higher" cognitive phenomena are emergent properties of neural activities.

  - *computer science* is reduced to *Parallel Distributed Processing* (PDP, connectionism, neural computing, *Rumelhart & McClelland* [44, 30, 31], etc.); we will shortly discuss this approach in section 3. It is obvious that it has a neuroscientific background.

- *epistemology* is understood as naturalistic epistemology as it is suggested by the *Churchlands* [5, 6, 7, 8, 48] or by *Oeser* [37] – we could call it a neuroepistemological or neurophilosophical approach.

- if we are assuming scientific cooperation to be a *cybernetic system* (i.e. the flow of information, knowledge, etc., *W.Krohn & Küppers* [20, 21, 22]) we can understand why most interdisciplinary research so often is doomed to failure; the reasons have been discussed above. From a cybernetic point of view we can look at interdisciplinary research as a *feedback system* trying to find a state of *equilibrium*. This feedback process can only be successful if this loop is closed; i.e. if the interfaces between the disciplines are compatible a closure in this sense is possible and, thus, a closed (feedback) flow of knowledge is enabled. Such a system will sooner or later head for a stable state (representing a common result from the interdisciplinary cooperation). As has been shown above such a closure is possible in the case of computational neuroepistemology, because of the common level of discourse.

- another justification for the reduction to these three disciplines is that many other disciplines, such as linguistics or cognitive psychology are *emerging* from this approach; of course, they would have completely different methods, results, etc., if they were based on this approach – if we are interested in adequate models, descriptions and explanations of cognition, language, etc. we have to start with investigations at the "roots" of these processes. In our age we do *not* have the excuse of having not enough (empirical neuroscientific) knowledge. Sure, many problems are still unsolved in neuroscience, however, psychology has not really made use of the present knowledge being provided by neuroscience so far.

• The aim of this alternative approach to cognitive science is *not* to build commercial systems – rather computational neuroepistemology investigates *basic* problems concerning questions of knowledge and knowledge representation. In the long run, however, these results will become interesting for commercial applications too, because the demands are continuously increasing: more complex knowledge, more common sense knowledge, etc. for more "intelligent" and "smarter" systems.

This short overview of an alternative approach to cognitive science illustrates the background of the aims and models to be presented in sections 4 and 5. It has turned out that we understand cognitive science as the *continuation* of traditional epistemology in the context of modern (natural) scientific knowledge and of modern (computer) technology. In the following sections the contributions and the role of each of the three disciplines taking part in the computational neuroscience approach will be sketched.

1.1   Epistemology

As has been mentioned above epistemology is understood as neuroepistemology or neurophilosophy in the sense of *Churchland* or *Oeser*. Epistemology represents the science of *knowledge* – knowledge is not restricted to certain forms, such as scientific knowledge or linguistic knowledge, in this approach. It rather reaches from very simple and "primitive" forms (for instance reflexes or knowledge having been learned by conditioning) over more complex forms of knowledge, such as communicative behaviour, language) up to very complex constructs such as scientific knowledge. Of course, this is a very wide range; if we do not consider such a range we run the risk – as orthodox AI or traditional cognitive science did – of losing the context we are researching in and of becoming specialists for a very restricted problem domain.

On the one hand epistemology plays the role of contributing *speculative* knowledge. It has a very long and old tradition providing much knowledge on the questions of knowledge representation, human knowledge, etc.. Up to our time this knowledge has not been affected by modern scientific (neuroscience) knowledge, because empirical results were quite poor compared to the complexity of (cognitive) phenomena which have been investigated, described, explained by philosophy (of mind, of language) and epistemology. In our time, epistemology's knowledge can*not* be seen as isolated any more. As it is discussed in section 1.2 there is no excuse of the lack of empirical data any more; thus neuroscience places *constraints* on the speculative ideas of epistemology. In this approach to cognitive science (CNE) epistemology has to accept these constraints as an input for the developments of further theories, hypotheses, etc.. In computational neuroepistemology the epistemological concepts of *constructivism* are introduced; they are discussed in section 2.

On the other hand epistemology plays the role of *reflecting* the results, methods, the paradigm, etc. of neuroscience and computer science. Thus, it has not only a "speculative" contribution, but also a *regulating* function in this feedback process of interdisciplinarily developing knowledge. By making explicit the implicit assumptions being made in the neuroscience's paradigm it has to articulate the problems and misleading assumptions.

1.2   Neuroscience

Neuroscience provides this "cybernetically organized system of interdisciplinary cooperation in cognitive science" (i.e. computational neuroepistemology) with *empirical data*. These data are the result of very sophisticated methods and theories. As it is well known there are several levels of discussion within neuroscience (*Sejnowski & Churchland* [48]) reaching from molecular biology over the investigation of processing taking place in synapses and cells up to the investigations of more or less complex (natural) neural networks. Each of these levels is of interest and relevance for our approach – for the interdisciplinary discussion, however, they have to be "(pre)processed" so that they can be understood and modelled by computer scientists as well as by epistemologists.

The aim of the investigations is to achieve a deeper understanding of *knowledge representation* in natural nervous systems – this aim is often forgotten in the course of the very detailed results. In this context the process of *learning* seems to be one of the most important aspects; there are, however, many different opinions and hypotheses on these problems ([9, 23, 1, 8, 49], etc.).

In this approach neuroscience *always* has to reflect on its methods and the paradigm it is working in. As is shown by *P.Feyerabend* [10] (so called "objective") empirical results are always contaminated by the theories and the methods of the chosen paradigm (in the sense of *T.S.Kuhn* [24]). The requirement for a successful interdisciplinary cooperation is the willingness to *reflect* and – if required – to change the paradigm.

1.3 Computer Science

The contribution of computer science will be discussed in section 3 in detail. In computational neuroepistemology computer science is *not* the dominating discipline. It is only interesting because of the possibilities being offered by the computer's very high flexibility, plasticity and speed. As has been discussed we do not follow the symbolic paradigm of modeling cognitive processes, but rather apply the alternative form of *neural computing*.

Computer science plays the role of *verifying* empirical data as well as of "*triggering*" new or alternative models, investigations, theories, etc.. The *simulation* and modeling of cognitive phenomena represent the central task for this discipline.

2 Constructivist Aspects

Having discussed the methodological background of this project I am going to focus in this section on some *epistemological* issues which are of interest for our cognitive model: a *constructivist* perspective and its implications are presented. As we have seen above this aspect represents only a part of the contributions being made by epistemology, philosophy and philosophy of science.

Knowledge is understood as the result of a process of *construction*. What does this mean? As has been mentioned above I do not have a restricted view of what knowledge is – knowledge in this sense reaches from very simple forms of behaviour up to complex scientific knowledge. This assumption of constructing knowledge is not new and has become even more relevant in the context and in the light of our modern (natural) scientific knowledge (especially in neuroscience and computer science). Hence, we understand knowledge not only as the result of a process of construction but also as a result of a process taking place *physically* in natural (or artificial) cognitive systems (nervous systems). *H.Maturana* [26, 27, 28, 29], *H.v.Foerster* [11, 12], *E.v.Glasersfeld* [14, 15, 16], etc. are the most important exponents of this approach to understanding cognition.

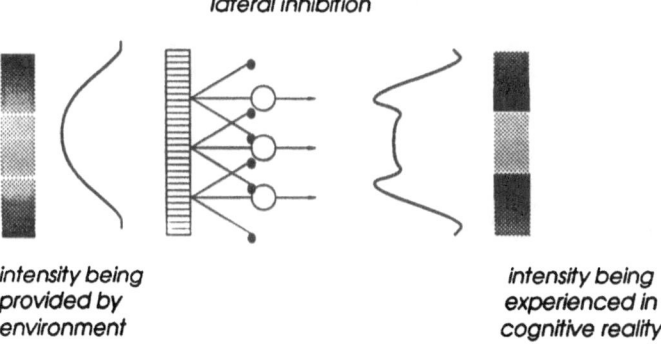

Figure 1: An example of the *construction* of cognitive reality: lateral inhibition increases the contrast.

Figure 2: Black rectangles with grey crosses.

For the reason of motivation I want to give an example of the process of construction in our everyday lives; it is well known and understood from investigations of the visual system: *lateral inhibition* plays an important role for increasing the contrast not only in the domain of vision but also for auditory or tactile modalities. If we look at fig. 1 we can see how such an increase of contrast is "realized" in "wet-ware". Each neuron locally inhibits its neighbours – an alternative kind of a "winner-takes-all" mechanism is responsible for the global increase of contrast.

This process of construction takes place even in peripheral parts in the first steps of processing a stimulus – it "produces" the impression of a reality with high contrast. Other examples which are well known are the receptive fields which were investigated first by *Hubel & Wiesel* [17, 18, 19]. A further example of constructing knowledge can be seen in fig. 2 where the black rectangles cause a grey glimmer in the white crosses. What conclusions can be drawn from these examples?

- In each case the construction is realized in *physical* processes, in the architecture of the (natural or artificial) nervous system. We are assuming with Maturana and *v.Foerster* that the nervous system is a closed system; its ar-

216

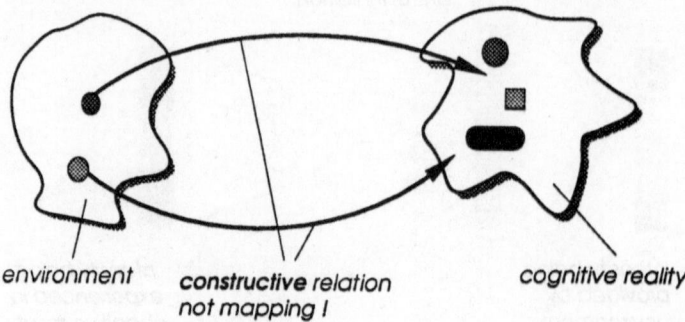

environment   **constructive** relation   cognitive reality
not mapping !

Figure 3: A process of construction is the relation between environment and cognitive reality.

chitecture is *highly recurrent* and, thus, has its own dynamics triggered by external (and internal) stimuli ("perturbations"). As *H.v.Foerster* puts it, the nervous system is a self-organizing system computing its own stable (cognitive) reality ([11], p 57). From a cybernetical point of view we can look at it in the sense of a *homeostatic* system.

- As an implication of the examples discussed above we have to accept that there seems to be a difference between the "real" stimulus and the way we experience it. In the case of fig. 2, for instance, we can be quite sure that the paper on which the black rectangles are printed is white (and not grey). Hence, we have to give up the assumption that the experienced world is the mapping of "real" world – one aim of this project is to show that it is wrong to assume a trivial isomorphic relation between the "real" world which I will call "*environment*" and the experienced world, which will be called "*cognitive reality*". We have to clearly differentiate between these two domains; the relation between them is rather an *active* process of *construction* than a passive process of mapping (see fig. 3).

- This concept has very important implications for our assumptions of knowledge, knowledge representation, cognition and of language: knowledge cannot be understood as something "objective" – it is rather the result of very private construction processes of construction, which can be accepted, if a group of organisms "agrees" to it, by a population of cognitive systems as "public" or "objective" with regard to this group. The implications for science or other domains cannot be discussed here, because this discussion fills many books in philosophy of science. As we are capable of dealing only with quite simple forms of knowledge and behaviour which represent the foundation for more complex forms we are going to restrict these problems to very basic questions, such as how knowledge or meaning arises in a population) of (natural or artificial) cognitive system(s).

It has become clear from the discussion that I am proposing a *bottom-up* approach to these problems, as I am giving up the.assumption of a "given" (natural) language or of cognition being the result of manipulating symbols, etc. – we are rather interested in the questions *behind* these assumptions; they are completely ignored by traditional approaches of Artificial Intelligence or cognitive science. They become evident and important, however, if the problems and contributions from epistemology/philosophy are taken seriously, as is suggested by the computational neuroepistemology approach (see section 1).

- As another implication we cannot assume a "public" meaning of a symbol, word, etc. any more. Rather we have to look at it as the result of the private history and experiences, of the state and of the structure (architecture) of the nervous system. Communication arises as the result of a process of mutual adaptation to the use of a symbol; *H.Maturana* calls this the establishment of a *consensual domain* [26, 28]. One aim of these investigations and simulations is to find out how a symbol gets its meaning.

To conclude this section: knowledge is the result of a process of construction. How can we model such a process? As we will see in section 3 the *Parallel Distributed Processing* paradigm offers such mechanisms which are not based on (natural) language or symbols. If we are interested in the phenomenon of meaning and symbolization we can*not* make use of symbols themselves (as orthodox AI or traditional cognitive science pretends when investigating "semantics")! We will see that the process of *adaptation* plays an important role (*Beer* [3]) in the construction of knowledge – cognitive reality (knowledge) has to adapt in such a way to the environment that it *fits* into it (*v.Glasersfeld* [15]). This understanding of knowledge represents the basis for all considerations in the following sections. Hence, knowledge (knowledge representation) is responsible for *acting adequately* in our environment (*H.Maturana*, [26]).

We will see that the process of *association* is closely related to the construction of knowledge. It links together impressions, experiences, phenomena, etc. which coincide in time or space. This causes a first step towards the construction of complex knowledge and towards *inductively* building up hypotheses. It seems that the meaning of a symbol is also one result of such a process of association (or construction).

3   Neural Computation – an Alternative?

In this section I want to focus on the alternative paradigm of connectionism (Parallel Distributed Processing, neural computing, etc.). It cannot be discussed in detail (for a detailed overview I can recommend the "bible" of PDP: the three volumes by *Rumelhart & McClelland* [44, 30, 31]; a short introduction is given in *Rumelhart* [45]). To give a short characterization: the aim is to simulate *neurons* which are assumed to be the functional "atoms" of our brain and cognitive processes. As

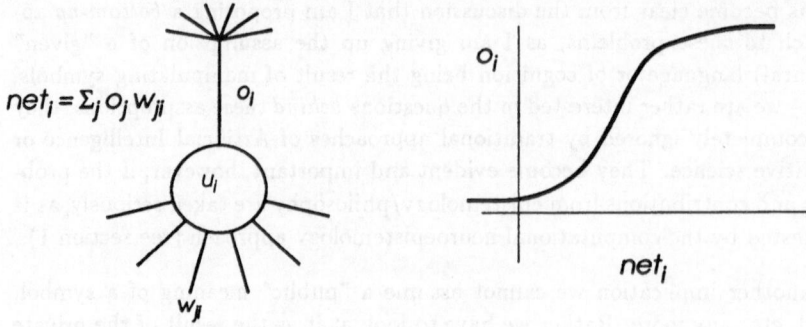

$net_i = \Sigma_j o_j w_{ji}$

Figure 4: One unit and typical output functions

is well known these (natural) neurons form a huge network containing about 10^{11} neurons which are highly interconnected and all work in parallel. These connections are used for exchanging "information" in the form of pulses or *spikes* which are spreading through the whole network; they may be caused by an internal neuron or by an external stimulus being transformed into such a spike by a sensor neuron. As these neurons are all working in *parallel* each of them has a certain activity ($a(t)$ at a certain time t. This activity is – put very simply – the result of the others neuron's (being connected to the neuron) activities and the activity at time $t - \varepsilon$. On the output side of this (highly recurrent) network these activities are transformed into action by effectors. We will see that this forms a feedback loop.

The aim of artificial neural networks (PDP networks) is to simulate these natural neural networks. Of course, they represent only very (mathematically) *abstract* and rough models of their natural originals; the relation between these artificial and natural neurons is discussed, for instance, by *P.M.Churchland* [8]. Many attempts have been made to model the function of a single neuron in detail (e.g. [4, 39, 32, 46, 47, 51], etc.). The connectionist approach, however, is not so much interested in the function of a single neuron, but rather in the global behaviour of quite simple processors called "*units*". One unit represents a neuron in a very abstract manner: it is connected by "*weights*" (they could be compared to the synaptic connections) to some other units and computes its own activation by summing up the product of the weights and the other units' activations. As shown in fig. 4 the output of the unit is computed by a sigmoid or linear function mapping this sum of products ("net input") to a certain range which is typical $[0, 1]$ or $[-1, 1]$, etc.. This output again acts as an input for other units doing exactly the same in parallel. In most cases a synchronous and discrete timing is assumed.

The interesting thing about connectionism is that these networks are capable of *learning*. This process of learning has a strong influence on how *knowledge* is *represented*. This is very closely related to the problem how this knowledge is acquired by the system. Put simply spoken, PDP networks learn by *associating* two or more patterns (of activity); i.e. if we assume a feedforward network[3] a certain pattern of activations is put to the input units and represents the "input vector". These activations are propagated through the network and produce a vector of activations in the output layer. We can look at this as a process of *association*; i.e. the input vector is associated with the output vector. *Learning* means that we can change this process of mapping of input vectors to output vectors (either by an external teacher determining the input-output relation or by an process of self-organization). This is realized by slightly changing the weights of the network – very small *local* changes cause a change in *global* behaviour of the network. These changes in the weights are computed by a *learning rule* adding or subtracting a small increment to/from the weight. I cannot go into detail here – put simply, these learning rules represent a kind of self-organizing mechanism trying to minimize an error or trying to find a minimum. Thus, we can PDP networks regard as a very *general information processing concept* transforming any input vector of activations to any output pattern of activations by propagating patterns of activations.

We will see, however, that this simple assumption of a feed forward PDP network is *not* adequate for modeling cognitive processes; this would reduce the cognitive system to a "stimulus-response machine". A *recurrent* network architecture avoids this simplification, because it has its own *dynamics* and has not to be stimulated by an external stimulus. Hence, the system has a kind of "intentionality" which is determined by the architecture of the PDP network, by the current state of activations, by the experiences (i.e. by the changes in the weights) as well as by the stimuli. In the following section we will have a closer look at this suggested model which is based on such a recurrent PDP network being connected to the environment via sensors and effectors.

4 A Suggestion for a Cognitive Model

The whole simulation takes place inside the computer; i.e. the cognitive model as well as its environment are simulated in software. This cognitive model (it is called $\kappa o\gamma\mu o\delta$) has the shape of a rectangle and moves around in its environment which is reduced to two dimensions. The reason for this reduction lies in limitation of computational power. As has been mentioned one of the most important points of this approach is that *no* symbolic interaction between the environment and $\kappa o\gamma\mu o\delta$ takes place. How is this realized? For more technical details see *Peschl* [40, 41]; $\kappa o\gamma\mu o\delta$ is equipped with sensors for different modalities and with effectors for different actions (a simplified sketch can be seen in fig. 5):

- effectors for *moving* around: $\kappa o\gamma\mu o\delta$ is capable of moving around in any direction; this is realized by four effectors each being responsible for an acceleration in one direction (each 90°). Each of these effectors is linked to a unit of the PDP network – this unit's activity is transformed into an acceleration. By

---

[3]The activations are propagated only in one direction: from an input layer over one or more "hidden" layers to the output layer. There is no recurrent flow of information.

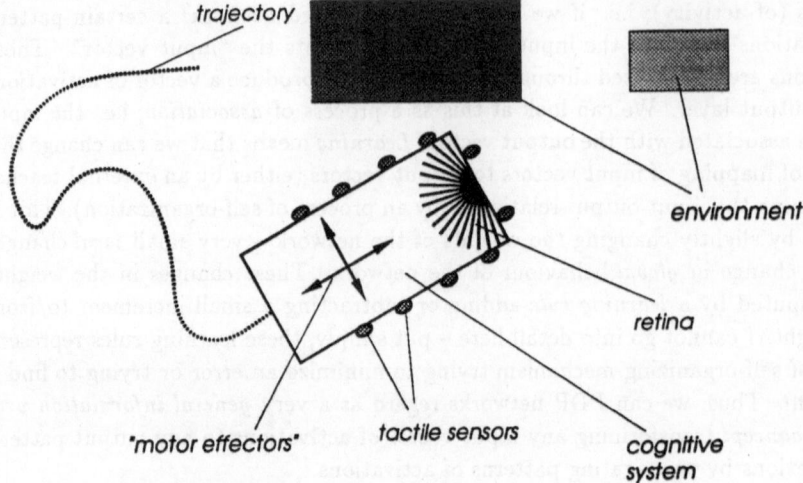

Figure 5: κογμοδand its environment

summing up these four accelerations a new direction and speed are computed; the result is a new position in the environment.

- effectors for *sending out signals*: the artificial neural network can be coupled to an effector which is capable of sending out acoustic signals. These signals will become more interesting if two or more cognitive systems are interacting.

- *optical sensors*: κογμοδis equipped with one or more artificial retinas which are capable of perceiving optical stimuli from their environment. By making use of techniques from computer graphics (ray tracing, etc.; [13, 36]) the optical impression of the environment can be transformed into neural activities.

- *tactile sensors*: these transform any collision with the environment into neural activities perturbating the recurrent architecture and dynamics of the PDP network of κογμοδ.

- *acoustic sensors*: acoustic signals are transformed into neural activities by these sensors.

If we look at the sensors, the constructivist aspect of perceiving the environment becomes clear: there is *no* passive mapping from environment to cognitive reality (see fig. 3). The optical impression determined by the artificial retina, for instance, very much depends on the density of the photo receptors and on their sensitivity. We have seen that the first steps of processing have rather a construction character than isomorphically mapping environment to cognitive reality. Cognitive reality has a completely different quality; it is the domain of meaningless patterns of activations which dynamically change. The nervous system's architecture (connectivity, etc.) brings together categories which may appear in the environment independently. Knowledge has been characterized as the result of a process of construction – the aim of this construction is to act adequately in the environment; knowledge is the association of phenomena, modalities, experiences, etc. which may or may not belong together. The aspect of *acting* in the environment, actively *changing* it shows the success of these constructions.

These simulations are computed on two SUN "sparc stations" – one for the "pure simulation" and the other for interaction and graphical presentation of the results. The software is programmed very flexibly: the environment consisting of boxes of different size and colour as well as the cognitive model can be designed; this implies that we can study the effect of a change in the architecture of the PDP network on the observed behaviour or on the knowledge representation. It is possible to interactively communicate with the simulation part; at each time step the pattern of activations or of weights as well as, for instance, the state of sensors and effectors can be inspected and graphically displayed, etc.. These functions are very important as we are interested in the relation between knowledge representation, learning, environment and the structure of the network. For the future the following things are planned:

(i) to increase the number of cognitive systems to two or more; the reason for this is an interest in how *communication* between two or more cognitive systems can be established; how a (very simple form) of a symbol system can be set up; how symbols get their meaning, etc.;

(ii) to build a physical model which moves around in a physical environment – as we will see this would not make much difference as sensors and effectors are connected to an artificial nervous system which is in both cases simulated by software. The only difference is in the way the environment is perceived or how it is changed; i.e. by which means stimuli are generated and by which means neural activities are transformed into actions changing the environment.

5 Discussion and Conclusions

What conclusions can be drawn from this approach? The original idea was to find and construct an adequate model of cognition. Traditional cognitive science, however, could not achieve this aim, because it has chosen a processing paradigm which is too naive and inadequate for the complexity of the questions concerning cognition, language, etc.; hence, an alternative methodological conception of cognitive science and its implications (in the form of an alternative model of cognition) were presented:

• the *computational neuroepistemology (CNE)* approach represents the methodological foundation for our investigations: computer science, its methods and concepts do not play such an important role any more, because of the *consequent integration* of the knowledge of epistemology, neuroscience and computer science. Each of these disciplines has "equal rights" and *has to* integrate the results of the interdisciplinary discourse in its own work.

- *constructivist* ideas (coming from epistemology) and the *Parallel Distributed Processing* paradigm (coming from computer science and neuroscience) play a central role in this approach to modeling and understanding cognition: the process of learning in natural as well as in connectionist networks can be understood as a process of *constructing* knowledge, of constructing a private cognitive reality which is triggered by an external environment. This approach to representing knowledge goes far beyond the borders of linguistic categories and is rather behavioural.

- as an implication a model of cognition has been presented which tries to come up to these epistemological (and neuroscientific) claims. It does *not* focus on computer science problems but rather on basic questions of natural as well as artificial knowledge representation. As this model is embedded in an environment and is linked to this environment over sensors and effectors threre is *no* symbolic instance between the perception of this environment and the "internal" representation of it.

The implications of this approach in cognitive science are diverse:

(i) *bottom up vs. top down*:

we are trying to understand cognitive processes in a bottom up manner; i.e. this approach is not so much interested in very complex and formal problem solving strategies or reasoning mechanisms – this is done quite well by AI and traditional cognitive science which reduce cognitive processes to manipulating symbols. Their top down strategies have brought about some quite spectacular results, such as computers playing chess or making a medical diagnosis; these results, however, must not mislead us that this has anything in common with cognitive processes. We have to be aware of the fact that cognition is "neurally grounded" (*Churchland* [7]) and, thus, has another basis apart from the manipulation of propositions.

We have to understand cognitive processes rather as processing from one activation pattern to another than from one sentence to another. The aim is to achieve a deeper understanding of basic processes, such as physical interaction with the environment or dynamics of activation patterns. "Higher" cognitive processes are embedded in these "low level" activities.

(ii) *new understanding of cognition*:

as an implication of (i) we have to *rethink* and reformulate our assumptions of what cognition is; commonly it is assumed that cognitive processes take place in (natural) language categories. From this alternative perspective we have to give up this assumption in favour of a more *behavioural* view: language, thinking, acting, etc. is an expression of *behaviour*, *not* of a linguistic process – hence, we have to give up

  (a) the *natural language* level: i.e. the level of describing cognitive processes as results of manipulations of linguistic descriptions.

(b) the *syntactic* level: cognition cannot be reduced to the manipulation of meaningless syntactic structures,

if we want to model and understand cognitive processes.

(iii) *"cybernetics and behaviour"*:
as has been mentioned above, "higher" cognitive processes, such as language, formal reasoning, etc., have to be understood as a very complex form of behaviour in this approach. They are embedded in these low level processes of spreading activations, local changes of connectivities, etc.. This model can be seen as a cybernetical system having its own complex dynamics – the flow of activation, the feedback loops, the recurrent architecture, etc. are responsible for the (externally) observed global *behaviour* of the cognitive model. The aim is, however, not only to observe the behaviour, but rather to investigate the relations between the architecture, the construction and structure of the knowledge representation, the external environment and how this organism behaves in it.

(iv) *knowledge representation*:
Knowledge representation is understood as the result of a process of *construction* and adaptation; it is not restricted to (natural) linguistic categories ("subsymbolic[4]", *P.Smolensky* [50]). This construction is realized in the process of learning and building up associations. It is an inductive process and is not logically justified. The feedback over the environment, however, could be compared to a "deductive" instance in the following sense: by acting in the environment the constructs of knowledge representation are "tested" if they are suitable, if they "fit", if they are adequate at all. This "experience" acts as an input for the next step of construction, change, etc. of the cognitive reality (i.e. of knowledge representation). Thus, we can look at it as a sort of "spiral" developing over time: induction and deduction are alternately applied for the development of knowledge; this is a well known strategy in philosophy of science (*Oeser* [38]).

(v) *applications*:
Of course, this suggestion of an alternative approach to cognitive science and the suggested model do not have any commercial application at the moment, because the investigation of basic research problems, such as of *common sense knowledge*, is necessary. If (commercially) reasonable mechanisms and strategies for the processing of, for instance, common sense knowledge can be found, these investigations could be the theoretical background for the development of:

  − more intelligent *human-computer interfaces*: we could learn a lot about the representation and processing of various forms of knowledge; this

---

[4]This term is still quite vague and it will have to be discussed in this approach – it will even change its meaning from a constructivist point of view.

could help us in designing user interfaces which are, for instance, "cognitive adaptive", i.e. the computer adapts itself to the nuances of the user's knowledge, use of language (meaning), etc..

- the integration of *real common sense knowledge* and not only of "pseudo common sense knowledge" of a very restricted and formal domain. If we are going beyond the borders of (natural) language and symbols *implicit or tacit knowledge* (in the sense of *M.Polanyi* [43]) could be included in "intelligent" systems as well. As has been shown this would imply that a pragmatic as well as (real) semantic aspect (not being reduced to syntax) would be introduced in the processing of knowledge. One could think, for instance, of hybrid systems making use of the advantages of both formal symbol manipulation and subsymbolic processing [42][5].

- this approach represents an interesting alternative for the domain of traditional *language processing and understanding*. The discussed understanding of language being ("only") a very complex form of behaviour making use of a medium for exchanging signals (patterns) which are based on a consensual code could bring forth a new concept of semantics and processing of semantics which is based on a *"pre-symbolic"* level.

- another application for this approach lies in the field of robotics: *autonomous vehicles* which operate in a more or less common sense context which is not restricted to blocks (world) and which is not predetermined could make use of the ideas and concepts presented in this paper as the aim is to *autonomously construct* a cognitive reality which fits into the environment.

There are, however, clear *limits* of application: as we have discussed above we assume that each (artificial as well as natural) cognitive system has its own, very private cognitive reality (i.e. its view of the world) determined by the experiences, structure of the nervous system, etc.; the implication of this assumption is that we can *not* fall back upon (the illusion of) an "objective" meaning or use of language. This may cause some problems in very formal domains.

What are the aims of this project? What is the use of observing one or more artificial cognitive systems moving around in an environment? What is the use of such a methodological approach to cognitive science?

(a) as has been mentioned, the most important aim is to achieve a deeper understanding of an alternative form of *knowledge representation*, a form which

---

[5]This approach, however, *must not* be misunderstood as a model for cognition!

is not restricted to linguistic or symbolic constraints. The syntactic representation is given up in favour of a "pre-symbolic" (cybernetically inspired) knowledge representation ($\Rightarrow$"embodiment of environment");

(b) the investigation of the *relation* (in the context of (a)) of: environment – behaviour – structure & architecture of the (natural or artificial) nervous system – knowledge representation – process of learning & construction.

(c) finally I want to ask a question in reply: what is the aim of a natural cognitive system?; does it have any explicit aims (besides surviving, etc.) as, for instance, an expert system?

One of the most important points of this approach is the aspect of a cognitive system actively *acting* in its environment. This perspective is often neglected or even completely ignored. This seems to be due to the failure of computer science to provide the required concepts. Hence, the approach of *computational neuroepistemology* has been suggested, an approach which addresses these problems and provides solutions and paradigms. The aspect of acting is of great importance because of the following reasons: it closes the feedback loop between the effectors and sensors over an external environment, it claims for an *direct* relation to the environment ("being-in-the-world"), and it enables a *verification* whether the constructs of cognitive reality fit in the environment. From a philosophy of science perspective this is a very interesting process as it is very important for the theory of the development of new theories, hypotheses, etc.. Of course, much work has to be done until we reach this level of complexity, the foundation of these processes, however, can be studies in this approach. So, things have come full circle – the aim was to find a model, approach, concept, suggestion, etc. for the investigation of various forms of knowledge, how they can be integrated, what the "interface" to (natural or artificial) neural activities could look like, etc.; the computational neuroepistemology approach presented in this treatise could be a step towards the solution of these problems and could (in principle) fulfill most of these claims.

Open Questions & Problems for Further Investigation

Many questions and problems are left unanswered, many problems will have to be investigated more closely. In the following I want to present only a small selection:

- Can a more general theory of *knowledge representation* be found which is based on the assumptions being suggested in this paper (behavioural, "pre-symbolic", etc.)?

- What could the "interface" between "higher cognitive processes" and lower processes look like? What role is the concept of *emergence* (F. Varela [52]) playing in this context?

- How do symbols get their meaning (investigation of the process of symbolization)?

- What is the relation between language and the dynamics of neural activation patterns?

References

[1] Alberts B. (1987): Molekularbiologie der Zelle (deutsche Übersetzung von: Molecular Biology of The Cell); *VCH Verlagsgesellschaft, Weinheim (D), Deerfield Beach (FL), 1987.*

[2] Anderson J.R. (1988): Kognitive Psychologie; *Heidelberg, Spektrum der Wissenschaft Verlag, 1988.*

[3] Beer R.D. (1990): Intelligence as Adaptive Behavior. An Experiment in Computational Neuroethology; *Academic Press, Boston, New York, 1990.*

[4] Caianiello E.R. & de Luca A. (1965): Decision Equation of Binary Systems. Application to Neural Behavior; *Kybernetik (Bilogical Cybernetics)' 3, pp 33 – 40 (1965).*

[5] Churchland P.S. (1986): Neurophilosophy. Toward a Unified Science of the Brain; *MIT Press, 1986.*

[6] Churchland P.M. (1989): A Neurocomputational Perspective – The Nature of Mind and the Structure of Science; *MIT Press, 1989, Cambridge, MA.*

[7] Churchland P.M. & Churchland P.S. (1990): Could a Machine Think?; *Scientific American, January 1990, pp 26-31.*

[8] Churchland P.M. (1990): Cognitive Activity in Artificial Neural Networks; *in Osherson et al. (eds.), An Invitation to Cognitive Science, MIT Press, Massachusetts, Vol. 3, pp 199-227, 1990.*

[9] Dudai Y. (1989): The Neurobiology of Memory. Concepts, Findings, Trends; *Oxford University Press, Oxford, 1989.*

[10] Feyerabend P. (1983): Wider den Methodenzwang (Against Method); *Suhrkamp, stw 597, Frankfurt/M., 1983.*

[11] von Foerster H. (1973): Das Konstruieren einer Wirklichkeit; *in Watzlawick (ed.), Die erfundene Wirklichkeit, Piper Verlag 1981, pp 39-60, 1973.*

[12] von Foerster H. (1984): Erkenntnistheorien und Selbstorganisation; *in Schmidt, Der Diskurs des Radikalen Konstruktivismus, pp 133-158, Suhrkamp, stw 636 (1987).*

[13] Foley J.D. & van Dam A. (1982): Fundamentals of Interactive Computer Graphics; *Addison-Wesly Publishing Company, 1982.*

[14] von Glasersfeld E.v. (1976): Sprache als zweckorientiertes Verhalten: zur Entwicklungsgeschichte; *in E.v.Glasersfeld, Wissen, Sprache und Wirklichkeit, Vieweg 1987, pp 63-79.*

[15] von Glasersfeld E.v. (1981): Einführung in den radikalen Konstruktivismus; *in Wat-zlawick P. (ed.), Die erfundene Wirklichkeit, Piper Verlag, 1981, pp 16–38, 1981.*

[16] von Glasersfeld E.v. (1987): Siegener Gespräche über den Radikalen Konstruk-tivismus; *in Schmidt, Der Diskurs des Radikalen Konstruktivismus, pp 401–440, Suhrkamp, stw 636 (1987).*

[17] Hubel D.A. & Wiesel T.N. (1962): Receptive Fields, binocular interaction and func-tional architecture in the cat's visual cortex; *Journal Physiol. 160, pp 106–154.*

[18] Hubel D.A. & Wiesel T.N. (1965): Receptive Fields and functional architecture in two non-striate visual areas; *Journal Neurophysiology 28, pp 29–289.*

[19] Hubel D.A. & Wiesel T.N. (1968): Receptive Fields and functional architecture of monkey striate cortex; *Journal Physiol. 195, pp 215–243.*

[20] Krohn W., Küppers G. Paslack R. (1987): Selbstorganisation – Zur Genese und En-twicklung einer wissenschaftlichen Revolution; *in Schmidt, Der Diskurs des Radikalen Konstruktivismus, pp 441–465, Suhrkamp, stw 636 (1987).*

[21] Krohn W. & Küppers W. (1988): Die Selbastorganisation der Wissenschaft; *Suhrkamp, Frankfurt/M., stw 776, 1988.*

[22] Krohn W. & Küppers G. (1990): Science as a Self-Organizing System. Outline of a Theoretical Model; *in Krohn et al (eds.), Selforganization. Portrait of a Scientific Revolution, Kluwer Academic Publishers, Netherlands, 1990, pp 208–222.*

[23] Kuffler St.W., Nicholls J.G. & Martin A.R. (1984): From Neuron to Brain, A Cellular Approach to the Function of the Nervous System (second edition); *Sinauer Associates Inc. Publishers, Sunderland, Massachusetts, 1984.*

[24] Kuhn T.S. (1967): Die Struktur wissenschaftlicher Revolutionen; *Suhrkamp Taschen-buch, Frankfurt, stw 25.*

[25] Mandl H. & Spada H. (eds.) (1988): Wissenspsyochlogie; *Psychologie Verlagsunion, München - Weinheim 1988.*

[26] Maturana H.R. (1970): Biology of Cognition; *in H.R.Maturana & F.J.Varela, Au-topoiesis and Cognition; pp 2–60, D.Reidel Publishing Company, Dordrecht, Boston (1980).*

[27] Maturana H.R. & Varela F.J. (1975): Autopoiesis: The Organization of the Living; *in H.R.Maturana & F.J.Varela, Autopoiesis and Cognition; pp 63–134, D.Reidel Publishing Company, Dordrecht, Boston (1980).*

[28] Maturana H.R. & Varela F.J. (1980): Autopoiesis and Cognition. The Realization of the Living; *D.Reidel Publishing Company, Dordrecht, Boston, London (1980).*

[29] Maturana H.R. (1983): What is it to see?; *Arch. Biol. Med. Exp. 16 (1983), p 255–269, (Chile).*

[30] McClelland J.L. & Rumelhart D.E. (1986): Parallen Distributed Processing, Explorations in the Microstructure of Cognition, Volume II: Psychological and Biological Models; *MIT Press, Cambridge, Massachusetts 1986.*

[31] McClelland J.L. & Rumelhart D.E. (1988): Explorations in Parallel Distributed Processing; *MIT Press, Cambridge, Massachusetts.*

[32] Nagumo J. & Sato S. (1972): On a Response Characteristic of a Mathematical Neuron Model; *Kybernetik (Biological Cybernetics) 10, pp 155–164.*

[33] Newell A. & Simon H.A. (1976): Computer Science as Empirical Inquiry: Symbols and Search; *Communications of the ACM, March 1976, Vol. 19, Number 3, pp 113–126*

[34] Newell A. (1980): Physical Symbol Systems; *Cognitive Science 4 (1980), pp 135–183.*

[35] Newell A. & Simon H.A. (1981): Computer Science as Empirical Inquiry: Symbols and Search; *in J. Haugeland (ed.): Mind Design. Cambridge/Mass., London(GB): The MIT Press, pp 35–66.*

[36] Newman W.M. & Sproull R.F. (1979): Principles of Interactive Computer Graphics; *McGraw Hill International Book Company, 1979.*

[37] Oeser E. & Seitelberger F. (1988): Gehirn, Bewußtsein und Erkenntnis; *Wissenschaftliche Buchgesellschaft Darmstadt, 1988.*

[38] Oeser E. (1990): The evolution of scientific methods; *Fresenius' Journal of Analytical Chemistry (1990) 337, pp 150–154.*

[39] Perkel D.H., Schulmann J.H., Bullock T.H., Moore G.P. & Segundo J.P. (1964): Pacemaker Neurons: Effects of Regularly Spaced Synaptic Input; *Science Vol. 145 (1964), pp 61–63.*

[40] Peschl M.F. (1990): Cognition and Neural Computing – an Interdisciplinary Approach; *in Proc. of International Joint Conference on Neural Networks, Washington (IJCNN '90), Lawrence Erlbaum Associates, Publishers, pp 110–113, 1990.*

[41] Peschl M.F. (1990): A Cognitive Model Coming up to Epistemological Claims – Constructivist Aspects to Modeling Cognition (1990); *in Proc. of International Joint Conf. on Neural Networks 1990 (IJCNN '90), San Diego, CA, IEEE, Vol. III, pp III-657–662.*

[42] Peschl M.F. (1990): Some Critical Reflections on Symbolic Knowledge Representation; *in Proc. of Second Inernational Congr. on Terminology and Knowledge Engineering (TKE '90), Trier (FRG), 1990, pp 131–139.*

[43] Polanyi M. (1966): The Tacit Dimension (Implizites Wissen); *Doubleday & Company, Inc. (1966), Garden City, New York (Suhrkamp-Taschenbuch Wissenschaft, stw 543 (1985), Frankfurt/M.).*

[44] Rumelhart D.E. & McClelland J.L. (1986): Parallel Distributed Processing, Explorations in the Microstructure of Cognition, Volume I: Foundations; *MIT Press, Cambridge, Massachusetts 1986.*

[45] Rumelhart D.E. (1989): The Archtecture of Mind: A Connectionist Approach; *in Posner M.I. (ed.), The Foundations of Cognitive Science, MIT Press, Massachusetts, pp 133-159, 1989.*

[46] Sato S. (1972): Mathematical Properties of Responses of a Neuron Model; *Kybernetik (Bilogical Cybernetics) 11, pp 208-216 (1972).*

[47] Sato S., Masaaki H. & Nagumo J. (1974): Response Characteristics of a Neuron Model to a Periodic Input; *Kybernetik (Biological Cybernetics) 16, pp 1-8 (1974).*

[48] Sejnowski T.J. & Churchland P.S. (1989): Brain and Cognition; *in Posner M.I. (ed.), Foundations of Cognitive Science, MIT Press, Massachusetts, 1989, pp 301-356.*

[49] Sejnowski T.J., Koch C. & Churchland P.S. (1990): Computational Neuroscience; *in Hanson et al.(eds), Connectionist Modeling and Brain Function, MIT Press, Massachusetts, 1990, pp 5-35.*

[50] Smolensky P. (1988): On the proper treatment of connectionism; *Behavioral and Brain Sciences (1988) 11, pp 1-74.*

[51] Torras C. (1987): On the Relationship between two models of Neural Entrainment; *Biological Cybernetics 57, pp 313-319 (1987).*

[52] Varela F.J. (1990): Kognitionswissenschaft – Kognitionstechnik. Eine Skizze aktueller Perspektiven (Cognitive Science); *Suhrkamp, stw 882, Frankfurt/M., 1990.*

Section 6:

Natural Language

Natural Language Dialogs for Knowledge Acquisition

David N. Chin

ABSTRACT

The state of the art in Natural Language Processing (NLP) precludes understanding of unconstrained text. However, NLP can still be applied successfully to knowledge acquisition if the problem domain is limited in scope and/or if the NLP task does not require complete NL understanding. These principles are demonstrated in the MOANA system which applies NLP to acquiring formal software specifications. By controlling the dialogue initiative and asking very specific questions, MOANA avoids unconstrained textual input. It uses stereotypical models of software systems to help interpret the user's input and produces graphical conceptual process models and function models.

INTRODUCTION

In general, artificial intelligence is knowledge-intensive and one of the hardest problems in AI has been how to get that knowledge into an AI system in an easy manner. Approaches to solving the problem of knowledge acquisition (KA) have included applying machine learning, incorporating large outside knowledge bases, building tools for knowledge programmers, and natural language processing (NLP). The last is problematic since current NLP technology is not yet very advanced. Ideally, an NLP-KA system would be able to acquire knowledge by reading and library research. Unfortunately the state of the art in NLP does not allow the understanding of unconstrained text.

The main limitation of current NLP is in understanding unconstrained text. Parsing arbitrary text is possible using currently available grammars which have tens of thousands of rules and fairly complete coverage. However understanding the meaning of the text is still problematic due to the inherent ambiguity of natural language (NL) text, the need for fairly complex inference processes for understanding, and the need for very large amounts of knowledge.

Ambiguity

NL ambiguity occurs in many forms. For example, individual words often have more than one sense, so there is ambiguity as to which sense of a word is meant. In sentence (1), the word "bank" is ambiguous and could be interpreted either as referring to a commercial institution as in sentence (2) or as referring to the ground beside a river as in sentence (3). In general, a NLP system needs to refer to the context of a word to determine which sense of the word was meant. The necessary context is provided by modifiers in sentences (2) and (3), but for sentence (1), the context could be mentioned many sentences earlier or it may even be implicit and hence must be inferred from the general situation.

I went to the bank.	(1)
I went to the bank of Hawaii.	(2)
I went to the bank of the Seine.	(3)

Another form of NL ambiguity is found in the scoping of modifiers such as prepositional phrases, adjectives, or multiple noun collocations. For example, in sentence (4), the prepositional phrases, "on the hill" and "with the telescope" can be interpreted with different modifier scoping:

I saw the man on the hill with the telescope.	(4)

In particular, "with the telescope" can be interpreted as modifying "the hill," "the man," or "saw." Determining which interpretation is meant depends on the context of the conversation. For example, if there are two hills, one with a telescope observatory, and the other without, then "with the telescope" might be interpreted as modifying "the hill." On the other hand, if there are two men being discussed, one holding a telescope, and the other not, then the right interpretation might be as a modifier of "the man." Finally, if the speaker had previously mentioned possessing a telescope which could be used for sighting a man on some distant hill, then "with the telescope" might modify "saw."

A final example of NL ambiguity is pragmatic ambiguity as shown in sentence (5). This sentence can be interpreted as a direct or indirect speech act. In the direct interpretation, the speaker wishes to know whether the listener knows how to delete a file. In the indirect interpretation, the speaker is politely asking the listener to explain how to delete a file.

Do you know how to delete a file? (5)

Complex Inferences

Besides needing to disambiguate words, scoping, and pragmatic usage, NLP systems often must make complex inferences in order to understand even the simplest texts. Typically these inferences are not logically proper deductions or inductions, but rather are only default or plausible inferences.

To illustrate a few of these inferences that are required for understanding even simple texts, consider the following excerpt from a children's fairy tale with analysis originally provided by Peter Norvig [1].

> In a poor fishing village built on an island not far from the coast of
> China, a young boy named Chang Lee lived with his widowed mother. Every
> day, little Chang bravely set off with his net, hoping to catch a few fish
> from the sea, which they could sell and have a little money to buy bread.

One of the inferences required for complete understanding of the text is the fact that there is one sea that is used by the villagers for fishing, that surrounds the island, and that forms the coast of China. This inference is obvious, yet it is only a plausible inference since it is conceivable that the village is on an island in the middle of a lake near the coast of China and that the villagers fish on a river that flows through the middle of their island. Naturally no human reader would make such unwarranted assumptions, but it is problematic to get a NLP system to make the correct plausible inferences.

Another required inference is that Chang intends to trap fish in his net, which is a fishing net as opposed to a butterfly net, hair net or a computer net. Conceivably, Chang could use a computer net as a deposit on a fishing boat and go trolling for fish with rod and reel. However, this is not as nearly as plausible as the normal interpretation.

Other complex inferences involve resolving referents of pronouns. For example, the word "which" in "which they could sell" refers to the fish and not to the sea. Conceivably, sea water could be evaporated and sold as salt as is done in India and elsewhere. Also, the word "they" in "they could sell" refers to Chang and his mother and not to the fish or the sea. This is problematic for most pronominal reference disambiguation algorithms since nowhere in the text are Chang and his mother mentioned together. An NLP system must infer from the phrase, "lived with his widowed mother," that Chang and his mother form a conceptual group and this group may later be referenced with a plural pronoun such as "they."

This example shows that even in very simple texts which even children would have no trouble understanding, there are many complex inferences that are problematic for NLP systems to make. The great variety of such inferences and the large amounts of world knowledge needed for such inferences makes the complete understanding of even simple unconstrained text beyond the capability of current NLP technology.

Successful NLP

Although NLP technology is not yet advanced enough to be able to understand unconstrained text, NLP can still be successfully applied to a variety of interesting areas. An analysis of successful NLP systems shows that many successful systems have very limited domains, such as consultants for the UNIX file system, the analysis of terrorist bombing newspaper stories, or front-ends for the blocks world or chess world. Severely restricting the domain allows the intensive use of built-in knowledge, since a limited domain will require less knowledge to be effective.

Other successful NLP applications include NL front-ends to databases. These commercially successful NLP systems work because they do not need to completely understand the meaning of their input; they only need to understand the input relative to their database models. For example, the query, "How many widgets were sold in 1989?" can be translated into a database query language by either a human or a machine without the need to understand either what is a widget (which has no definite meaning) or what it means to sell something (involving the mutually agreed exchange of goods for money between merchant and customer). So long as there is a sales relation which lists widgets and sales per year in the database and the NLP front-end knows that sell refers to the sales relation, the transformation is straightforward.

These examples of successful NLP systems show that in order to apply NLP successfully, one must choose problems that either do not require complete understanding of their natural language input and/or have very limited domains. These two NLP application principles are most easily achieved in a structured NL dialogue. But by controlling the initiative in the dialogue, and by asking very specific questions of the user, dialogues can be designed to avoid (as much as possible) unconstrained textual input. Applying this to knowledge acquisition means that NLP-KA systems should be designed to interview people in order to acquire knowledge.

Past Research

In the past, several systems have acquired knowledge by querying the user in natural language. The GUS system of Bobrow et al. [2] acquired knowledge about traveler's flight plans. It used a frame to store the information that it needed to make flight reservations. When it found a slot that did not have a necessary value, GUS executed the attached procedure which queries the user for the information needed to fill in the slot value. Another system, Tereisias, by Davis [3] acquired expert-system rules. Although Tereisias used a primitive key-word parsing approach to NLP, it was still successful in interpreting rules because the context and domain were highly focussed and it only needed to understand the text enough to restructure it into the syntax of an expert-system rule. Other expert system NLP-KA systems with more advanced NLP aspirations are currently being developed by Arinze [4] and Gao and Salveter [5].

The NANOKLAUS system of Haas and Hendrix [6] acquired NL grammar and concept categories by querying users following a predefined script. This is the original model upon which most current NL front-ends to data-base are based. The most important advance demonstrated in NANOKLAUS is the effort it made to phrase its queries so that people who were not linguists could easily understand its queries. For example, instead of asking whether a verb like ASSIGN can be passivised, it instead asks, "Could an UNIT be ASSIGNED a TASK GROUP by an OFFICER?" Not only does the phrasing of queries have a major impact on the understandability of the query, but also on the kinds of replies received from users. Many problems in understanding the user's answers can be avoided by proper phrasing of the knowledge acquisition queries.

NLP-KA FOR SOFTWARE SPECIFICATIONS

The two principles of successful NLP application, choose limited domains and avoid the need for complete understanding, can be applied to the domain for acquiring knowledge about the specifications of new software systems.

The Software Design Cycle

Figure 1 shows the typical software engineering design cycle. A manager who usually is not a software engineer writes the informal specifications for the new software system in a natural language like English. Next, this is transformed into some form of formal specifications using textual and/or graphical formal specification languages. Then the formal specification is transformed into a modular software design by a software engineer. Finally the design is implemented in programme code by software programmers. Areas of the software cycle which are amenable to artificial intelligence techniques include using NLP to transform the informal software specifications into formal specifications, and applying automatic programming to producing code from the software design.

238

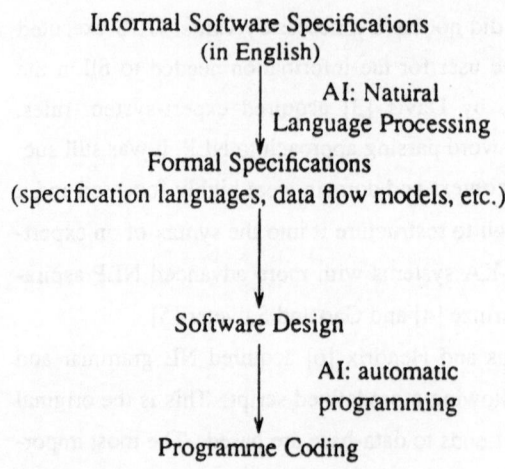

Figure 1. Typical software design cycle.

Unfortunately, AI techniques are not yet advanced enough either to automatically generate code or to understand natural language software specifications in the general case. The latter is as difficult as the general NL understanding problem since NL specifications, like other NL texts, are ambiguous and require complex inferences for understanding.

Two different approaches have been taken to attack the problem of producing formal software specifications. One approach is to build semi-automated systems where the system is meant to assist a human in the transformation process. For example, the system developed at the Tokyo Institute of Technology and Fujitsu by Saeki et al. [7] includes a process and the necessary tools to incrementally derive formal specifications from NL text with human guidance. The human user first selects the important nouns and verbs in the NL text for analysis by the system. Then the system guides the user through the process of incrementally elaborating the formal specifications. As the authors point out, it is not their goal to automate this process which would be impossible using current technology.

Another approach is used at the Software Engineering Research Laboratory at the University of Hawaii in the MOANA (MOdel Acquisition using NAtural language) system [8, 9]. Rather than try to understand the unconstrained text of informal software specifications, MOANA interviews users, who typically are not software designers, to acquire the knowledge needed to build formal specifications of the new software system. MOANA applies the two principles of successful NLP application to acquiring software specifications: limit the domain and avoid the need for complete understanding. Within the limited domain of business software systems, MOANA contains considerable knowledge about typical software systems and applies this in its dialogue with the user.

Also, MOANA does not need to completely understand all of the user's input. For the purposes of building software specifications, it suffices for the system to just understand the general category of any unknown terms. Usually, this information can be acquired through the context or through short clarification sub-dialogues. For example, a user might specify that the software should produce a "discontinued-items report." A complete understanding of this term is not needed so long as the system (or software engineer) understands that this is a type of report. The semantics of the report is defined not by understanding its name, but rather by later specification of what data serves as input to the process that produces the report and by specification of the report's format and processing. The fact that it is called a "discontinued items" report rather than a "gobblygok" report is unimportant.

THE MOANA SYSTEM

The MOANA system interviews users, who typically are not software engineers, to acquire information about the specifications of their proposed software system. The knowledge acquired by MOANA is used to build a series of models (including conceptual process and function models), which are then used as a guide for the design of the software by human software engineers. The current prototype of MOANA is implemented in Common LISP and runs on a TI microExplorer LISP machine.

Organisation of MOANA

As shown in Figure 2, MOANA is organised into two main sub-systems, a natural language processor and a software modeller. The NL processor includes components such as a spelling corrector, a natural language parser/analyzer, a natural language disambiguator, a dialogue manager, and a natural language generator. The software modeller's components include a model builder, a model analyzer, and a model grapher.

Knowledge bases include a linguistic knowledge base, a dialogue model, a user model, and stereotypical software models. The linguistic knowledge is used in the analysis of the user's input and for natural language generation. The dialogue model contains information concerning the planned dialogue as well as the actual course of the dialogue. It provides context for disambiguating the user's input as well as information to guide the dialogue manager. The user model contains information about the user's linguistic usage and understanding of software models. It is used by the natural

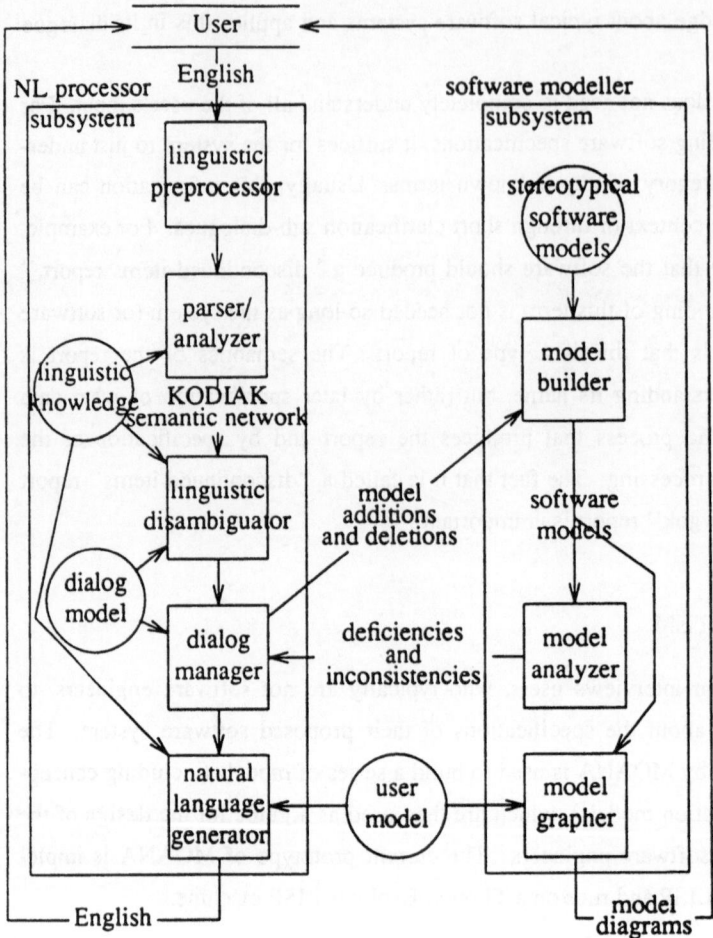

Figure 2. Organisation of MOANA.

language generator and the model grapher to produce output which fits the user's linguistic usage and understanding of software.

Sample MOANA Dialog

Figure 3 shows a dialogue from a MOANA session that demonstrates some of the current capabilities of the system. The sample new software system is a programme which produces a microfiche summary of warehouse stock movement for customers, and is taken directly from Cameron's textbook [10] on the Jackson Structured Programming (JSP) and Jackson System Development (JSD) software design methodologies.

```
Welcome to MOANA.

Please answer the following questions regarding your proposed software
system.

What type of software system do you wish?
a.  point-of-sale and inventory control business system
b.  other (use Jackson Design Methodology)
# b

What is the name of this software system?
# microfiche problem

Please list all of the input files of the program.
# smf

Just one input file?
# yes

Please list all of the output files of the program.
# of1, of2

What is the media_type of SMF?
# computer file

What is the media_type of OF1?
# microfiche

What is the media_type of OF2?
# microfiche

What is  the frequency of execution of program?
# Once a month

Please list all of the elements in each record of SMF.
# warehouse id, item id, customer id, date of the stock movement,
# quantity of the item moved,  type of movement, id of the
# responsible employee, additional description

Please list all of the elements in each record of OF2.
# item id, word "warehouse", warehouse id, symbol "#", fiche number

Please list all of the elements in each record of OF1.
# fiche header, frame header, frame index, frame detail, end of frame
# mark, end of fiche mark

Thank you for using MOANA
```

Figure 3. MOANA session showing use of Jackson Structured Design methodology.

The dialogue in Figure 3 shows some of the basic capabilities of MOANA. The first step in MOANA's knowledge acquisition script is to identify what type of software system the user wishes to have built. The user's answer allows MOANA to activate its stereotypical models of software systems. The current MOANA prototype only has stereotypical models for JSP and for point-of-sale and inventory control software systems. JSP assumes that the software system performs a transformation of input to output files. Future versions of MOANA will need many more stereotypical models, perhaps organised hierarchically.

After determining that the new software system falls under the JSP stereotype and identifying a name for the new software system, MOANA proceeds to identify the input and output files of the system. Screen images of MOANA at this stage are shown in Figures 4 and 5.

The screen image is divided into three panes. The Dialog Manager pane at the top left shows MOANA's dialogue with the user. The Graph pane at the bottom left shows the current state of the system's understanding of the user's specifications represented using a graphical conceptual process model. As the system acquires more information about the user's requirements, this pane changes interactively to display MOANA's current understanding of the requirements. The third pane on the right side, Trace Output, shows the results of parsing and the information flow between MOANA's NL processor and software modeller subsystems.

The next step in MOANA's JSP dialogue script is to fill in any unknown attributes of the files (in this case, their media type) and of the transformation programme (in this case, its frequency of execution). Finally, MOANA identifies all of the elements in each record of the input and output files. A screen image after the user identifies the elements in each record of OF2 (Output File 2) is shown in Figure 6.

The final result of the MOANA dialogue is a series of software models which provide information to a software designer who will use the models in designing the actual software.

NL Processor Subsystem

The NL processor subsystem of MOANA is responsible for interacting with the user. The user's English input is first processed by the linguistic preprocessor which corrects spelling and does morphological analysis. The result is parsed and analyzed by the parser/analyzer which is based the the phrasal analysis techniques described by Wilensky, Arens, and Chin [11]. The parser/analyzer uses a phrasal grammar whose elements include both syntactic and semantic categories. A chart data structure is used to keep track of the process of the parse which proceeds in parallel along all possible parses without backtracking. The parser/analyzer produces a semantic interpretation of each parse element and of the complete sentence in the form of a KODIAK [12] semantic network.

After the parser/analyzer has produced a context-free interpretation of the user's input, the linguistic disambiguator uses contextual and pragmatic information to reinterpret the result. It is responsible for handling ellipsis and anaphora resolution. Its information sources include MOANA's dialogue and user models. Its main task is to

243

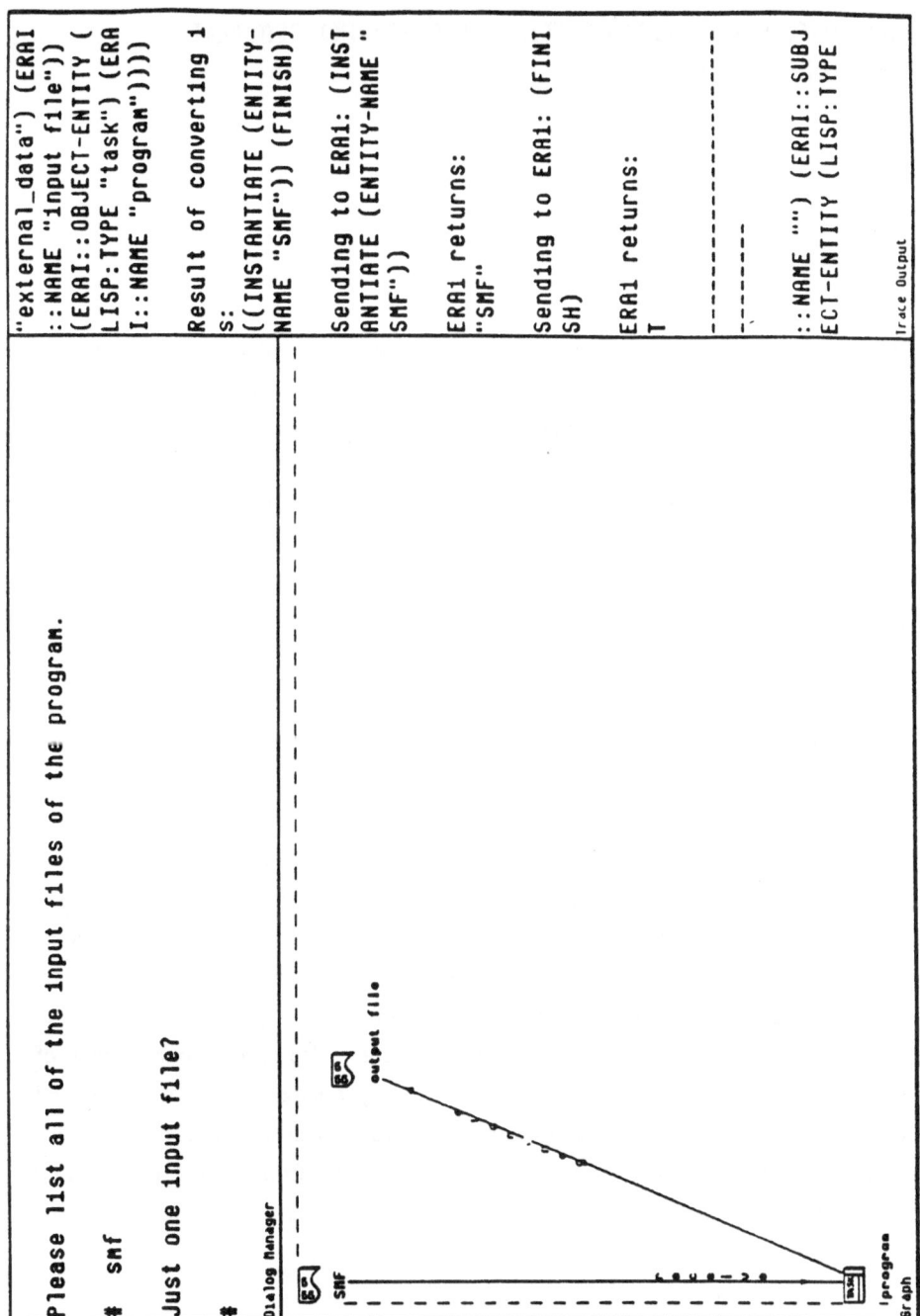

Figure 4. MOANA screen image showing acquisition of input files.

PARSING ENGLISH DIRECTLY INTO PROPOSITIONAL REPRESENTATIONS

Jerry T. Ball

1. ABSTRACT

By making a strong commitment to the basic relational structure of English, it is possible to conceive of parsing English text directly into propositional representations without the need for any intermediate constituency based syntactic analysis. In fact, constituent grammars distort the basic relational structure of English and make the construction of propositional structures more difficult. This paper presents a propositional system of representation and a processing mechanism which demonstrates the viability of a direct interpretation of English text into propositional structures.

2. INTRODUCTION

This paper describes a processing algorithm which constructs propositional representations directly from English text. No intermediate syntactic representation is constructed nor is such a structure needed. This is in accord with Johnson-Laird's psychological claim that 'the [mental] parser delivers an almost immediate propositional representation of a sentence constituent by constituent, or word by word. (It does not set up a representation of syntactic structure.) [1]' and follows from the Preference Semantics [2,3] roots of this research. The propositional system of representation adopted in this work is most closely related to that of Miller and Johnson-Laird [4]. The processing mechanism presented in this paper demonstrates the viability of constructing such representations directly from English text without the need for an intermediate syntactic representation of constituent structure.

3. A PROPOSITIONAL SYSTEM OF REPRESENTATION

The propositional system of representation described in this section is a generalization of work of Miller and Johnson-Laird [4] and Miller [5]. Miller and Johnson-Laird [4] considered the relational status of a large number of English verbs, prepositions and adverbs. This work builds on the results of their analysis and identifies the basic propositional forms which occur in English. Based on an examination of relational terms in English, eight propositional structures have been identified. Examples of each type of propositional structure are provided below:

He went	pred(arg)	went(he)
He kissed me	pred(arg,arg)	kissed(he,me)
He gave me it	pred(arg,arg,arg)	gave(he,me,it)
He quickly went	pred(prop)	quickly(went(he))
He believes you like me	pred(arg,prop)	believes(he,like(you,me))
He kissed me by it	pred(prop,arg)	by(kissed(he,me),it)
He told me you like me	pred(arg,arg,prop)	told(he,me,like(you,me))
I like you and you like me	pred(prop,prop)	and(like(I,you),like(you,me))

In terms of the notation, **pred** must be a relational term, **arg** is a nonpropositional argument (the internal structure of nonpropositional arguments is not discussed in this paper), and **prop** is a propositional argument. It should be noted that the sentences above were specifically selected because they directly reflect the underlying propositional forms. Of course, English has many other surface forms which do not directly reflect the underlying propositional form. For example, the surface form **subject verb infinitive** typically corresponds to the propositional forms **pred1(arg1,pred2(arg1))** or **pred1(arg1,pred2(arg1,arg2))**. Thus, the sentence

I want to go

has the underlying propositional form

want(I,to_go(I))

where **to_go** is a compound predicate which takes the argument **I** and forms aproposition which in turn functions as the second argument of **want**. Despite such variation, it is a basic assumption of this work that the propositional structure of English is more directly reflected in surface structure than is generally accepted. This assumption is in accord with Jackendoff's Grammatical Constraint [6]. It benefits from insights which follow from the system of representation called the predicate calculus. It contrasts with the Aristotelian system of representation in which sentences are divided into subjects and predicates. Unfortunately, the highly influential linguistic theory developed by Chomsky [7,8,9,10] and his followers (see Sells [11] for a description of Government-Binding Theory) adopts the Aristotelian system, and it is fallout from this historical fact which clouds the validity of the above assumption. According to Chomsky, sentences in English have the basic form **noun-phrase verb-phrase** , with the **noun-phrase** corresponding to the subject and the **verb-phrase** corresponding to the predicate. But, this structure dissociates the verb from the subject and creates an asymmetry in status between the subject and any objects which exist. In so doing, it destroys the relational status of the verb. Transitive verbs are relational entities which establish a relationship between two other entities, the subject and the object. Chomsky's system of representation fails to capture this semantic relationship. Prepositions fall victim to a similar fate. In Chomsky's system of representation, prepositions take a single object. However, prepositions, like transitive verbs, are relational entities which establish a relationship between two other entities. Chomsky's system of representation fails to capture this semantic relationship as well. That Chomsky's system of representation completely ignores the relational status of English words is reflected in his X-bar Theory. According to X-bar Theory, each of the four categories of words (Verb, Noun, Preposition, Adjective) is associated with one and only one complement. In relational terms, this is clearly an overgeneralization. Verbs may take from one to three arguments, prepositions take two arguments, predicate adjectives take a single argument, and nouns are not even relational. Reliance on X-bar Theory results in serious problems of overgeneration which are only overcome by the introduction of the Case and θ constraints. A relationally based system of representation provides the representational equivalent of X-bar Theory, Case Theory and the θ principle while at the same time providing structural representations which are semantically based and obviating the need for separate syntactic representations.

Intermediate between the system of representation described in this paper and that of Chomsky are systems like Case Grammar [12,13,14] and Valency Grammar (see Somers [15] for a survey of Case and Valency based systems). These systems of representation recognise the basic relational status of verbs, but generally fail to adequately consider the status of other relational categories. Thus, Case Grammar allows prepositional phrases to function as arguments to verbs and Valency Grammar allows them to be complements or adjuncts of verbs. In so doing, they ignore the basic relational status of prepositions as relational entities which determine a semantic relationship between two other entities. Prepositional phrases like verb phrases are not well-formed semantic units.

Providing a slightly different contrast is the system of representation which forms a part of Lexical-Functional Grammar [16,17]. Actually, LFG has two separate systems of representation, one syntactic and one functional. The syntactic system of representation is described by a phrase structure grammar which generates subject-predicate structures, and the functional system of representation follows along lines similar to Case Grammar. Syntactic representations mediate between surface text and functional representations. According to the system of representation of this paper, these syntactic representations are unnecessary. In fact, since they are based on the generation of subject-predicate structures they actually make the mapping from surface text to propositional structures more difficult than would otherwise be the case.

The system of representation described in this paper is essentially concerned with the identification and classification of relational terms and with the determination of how relational terms combine together with propositional and nonpropositional arguments and with each other to form propositions. The principle determinants of the structure of a piece of text are the relational units which occur in that text. A further analysis of relational terms in English leads to the following more extensive classification:

syntactic category	surface form	propositional form
predicate adjective	He is *sad*	**pred**(*pred*(**arg**)) = is(*sad*(he))
intransitive verb	He *ate*	*pred*(**arg**) = *ate*(he)
auxiliary verb	He *is* sad	*pred*(**pred**(**arg**)) = *is*(sad(he))
modal auxiliary	He *may* go	*pred*(**pred**(**arg**)) = *may*(go(he))

negative	He did *not* go
	$pred(pred(pred(arg))) = did(not(go(he)))$
adverb	He *already* went
	$pred(pred(arg)) = already(went(he))$
verb particle	He held it *up*
	$pred(pred(arg,arg)) = up(held(he,it))$
transitive verb	He *hit* it
	$pred(arg,arg) = hit(he,it)$
bitransitive verb	He *gave* you it
	$pred(arg,arg,arg) = gave(he,you,it)$
preposition	He is *in* the store
	$pred(pred(arg,arg)) = is(in(he,the\ store))$
preposition	He prays *on* Sunday
	$pred(prop,arg) = on(prays(he),Sunday)$
propositional attitude	He *believes* you like him
	$pred(arg,prop) = believes(he,like(you,him))$
conjunction	I like you *and* you like me
	$pred(prop,prop) = and(like(I,you),like(you,me))$
double complement verb	He *told* me you like him
	$pred(arg,arg,prop) = told(he,me,like(you,him))$

In addition to identifying and classifying relational terms, the system of representation must determine how such terms combine together to form well-formed propositional structures. Some of the examples above include multiple relational terms. The example containing a predicate adjective also contains an auxiliary verb since adjectives do not normally occur as predicates in English without being accompanied by such an auxiliary. Of course more complex structures are possible. For example, consider the sentence,

(3) He gave me the ball on Tuesday in the park

which contains the relational units **gave, on,** and **in** and the nonrelational units **he, me, the ball, Tuesday,** and **the park. Gave** is a relational unit which takes three nonpropositional arguments. **On** is a relational unit which takes two arguments, the first of which is propositional and the second of which is nonpropositional. **In** is a relational unit which has the same relational status as on. How can these relational units combine with the nonrelational units to form a well-formed structure in which each relational unit is associated with the appropriate number and type of arguments? The verb **gave** combines with the nonpropositional arguments **he, me** and **the ball** to form the well-formed propositional structure

gave(he,me,the ball).

The preposition on combines with the propositional argument **gave(he,me,the ball)** and the nonpropositional argument **Tuesday** to form the well-formed propositional structure

on(gave(he,me,the ball),Tuesday).

The preposition in combines with the well-formed propositional argument **on(gave(he,me,the ball),Tuesday)** and the nonpropositional argument **the park** to form the well-formed propositional structure

in(on(gave(he,me,the ball),Tuesday),the park).

In this last propositional structure each relational unit has the right number and type of arguments and the structure is well-formed. This representation captures the basic semantic information that the **verb** gave is an action involving a subject and two objects, that the preposition on takes the proposition **gave(he,me,the ball)** and determines the time at which the event occurred, and that the preposition in takes the proposition **gave(he,me,the ball)** which occurred **on Tuesday** and determines the location where the event occurred.

4. THE PARSING MECHANISM

The parse of a piece of text is driven by the relational entities contained in the text along with very basic rules of argument order which determine where the arguments to relational entities are to be found. Each relational entity has associated with it one or more relational structures in the lexicon. These structures determine the number and type of the arguments with which the relational entity is commonly associated. When a relational entity is encountered in a piece of text a relational structure is selected. The expectations of the relational structure in combination with the rules of argument order serve to drive the processing mechanism. For example, if a relational entity is encountered which commonly takes two nonpropositional arguments (e.g. a transitive verb), expectations are set up for the occurrence of the nonpropositional arguments. Based on typical argument order for English (i.e. SVO), one of the arguments should already be available in the memory structure where the representations of previously processed text are maintained. The memory structure is searched for this argument. If found, the argument is instantiated into the relational structure, and the resulting partially instantiated relational structure is added to the memory structure where it awaits processing of the second argument. After the second argument is processed it is instantiated into the partial relational structure to form a complete relational structure or proposition. This proposition is added to the memory structure as a unitary chunk of information. This proposition can then itself function as the argument of a subsequent relational entity which takes a propositional argument. If such a relational entity occurs, a new proposition will be constructed which is composed of the relational entity, the proposition as one of its arguments, and any other arguments which occur. This new and larger propositional chunk will be placed in the memory structure. The memory structure is structured so that it reflects both the temporality of processing (with more recently constructed representations being more accessible) and the organization of the representations themselves (with higher level entities being more accessible than more deeply embedded entities). Once an argument is instantiated into a relational structure it is no longer available in the memory structure as a separate unit of representation.

5. TWO SAMPLE PARSES

The parsing mechanism will be demonstrated using the following sentences:

The reporter saw her friend

and

The lecturer said the reporter saw her friend.

In the parsing of the first sentence, the processor begins by identifying the nonpropositional argument **the reporter**. Since there is no relational structure available for this argument to be instantiated into, the argument is made available in the memory structure. Next, the processor identifies the relational term **saw** and its preferred relational form **pred(arg,arg)**. Since the first argument of a transitive verb typically occurs to the left of the verb in the input, the memory structure is searched for the availability of a nonpropositional argument and **the reporter** is identified and instantiated as the first argument of **saw(the reporter,Arg2)**. This partially completed proposition is then placed in the memory structure to await its final argument. Next, the nonpropositional argument **her friend** is identified. Since the second argument of a transitive verb typically occurs to the right of the verb in the input, this nonpropositional argument is instantiated as the second argument of **saw(the reporter,her friend)**, and the processing of this sentence is complete.

In the parsing of the second sentence, the processor begins by identifying the nonpropositional argument **the lecturer**. This argument is made available in the memory structure as in the first sentence. Next, the processor identifies the relational term said and its preferred relational form **pred(arg,prop)**. The nonpropositional argument the lecturer is instantiated as the first argument of the **said(the reporter,Prop)**. Processing continues with the identification of the nonpropositional argument **the reporter**. This argument cannot be instantiated as the second argument of **said(the reporter,Prop)** , since a propositional argument is required. The nonpropositional argument is made available in the memory structure. Next, the relational term **saw** is identified along with its preferred relational form saw(arg,arg). The memory structure is searched and the nonpropositional argument **the reporter** is instantiated as the first argument of **saw(the reporter,Arg2)**. Processing continues with the identification of the nonpropositional argument **her friend** which is instantiated as the second argument of **saw(the reporter,her friend)**, forming a complete propositional argument. This completed propositional argument is then instantiated as the second argument of **said(the lecturer,saw(the reporter,her friend))**, forming a completed proposition and processing terminates.

6. SUMMARY

By making a strong commitment to the basic relational structure of English, it is possible to conceive of parsing English text directly into propositional representations without the need for any intermediate constituency based syntactic analysis. In fact, constituent grammars distort the basic relational structure of English text and make the construction of propositional structures more difficult. The direct construction of propositional structures from input text has strong psycholinguistic support. This paper presents a propositional system of representation and a processing mechanism which demonstrates the viability of a direct interpretation of English text into propositional structures.

REFERENCES

[1]   Johnson-Laird, P. Mental Models. Harvard Univ Press, Cambridge, Mass, 1983

[2]   Wilks, Y. Deep and Superficial Parsing. In: King M (ed) Parsing Natural Language. Academic Press, London, 1983, pp 219-246

[3]   Wilks, Y. Preference Semantics. In: Keenan E (ed) Formal Semantics of Natural Language. Cambridge Univ Press, New York, 1975, pp 329-348

[4]   Miller G A, Johnson-Laird P. Language and Perception. Harvard Univ Press, Cambridge, Mass, 1976

[5]   Miller G A. Semantic Relations among Words. In: Halle M, Bresnan J, Miller G A (eds) Linguistic Theory and Psychological Reality. The MIT Press, Cambridge, Mass, 1978, pp 60-118

[6]   Jackendoff R. (1983). Semantics and Cognition. The MIT Press, Cambridge, Mass, 1983

[7]   Chomsky N. Language and Problems of Knowledge. The MIT Press, Cambridge, Mass, 1988

[8]   Chomsky N. Lectures on Government and Binding. Foris, Dordrecht, 1981

[9]   Chomsky N. Aspects of the Theory of Syntax. The MIT Press, Cambridge, Mass, 1965

[10] Chomsky N. Syntactic Structures. Mouton, The Hague, 1957

[11] Sells P. Lectures on Contemporary Syntactic Theories. The Univ of Chicago Press, Chicago, 1987

[12] Fillmore C. The Case for Case Reopened. In: Cole P, Sadok J (eds) Syntax and Semantics, Volume 8. Academic Press, New York, 1977, pp 59-81

[13] Fillmore C. Some Problems for Case Grammar. In: O'Brien (ed) Monograph Series on Language and Linguistics 22nd Annual Roundtable. Georgetown Univ School of Languages and Linguistics, DC, 1971, pp 35-56

[14] Fillmore C. The Case for Case. In: Bach E, Harms R (eds) Universals in Linguistic Theory. Holt, Rinehart, and Winston, New York, 1968, pp 1-88

[15] Somers H. Valency and Case in Computational Linguistics. Edinburgh Univ Press, Edinburgh, UK, 1987

[16] Bresnan J (ed). The Mental Representation of Grammatical Relations. The MIT Press, Cambridge, Mass, 1982

[17] Bresnan J. A Realistic Transformational Grammar. In: Halle M, Bresnan J, Miller G A (eds) Linguistic Theory and Psychological Reality. The MIT Press, Cambridge, Mass, 1978, pp 1-59

SIMPR: Using Syntactic Processing of Text for Information Retrieval.

Alan F. Smeaton
Paraic Sheridan

ABSTRACT

This paper gives a brief overview of the work of Dublin City University in the SIMPR (Structured Information Management: Processing and Retrieval) project. One of the tasks of SIMPR is to apply the automatic morpho-syntactic analysis of English text to the problems of text indexing and text retrieval. We present here a brief review of how natural language processing is done in the SIMPR project. We then describe how we use the output of this text analysis to index texts by structured representations and we outline how we intend to do interactive matching or retrieval of texts using these structured representations.

INTRODUCTION

The SIMPR Project (Structured Information Management: Processing and Retrieval) is an ESPRIT II project which commenced in January 1989 and will run for 3½ years. It is targeted to lead to advances in information storage and retrieval, particularly the retrieval of textual information by using natural language processing techniques. The work being done as part of SIMPR includes research work on text and language analysis, automatic indexing, automatic classification methods and techniques, machine learning, and domain and task modelling. SIMPR will also include the design and development of software for indexing, analysis, classification and retrieval of text. The first prototype indexing system was produced in mid-1990.

With a total of 9 partners, a mixture of academic, commercial and research partners, there is a lot of disparate work going on in SIMPR. This paper, however, concentrates on just one aspect of that work. The task of Dublin City University within SIMPR is to research methods of using morphological and syntactic analysis of text to provide a <u>rich</u> indexing of text, and to develop a method of providing <u>effective</u> phrase matching based on that rich indexing. In processing text at this level we will have to tolerate and handle any syntactic ambiguities inherent in our indexing and phrase matching processes. This paper presents an outline of SIMPR from the point of view of our work, which is to design an internal representation for text phrases derivable from a parse and to design a matching operation to function on this internal representation in an effective manner for use in both interactive retrieval and automatic text classification.

THE SIMPR ARCHITECTURE ACCORDING TO DCU

The following diagram illustrates the architecture of the SIMPR system as seen from the point of view of Dublin City University's work. The SIMPR prototype is being implemented on SUN 3/60 workstations with 12 Mb of main memory, in SUN Common LISP and using a set of Application Interface Generation Tools (AIGT) supplied by the NOKIA Research Centre.

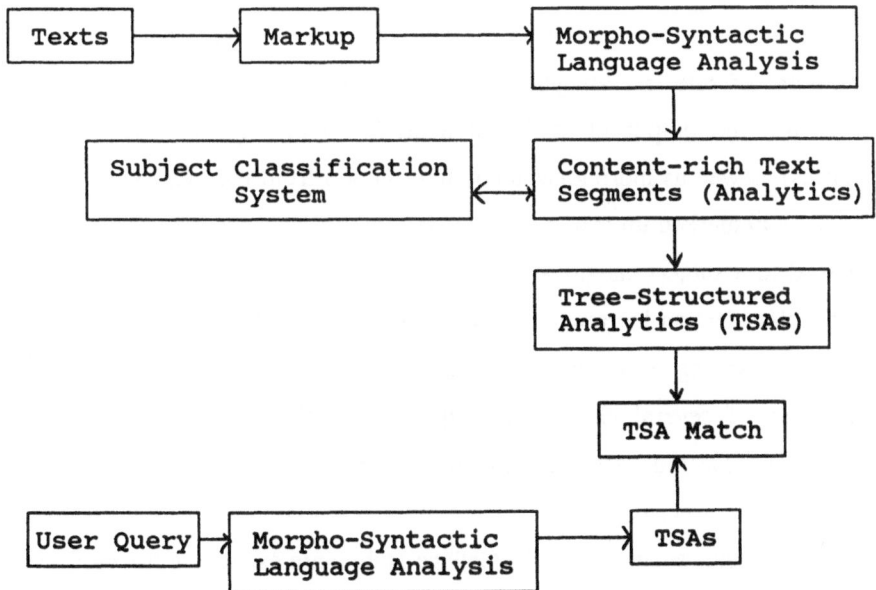

Fig 1: Architecture of SIMPR Prototype System

As we can see from the diagram, newly-created documents are entered into the system and go through a document markup process. This phase of the indexing process is used to reveal the structure of the document by marking sentence, paragraph, sub-section, section, chapter, heading, etc. boundaries. A cleaning process is first used to identify word boundaries, punctuation marks, etc. The marked up document is then passed to a morpho-syntactic analysis phase which implicitly does spelling error detection and which is being implemented by the Research Unit for Computational Linguistics (RUCL) at the University of Helsinki. Since this is such an important part of the SIMPR indexing and retrieval process for the work at Dublin City University, we will outline it in more detail in the following section of this paper.

The morpho-syntactically analysed text is then passed to a module which identifies <u>analytics</u> or potentially rich word sequences which may be one word, or a sequence of words not necessarily in the same sequence as in the original text. Analytics could in fact be as large as sentence clauses, which could be normalised for consistency. This component of the system is being researched by Strathclyde University and the research component of this is outlined in [1]. In the prototype developed by SIMPR this year, this process involves the author or editor of the original document in the role of validating the analytics chosen, but eventually this will be an entirely automatic process. The process is being implemented by defining a set of rules for analytic identification, using forward chaining. These rules will use the output of statistical analysis routines (frequencies and co-occurrences) as well as morpho-syntactic information from the language analysis.

The original marked up text, and the generated analytics mentioned above, are then passed to a subject classification system which classifies input documents in terms of a classification hierarchy that it maintains. This classification hierarchy is dynamic and changing as new texts are processed by the system. Having evaluated the suitability of several classification approaches, Strathclyde University has decided to use a faceted classification technique and this work is described in [2].

The SIMPR project is also addressing other research issues within document classification. In particular, University College Dublin are examining the usefulness of machine learning techniques to automatic subject classification, as described in

[3]. Using this AI approach it is hoped that the subject classification system will learn how to classify texts itself. The machine learning component of the system would look at document texts and their representations, and generate rule(s) for distinguishing between texts on different subjects or topics.

The analytics which provide input to the subject classification system and are the size of sentence clauses, are also used to generate tree-structured analytics (TSAs), which will yield a richer representation of the text. These TSAs, their design and their matching algorithm, form one of the research tasks being done by Dublin City University within SIMPR, and are described in more detail later in this paper. TSAs are also used as input to the subject classification expert system.

When a user wants to retrieve a document he has a choice of retrieval strategies available to him. A user may retrieve by browsing through the heading hierarchy, by specifying a single keyword to start retrieval, by browsing through the subject classification hierarchy or by specifying a natural language statement of his information need as illustrated in the architecture diagram in Fig 1. Fundamental to this type of search, and also used in automatic document classification, is the facility to be able to match two texts, query v document texts or document v document texts. This matching can be done on several levels; word indexes, analytics, TSAs, or all three. The remainder of this paper will briefly review the RUCL morpho-syntactic analysis process, which is crucial to our work, and we will then describe the format of the TSAs and how the matching algorithm between phrases operates.

LANGUAGE PROCESSING IN SIMPR

The automatic language processing aspects of SIMPR are being researched and developed by the Research Unit for Computational Linguistics at the University of Helsinki. RUCL are initially developing a morphological analyser, a syntactic analyser and lexicon for the English language. The master lexicon presently contains some 54,000 entries or word forms. The morphological analyser is capable of correctly identifying all, and only the legal inflectional forms of lexicon entries, and of retrieving the proper base forms of words as they are inflected in text. Thus computational and computerised are correctly identified as lexical variants of the base form of the word computer. The analyser also supplies the parts of speech and other grammatical properties of input words, as needed by the disambiguation and syntactic analysis components.

The output from morphological analysis of an input phrase consists of a list of lexical interpretations of each input token. Fig 2. shows the analysis of the input phrase "controlled research techniques".

controlled

control	**V PCP2**	**Past Participle, 2nd person**
control	V PAST VFIN	Past Tense, Finite Verb

research

research	V SUBJUNCT VFIN	Subjunctive Finite Verb
research	V IMP VFIN	Imperative Finite Verb
research	V INF	Infinitive Verb
research	V PRES -SG3 VFIN	Present Tense, not 3rd person
research	**N NOM SG**	**singular Noun**

techniques

technique	**N NOM PL**	**Plural Noun**

Fig 2. Morphological Analysis of "controlled research techniques".

Having done simple morphological analysis on input tokens, a set of disambiguation and syntax rules are then applied to discard all ambiguities and ultimately leave each input word-form with one morphological interpretation [4]. At present[1], 900 such rules have been formulated and 90% of morphological ambiguities are discarded. These would be used to correctly assign the underlined morphological alternatives in the example in Figure 2.

The explicitly disambiguated morphological analysis is then input into the syntactic parser which will determine the clause boundaries, syntactically disambiguate tokens and then determine the syntactic functions of the tokens. A dependency syntax will be created which will identify for each input token, whether

---

[1] October, 1990.

it is a head or a modifier, and if it is a modifier then what is its head. For example, the syntactic analysis of the phrase used above is given in Fig 3.

Controlled	research	techniques
Noun	Noun	Noun
@an>	@nn>	@nphr

Fig 3. Syntactic Analysis of "controlled research techniques".

The interpretation of this analysis is that controlled is an adjectival form modifying a noun to the right, research is a noun modifying another noun to the right and techniques is a nominal phrase, the head of the clause. The rules for syntactic disambiguation are being further refined by RUCL at present and currently number about 350. Between 70% and 80% of all input words are assigned a unique syntactic category as a result of syntactic disambiguation and it is expected that this total will reach 85% with a fuller rule set.

It is worth pointing out here that these examples are of an analysis of a simple noun phrase. The processes outlined here operate on full sentential units as well as on the smaller clauses but for brevity these are not shown. The syntactic structures created by the RUCL processing is a flat structure, almost linear in nature. This allows it to encode ambiguity from any linguistic input which has inherent syntactic ambiguity. This is ideal for the purposes of text matching because the SIMPR project has decided to do natural language processing at the morpho-syntactic level only and not to resort to higher level semantic language processing for disambiguation purposes. This means that we must work with syntactic ambiguity which is already encoded in the syntactic structures derived from language analysis but we do get the benefit of being domain-independent in our language analysis.

USING NATURAL LANGUAGE PROCESSING FOR MATCHING TEXTS

As mentioned earlier, one of the types of retrieval that SIMPR supports is where an input query is analysed by the RUCL morpho-syntactic analysis process, and

TSAs are derived from that and matched against TSAs for the database of stored texts. When a document is being classified, one of the fundamental operations that the classification system must do is measure the similarity between the input text and other filed documents. This matching can also be done by comparing the TSAs for the two documents.

Recent years have seen other researchers try to incorporate syntactic analysis into document retrieval processes. Martin Dillon developed the FASIT system [5] which did grammatical tagging and searched for grammatical tag patterns as content indicators for indexing. The first author of this paper has done a syntactic analysis of user queries from which typed word-word dependencies were derived and piggy-backed on top of statistical retrieval [6]. Fagan at Cornell performed a similar analysis on document texts as well as user queries and used this information to index by phrases [7]. Again, Fagan used the syntactic structure and patterns of the texts as indexing criteria. The Siemens Research group in Munich has also used shallow syntactic analysis to construct dependency trees from noun phrases [8]. These trees are used in matching by fitting both keywords and dependencies together. The Siemens approach to language processing is to use a small lexicon as positive and negative representatives of word categories instead of creating a large lexicon as in SIMPR. For example, a word with an "-IC" suffix is an adjective unless it is found in a negative list containing words like "MUSIC" and "TRAFFIC".

Another significant application of syntactic analysis of text to information retrieval is being undertaken by Metzler at the University of Pittsburgh [9,10]. This approach is based on stressing the hierarchical structure of syntax in matching texts and operates on a hierarchical structure of dependency relationships in the Constituent Object Parser (COP). COP produces binary trees with the branch containing the head of the construct always marked as being dominant. This representation allows the matching algorithm to note ambiguities and match texts which have the same meaning but use a variety of linguistic constructs. A brief review and comparison of these approaches can be found in [11].

Of the previous approaches to using syntactic analysis in information retrieval processes, it seems that those which decompose parse tree information into phrases and store the phrases, like Dillon, Smeaton and Fagan, are losing a lot of information as well as creating unnecessary problems. The decomposition process must be heuristic and complex and there must also be a complex set of normalisation rules to allow matching between syntactic variants. On the other hand, storing intermediate representations of text as complex structures and performing some kind of graph matching as the retrieval operation does seem more inherently attractive as ambiguities in the parse can be encoded in a structure and these ambiguities taken into consideration during retrieval. The computational expense of any graph matching algorithm can be quite large but we have shown in SIMPR that graph matching problems can be simplified and made efficient when dealing with a specific application [12].

Our research work has been directed at finding exactly such a structured intermediate representation of text and a matching algorithm for that structured representation. This representation had to be simple enough to be derivable from the output of RUCL's morpho-syntactic analysis, flexible enough to allow the encoding of different types of ambiguity but rich enough to provide a better retrieval mechanism (through the matching of the structures) than simple phrase matching. It should be noted that we do not build this structured representation for complete texts but only for the analytics or content-rich text segments that are extracted from the text by the software developed at University of Strathclyde and validated by the user.

We have proposed to represent analytics in binary tree form and these trees have certain special characteristics listed below.

- Word information is stored at the leaf nodes. This information includes the word, as it appears in the text, and all linguistic information supplied by RUCL (base form, lexical category, morphological information and syntactic labels).

- Conjunctions are stored at the ancestoral node which covers all of the conjuncts.

- Complete modifier/head subtrees are specially marked (nominal compounds).

262

- Prepositions are stored on the branches to the parent node of the prepositional complement.

An example of the structured representation of "<u>techniques of information processing and classification</u>" is shown in Fig. 4. It is quite a simple matter to build these tree structures using the modification relationships provided by the linguistic processing. We see it as a process of reading left to right through the analytic, building trees and subtrees and connecting these together.

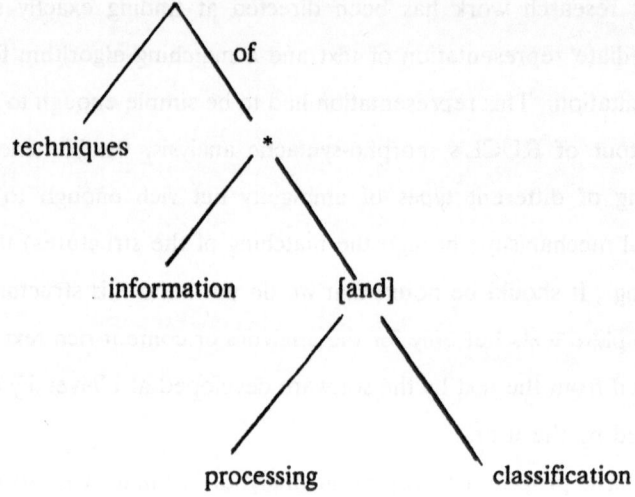

Fig 4: Structured representation of,
<u>techniques of information processing and retrieval.</u>

This binary tree structure is also flexible enough to encode all types of ambigutiy. We have carried out a quite comprehensive study of the different types of ambiguity that may occur and the likely frequency of these occurences [13]. At the level of syntactic processing problems of ambiguity can arise because of conjunctions, prepositions, adverbs, and ellipses. In most cases, too, it is impossible to disambiguate these without resorting to higher level (semantic) processing. Since, in SIMPR, we have excluded the use of any semantic processing, we had to find some way to <u>preserve</u> the ambiguity and so make each interpretation available for retrieval at matching time. For example, the above analytic should be matched to both "<u>information classification</u>" and "<u>classification</u>" since there is ambiguity involved with the distribution of the modifier over the conjunction. These two different interpretations are found by simply matching different sub-trees of the whole structured analytic. In general, ambiguity is implicitly encoded in the structure of the analytic and the different interpretations are then extracted at matching time by following different paths through the tree and matching different sub-trees together. If we take a much used example of prepositional ambiguity, "the man on the hill with the telescope",

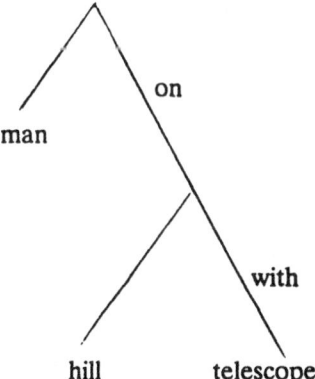

Fig 5: Structured representation of,
"the man on the hill with the telescope".

We see that because of the way the tree is structured we can find matches for both "<u>the man with the telescope</u>" and "<u>the hill with the telescope</u>" by following different paths through the tree. It does seem, however, that the way the tree is structured favours the interpretation of "the man on the <u>hill with the telescope</u>".

Indeed this area is one that we have already investigated; is it possible in cases of ambiguity like this to favour one interpretation over others. Our conclusion from this was that the answer to this would only become apparent after much empirical testing and evaluation of our system. This testing and evaluation will form a major part of our task in arriving at a final system.

INTERACTIVE RETRIEVAL IN SIMPR.

Part of the power of SIMPR's retrieval environment is that it allows the user to specify a request that is matched to the database of texts at several different levels - word/word, phrase/phrase and structure/structure matches and searching via the subject classification system. Once the user has found a relevant text as a starting point in the text database, s/he can broaden or narrow the subject of the texts by navigating through the heading hierarchy (like a table of contents). At Dublin City University we are concerned with the development of a structure matching algorithm to facilitate the "ranked" retrieval of texts based on the matching of structural representations of user queries and structured analtyics, in other words we are presently concerned only with matching phrasal-level constructs against each other.

We established at an early stage of our research that complex graph matching algorithms could be computationally efficient by looking at the graph isomorphism problem and its application in matching molecular structures in chemistry [12]. This served to remove any restrictions on the type of structures that we could propose for the internal representation of text. Having defined the binary tree structure above for the representation of analytics we then proceeded to specify how the matching of these structures would take place. The matching algorithm takes the form of a basic tree search algorithm but has some enhancements due to the special requirements of our application. Two major requirements of our application are:

1.   The need for inexact or relaxed matching in:

- The relaxed matching of nodes i.e. matching the noun "analysis" to the verb "analyse". This is required for matching, for example, the query "text analysis" to the analytic, "a system for analysing text". The matching algorithm has available to it, all information regarding the word as it appears in the text, its base form, and morphological and syntactic category in performing this match.

- The inexact matching of structures; i.e. matching the structure for "engine removal" to the structures for "removal of the engine", "remove the engine" and "removing the engine" (or indeed "carefully remove the big broken engine").

2. The manipulation of phrase/phrase scores in:

- The weighted matching of nodes (a measurement of the degree of "relaxation" as described above); i.c. scoring noun "analysis" matching noun "analysis" higher than noun "analysis" matching verb "analyse", (with probably only a slight difference in the score to reflect the fact that there is little difference between the match of "X-ray analysis" to "X-ray analysis" and "analyse the X-rays", except that the first is a perfect match and the second is not).

- The weighted matching of inexact structures; i.e. a weight reflecting the "exactness" of a match. For example, to reflect the difference in exactness between the match of "engine removal" to "engine removal" and "carefully remove the broken engine".

Having defined these basic requirements we carried out a survey of research that was directed at weighted structure matching and we found that two of the approaches taken [14,15] could be applicable to our own needs. In particular the work of Shapiro and Haralick presented a good mathematical theory of graph matching in general, defining different types of exact and inexact matching, introduced the idea of weighted structures and presented a theory for the inexact matching of these structures including the manipulation of weights and also went on to present several algorithms for tree searching including an evaluation of these. This has reassured us that the implementation of the type of complex matching algorithm that we require is feasible and that rather than having to develop such an algorithm from scratch, we have been able to take an existing one and expand and enhance it to suit our own particular needs.

CONCLUSION

The status of our match algorithm is that we are implementing it in LISP and will conduct experiments to measure its effectiveness as a phrase matching technique. This process will be subject to a number of iterations until we eliminate any possible inconsistencies with our algorithm and our representation for phrases. In the longer term we will work on incorporating this phrase match procedure into an overall text matching and ranking system.

There are many questions we have yet to address in our approach to indexing and retrieval and many details that have yet to be finalised. In particular, we are still quite unsure about the ranking of matches between queries and analytics and between analytics and analytics. We must also decide how such match scores can be combined to compute a relevance ranking for texts for retrieval purposes or a similarity measure for classification purposes. To date we have worked on TSA/TSA matching but we must now start to scale up to the document text level.

REFERENCES

1.   Gibb F, Karetnek D and Badenoch D, Knowledge-Based Extraction of Analytics from Text. SIMPR Document No. SIMPR-SU-1989-4.1e, December 1989.
2.   Sharif C, Gibb F, Karetnyk D and Badenock D, Subject Classification Research. SIMPR Document No. SIMPR-SU-1989-6.1e, December 1989.

3. Ward C, Kavanagh I and Dunnion J, Machine Learning in Subject Classification. In: McTear M and Creaney N (eds) AI and Cognitive Science 90. Springer Berlin Heidelberg New York, 1990.

4. Karlsson F, Syntactic Analysis for SIMPR Purposes. SIMPR Document No. SIMPR-RUCL-1989-13.1e, February 1990.

5. Dillon M and Gray AS, FASIT: Fully Automatic Syntactically-Based Indexing of Text. Journal of the ASIS, 34(2), 99-108, 1983.

6. Smeaton AF and van Rijsbergen CJ, Experiments on Incorporating Syntactic Processing of User Queries into a Document Retrieval Strategy. In: Chiaramella Y (ed) Proceedings of ACM SIGIR Conference on Research and Development in Information Retrieval, 31-51, Grenoble, 1988.

7. Fagan J, Experiments in Automatic Phrase Indexing for Document Retrieval: A Comparison of Syntactic and Non-syntactic Methods. PhD thesis, Department of Computer science, Cornell University, 1987.

8. Schwartz C, Content Based Text Handling. Information Processing and Management, 26(2), 219-226, 1990.

9. Metzler D and Haas SW, The Constituent Object Parser: Syntactic Structure Matching for Information Retrieval. ACM Transactions on Office Information Systems, 7(3), 292-316, July 1989.

10. Metzler D, Haas SW, Cosic CL and Weise CA, Conjunction, Ellipses and Other Discontinuous Constituents in the Constituent Object Parser. Information Processing and Management, 26(1), 53-72, 1990.

11. Sheridan P, Syntactic Processing for Text Analysis: A Survey. SIMPR Document No. SIMPR-DCU-1989-8.2e, November, 1989.

12. Smeaton AF, Searching Chemical Structures: Implications for Searching Parse Trees in SIMPR. SIMPR Document No. SIMPR-DCU-1989-8.2e, May 1989.

13. Sheridan P and Smeaton AF, Structured Analytics: A Method for Handling Syntactic Ambiguity. SIMPR Document No. SIMPR-DCU-1990-16.1e, March 1990.

14. Levinson R, A Self-Organizing Retrieval System for Graphs", in: Proceedings of AAAI-84, also MSc thesis, University of Texas at Austin, 1985.

15. Shapiro L and Haralick R, Structural Descriptions and Inexact Matching. IEEE Transaction on Pattern Analysis and Machine Intelligence, PAMI-3(5), 504-519, 1981.

Note: SIMPR Documents may be obtained from the SIMPR Prime Contractors, CRI A/S, Bregnerodveg 144, DK-3460 Birkerod, Denmark.

Very Large Lexicon Management System for Natural Language Generation

Gi-Chul Yang and Jonathan C. Oh

ABSTRACT

Minimal usage of storage space, maximal speed in retrieval of desired textual items, and ability to update with minimal side-effects on the lexicon are three major concerns in lexicon management systems. We propose an efficient mechanism for organization of a lexicon along with its retrieval, update mechanisms. The update operation is for all practical purposes side-effect-free; the retrieval operation is not closely dependent on the size of the lexicon nor adversely on the complexity of the internal representation of the lexical items, rather it has an interesting property of often allowing faster retrieval for more complex items; and the storage overhead is negligible.

1. Introduction

Fast retrieval from a very large lexicon is one of the major considerations for natural languageprocessing systems including machine translation and natural language interface systems. Even a severely limited system in its text domain requires a large lexicon if it is to serve a practical purpose at all. Much work is done on the organization of lexicons, their update and retrieval mechanisms [4, 7, 3, 2]. We propose in this paper some ways to improve on these organization, update, and retrieval mechanisms of lexicons for natural language processing. We will call the proposed system Efficient Lexicon Management System (ELMS).

ELMS is a part of machine translation system which is being implemented at the University of Missouri. The machine translation system is mainly based on lexical knowledge (syntactic knowledge is also used for efficiency) which is represented as a conceptual structure [9] and implemented as slot-filler structure (cf. ERKS structure [1]) using Common Lisp. We will review some literature on lexical item retrieval methods in section 2. In section 3, we will discuss our system ELMS in some detail. And we will conclude the paper in section 4.

2. Review of Some Previous Work

In this section we will review some earlier work on lexical item retrieval based on conceptual representations. Goldman introduces a retrieval method of lexical items for which he uses a 'discrimination net' [4]. It has a binary tree structure. Terminal nodes of the tree are lexical items. A nonterminal node consists of predicates which are used in determining the path to take. Nonterminal nodes in effect serve as a guide in picking out a terminal node. Goldman's approach is the first lexical item retrieval method based on conceptual structures.

Levelt, et al discuss some difficulties with Goldman's approach [5]. The problem with Goldman's approach is the employment of the serial arrangement of predicate tests at the nonterminal nodes. This makes a search in a large lexicon very expensive and creates a dependency on the complexity of the conceptual structure of a lexical item. The search for a more complex item takes a longer time than a simpler one. Also this approach is purely hierarchically organized, an obvious problem for a natural language lexicon, which often involves more than just a hierarchical structure.

Miller and Johnson-Laird propose an alternative method [7] based on a table -- known as 'decision table' --rather than a tree. Pollock shows that a decision table can be translated to a flow chart, truth table, or key [8]. The decision table provides a more abstract representation than other otherwise equivalent representation schemes. This is due to the fact that the ordering of the testing is immaterial in a decision table unlike in the other representation styles. This has a rather important consequence, that is, that the parallel operation is readily available for the decision table. A problem with this approach is the exponential growth of the table size as more conditions are considered. This problem may be alleviated somewhat by employing table linkage.

Cullingford and Joseph have developed a knowledge base organizing program called SOT (Self-Organizing Tree) [1, 2]. It is a variation of Goldman's work. SOT separates items into groups, on the basis of the path in the tree. Terminal nodes of SOT are lexical items as in a discrimination net. Each nonterminal node of a tree consists of a 'path' and an 'index'. The search always starts with the root node as is generally the case with slot-filler-based searches. Our approach departs from such traditional approaches in the search strategy, it employs a backward searching mechanism. An example of SOT tree is:

```
            -0-
      (b)< c 1 f 2 >
      _____|_____
      -1-                    -2-
      (d)< f 3 e 4 >         [(a b (f) h (nil))  (a b (f))]
      _____|_____
      -3-                  -4-
      [(a b (c) d (f))]    [(a b (c) d (e))]
```

The above tree contains the following five items:

    (a b (c) d (e))
    (a b (c) d (f))
    (a b (f) h (nil))
    (a b (f)).

In a SOT tree, terminal nodes are sets of lexical items enclosed in [ ]. Nonterminal nodes consist of a path enclosed in ( ) and an index in < >.

Such SOT trees are built recursively. Initially one starts with the total set of all internal representations of lexical items--ERKS structures explained below--to be discriminated. The nature of a tree depends crucially on the paths that are employed. A path is a sequence of slot-names that lead to a given value. The path for '(masc)' in the example below is '(to val partof gender)'. The only path all items have in the beginning is the null path '( )'. SOT creates a candidate path by adding a slot name to preexisting paths which distinguishes the items in the set under examination. The selection of an appropriate slot name in extending old paths will contribute to the efficiency of overall retrieval strategy[1]. Once the original set of items to be discriminated is divided on the basis of the new path, the same process recursively applies to each of the resultant subgroups.

Cullingford uses ERKS (Eclectic Representations for Knowledge Structures) structures for internal representation of lexical items. An ERKS structure consists of a header followed by zero or more pairs of a role and its filler. The filler itself is an ERKS structure. An example of a well-formed ERKS structure is:

(propel actor (person persname (Olivia) gender (fem))
         object (bpart bptype (hand) partof (person persname (Olivia) gender (fem))))

```
to  (physcont val (bpart partof (person persname (Muhammed)
                gender (masc)) bptype (nose) ref (def)))
    time (times time1 (:past))
    mode (modes mode1 (:t)))
```

Pointers may be used for slot-fillers. This example is a representation of the sentence 'Olivia punched Muhammed in the nose.' The syntax for ERKS is provided below, for the discussion in section 3 presupposes some knowledge of ERKS structure.

```
struct    --->   (header body)
header    --->   "atom" | nil
body      --->   slot  filler | (slot  filler)*
filler    --->   struct | <struct>   ; <> indicates the pointer
slot      --->   "atom"              ; symbolic atom,  not T & nil
```

3.   An Efficient Lexicon Management System

Three major concerns for large lexicon management systems are ability to update with minimal side-effects on the lexicon as a whole, maximal speed in retrieval of desired lexical items, and minimal usage of storage space. With the scheme outlined in the previous section, the update operation can be very expensive. The whole lexicon structure must be reconstructed everytime when an item is added or deleted. Otherwise the tree might go out of balance badly. Furthermore the retrieval operation is dependent on both the size of the lexicon and the complexity of the item to be retrieved.

ELMS meets all three criteria outlined in the previous paragraph. Lexical word entries are represented as slot-filler structures similar to Cullingford's ERKS structures. This slot-filler structure has compositional characteristics, so it is possible to have one conceptual structure for a whole sentence or a paragraph. From such an inclusive conceptual structure which contains more than one word, the language generator will select a word by recursive decomposition and search. We will not discuss the decomposition process in this paper. Our main concern is how efficiently we can select an appropriate target word from the conceptual structure which represents the word.

3.1 Organization and Update Mechanism of Very Large Lexicon

ELMS consists of a Lexical Item Table (LIT) for the lexical data itself and a Header Index Set (HIS) for the control of LIT. LIT is maintained for analysis of the source language and HIS is maintained for generation of the target language. LIT is a two dimensional array organized along the axes of item numbers, one for each item, and header numbers. A header index set provides for each header a list of indices each of which contains the slot-filler for the header in question and also all the item numbers of those items which have that value for that header.

We will illustrate the LIT organization with a simplified example. The following illustrates simplified conceptual structures for lexical items.

(p x (d))
(p x (b) y (c))
(p x (d) y (e))
(p x (d) y (c))
(p x (f) y (e) z (c))
(p u (d))

As the first item is added to LIT, item number[2]1 will be assigned to it together with header numbers for all the headers appearing in that item. There are two headers (p and d) in the item. The first header of the item is 'p' and the second header of the item is 'd'. Hence, ELMS assigns header number 1 (H1) to 'p' and header number 2 (H2) to 'd'. LIT at that point will have one item and two headers. The number attached to the item number is the number of headers in the lexical item called Total Header Number (THN). For instance, i1-2 means that the first lexical item (i1) has 2 headers (in this case 'p' and 'd'). LIT and HIS are given below:

LIT: item#   H1   H2
           i1   (p   x   (d))

HIS: HIS1 = {(p, i1-2)}   HIS2 = {(d,   i1-2)}.

When the second item, (p x (b) y (c)), is added, that item will have assigned item number 2 and one additional header (H3) since its header length is three. The Lexical Item Table (LIT) and Header Index Set (HIS) will be updated as below:

LIT: item#   H1   H2   H3

```
i1        ( p    x   (d))
i2        ( p    x   (b)    y    (c))
```

HIS: HIS1 = {(p, i1-2, i2-3)}   HIS2 = { (d, i1-2) (b, i2-3)}
       HIS3 = {(c, i2-3)}.

The final LIT and HIS are given below after all six items are added:

Item #	H1			H2		H3		H4	
i1	(p	x	(d))						
i2	(p	x	(b)	y	(c))				
i3	(p	x	(d)	y	(e))				
i4	(p	x	(d)	y	(c))				
i5	(p	x	(f)	y	(e)	z	(c))		
i6	(p	u	(d))						

Figure 1. Lexical Item Table

HIS1 = {(p, i1-2, i2-3, i3-3, i4 3, i5-4, i6-2)}
HIS2 = {(d, i1-2, i3-3, i4-3, i6-2) (b, i2-3) (f, i5-4)}
HIS3 = {(c, i2-3, i4-3) (e, i3-3, i5-4)}
HIS4 = {(c, i5-4)}.

The deletion is hardly expected but if it should happen then just delete the item from the LIT and its item number from the HIS. The cost for the latter would not be worse than $O(n)$.   Thus, the update operation does not cause any restructuring of the database, just addition or deletion of the item in LIT and of its item number in the header index lists.

Some simple software tool can be used here to order the role-filler pairs in a consistent way and also in such a way that the least frequently used role in the dictionary items with its filler come last. This order is fixed at first time when the structure is defined. This is a useful supplementary technique for reduced searching time, but it is not essential for ELMS.

Two or more items may have the ERKS structures in such a way that they have slot-fillers with the same header but their slots are different in the same position. Compare i1 and i6 in the above example. These two have the same headers but their second headers are for two different roles (i.e., x and u). In such a case, we may not be able to uniquely identify a lexical item. But we conjecture that whenever this happens the set will not be very large, and we can use some pattern matching algorithm (for each role) to find the particular item under search.

3.2 A Fast Lexical Item Retrieval Mechanism

The most important consideration in a very large lexicon management system is the speed of the retrieval operation. Furthermore, the complexity of the internal structure of an item should not adversely affect the efficiency of the retrieval operation. More complicated items (i.e., the lexical item which has more slot-fillers) should be accessed faster than the simpler lexical items from the lexicon, since more complicated lexical items are more specialized, i.e. more silent features which are not found in the majority of other lexical items (principle of greater specificity). In other words a large number of items may be eliminated from the set of items to be searched by using those extra salient features in comparison. Similar opinions have been expressed elsewhere. For example, Levelt et al states that more complex verbs take systematically less time to access than less complex ones do [6].

Assuming that the patterns for comparison and the dictionary entries are both maximally specified, or to put it differently that the retrieval operation is based on complete identity, retrieval can be accomplished by 1) getting the header index set HISj for the last header where j is the number of headers in the ERKS structure of the lexical item under search; 2) finding the header index list[3] (HIL) which contains the j-th header in the pattern. If it is a singleton list[4] (a list with exactly one item), then that item must be the entry we are looking for. If not, compare THN of each lexical item in the HIL with THN of the entry we are looking for and get the lexical item numbers which have the same THN with the lexical item under search. If it is a singleton list, then that item must be the entry we are looking for. If not, get the correct HIL from HIS(j-1) for the (j-1)th header and get the intersection list between these two lists. If the resultant list is a singleton list, the item therein is the entry we are looking for. If the inters ection list is still not a singleton, we repeat the above process until eventually we come to a singleton list , in which case we have succeeded in locating the lexical item we want, or hopefully to a small but not a singleton list, in this latter case we use some pattern matching algorithm to find the particular item under search.

For example, suppose that a generator tries to access the fifth item in Figure 1. First we get HIS4, which contains only one HIL, (c, i5-4), which is a singleton list. The item we are looking for is i5. For another example, let us consider a pattern identical to i3. We get HIS3 since the number of the headers in the structure is three. Since the last header (i.e., third header for this example) for this item is 'e', we get the HIL '(e, i3-3, i5-4)'. This is not a singleton set, hence, the system will check THN of each lexical item in the HIL. THN of i3 is 3 and THN of i5 is 4 in this case and THN of the lexical item we are looking for is 3. Therefore the system will retrieve i3. In order to retrieve the lexical item same with i4, get HIS3 as previous example with the same reason and get HIL, (c, i2-3, i4-3). This time the HIL is not a singleton list and THN of each lexical item in the HIL are same. So we go to HIL2 and extract the HIL '(d, i1-2, i3-3, i4-3, i6-2)' because the second header in the structure is 'd'. The intersection of these lists ((c, i2-3, i4-3)∧(d, i1-2, i3-3, i4-3, i6-2)) is the singleton list '(i4-3)', so the fourth item (i4-3) is selected before we consider the first header of the ERKS structure. Notice that the search in most cases need not exhaust the ERKS structures as these examples demonstrate. The complex items in internal representation are not necessarily penalized, rather often rewarded with a faster match. As the size of the dictionary increases, the individual HIL will grow longer, demanding a longer search time within a given HIL. However, since search might end sometime before we get to HIS1, the dependency of the efficiency of retrieval operation on the size of lexicon is not a close one.

The ELMS has been implemented on VAX11/785 using Common Lisp. It consists of a Lexical Item Table Organizer (LITO) and Fast Lexical Item Retriever (FLIR) with some additional routines. LITO builds the LIT and HIS with the user input. FLIR retrieves the correct lexical item quickly by using backward searching. Followings is the summary of the procedures discussed so far.

LITO
Extract all the headers from the input item
    & maintain Total Header Number (THN).
Build a Header Index Set (HIS) according to the position of each header.
    For all headers in the input lexical item:
        If HIS is empty then add (new_header, item_number) pair in the HIS
            else compare HIS with new header
            If any header in HIS is the same as the new header
                then append the (lexical) item number to correct HIL
                else add (new-header, item-number) pair on the HIS.

End for

FLIR

Extract all the headers from the input and maintain number of headers in THN
    for the input.

For THNth HIS:

Consider the last header (i.e., n-th header) of input & find the HIL
    (HIL is a subset of HIS) which contains the last header of the input.

    Let X = the number of items in the HIL

Loop:

    If X = 1, We are done. Exit with the item number in the HIL

    If X < 1, the item is not in the lexicon. Exit with an appropriate message

    If X > 1, get the lexical item which has same THN as input's THN.

    If there is only one lexical item then exit with the item number.

        If not, consider (n-1)th HIS & find the correct HIL which contains the
        same header as the (n-1)th header in the input.

    Let X = size of the intersection of the HIL here and the HIL of the (n-1)th HIS
        above. Current HIL = Intersection of two HILs. n = n - 1.

    Go to Loop.

4. Conclusion

We have outlined an efficient lexicon management system for a very large lexicon
with a fast retrieval operation, a side-effect-free update operation. The overhead for
organizing the lexicon for the system is favorably compared with other organizations.
Unlike the other lexicon ELMS employs a backward searching mechanism and it
allows to search the lexical item faster, since we often do not have to compare all the
headers in the ERKS structure (it does not travel from the root node to a leaf node all
the time). Whenever there is only one ERKS structure left for consideration the
searching process will be terminated. Each time when we compare headers,
numerous ERKS structures might be deleted from the next header comparison.

    Selection of an appropriate data structure in implementing the header index set
(these need not be lists) is an important consideration for the resultant efficiency of
the system. Also selection of the order of role-filler pairs used in lexical
representations can be used to minimize the searching time. Empirical studies need to
be conducted with a view to determining the relative frequency of the use of these
roles in the lexicon.

Notes

1. Cullingford works under the assumption that simpler lexical items tend to be used more frequently and therefore must come higher in the SOT tree.

2. This item number can be an actual word or pointer to the words in a real system. At least 2 words should be there, one for source language and the other one for target language.

3. Header Index List (HIL) is a subset of HIS. For example, (c, i2-3) is a HIL of HIS3.

4. For example, (c, i2-3) is a singleton list, since there is only one item (i2-3) in the list. We are not consider the header in the list as an element of the list.

References

[1] Cullingford, R. E. (1986) Natural Language Processing, A knowledge-Engineering Approach. Rownam & Littlefield.

[2] Cullingford, R. E. & Joseph, L.J. (1983) A Heuristically 'Optimal' Knowledge Base Organization Technique. Automatica, Vo. 19, No. 6

[3] Deering, M., Faletti, J. & Wilensky, R. (1981) PEARL-A Package for Efficient Access to Representation in LISP. IJCAI-81, Vancouver, British Columbia.

[4] Goldman, N. (1975) Conceptual generation. In: R. Schank, Conceptual information processing. Amsterdam: North-Holland.

[5] Levelt, W.J.M, & Schriefers, H (1987) Stages of lexical access. In: Gerard Kempen (Ed.) Natural language generation. NATO ASI Series. Martinus Nijhoff Publisher.

[6] Levelt, W.J.M, R. Schreuder & E. Hoenkamp (1978) Structure and use of verbs of motion. In: R.N. Campbell & P.T. Smith (Eds.) Research Advances in the Psychology of Language. New York: Plenum.

[7] Miller, G.A. & Johnson-Laird, P.N. (1976) Language and Perception. Cambridge, Massachusetts: Belknap/Harvard Press.

[8] Pollock, S.L. (1971) Decision table: theory and practice. New York: Wiley-Interscience. A Division of John Wiley & Sons, Inc.

[9] Schank, R.C. (1973) Identification of Conceptualizations Underlying Natural

Language. In: Schank, R.C. & Colby, K.M. (Eds) Computer Models of Thought and Language

Distributed Subsymbolic Representations for Natural Language: How many Features Do You Need?

Richard F.E. Sutcliffe

ABSTRACT

In a Natural Language Understanding system, be it connectionist or otherwise, it is often desirable for representations to be as compact as possible. In this paper we present a simple algorithm for thinning down an existing set of distributed concept representations which form the lexicon in a prototype story paraphrase system which exploits both conventional and connectionist approaches to Artificial Intelligence (AI). We also present some performance measures for evaluating a lexicon's performance. The main result is that the algorithm appears to work well - we can use it to balance the level of detail in a lexicon against the amount of space it requires. There are also interesting ramifications concerning meaning in natural language.

INTRODUCTION

Any computer program which aims to understand or process natural language to a significant degree must be able to represent the meanings of words, clauses, paragraphs and larger units or text. In the past, the problem of how to represent meaning within such Natural Language Understanding (NLU) Systems has normally been addressed by the use of list-based symbolic structures. In many cases, these structures have been based around a set of semantic primitives such as those proposed by Schank [1] or Wilks [2]. These primitives were often combined into semantic structures for representing sentence contents by means of various case

I would like to thank Paul Mc Kevitt, Adrian Baldwin, Jon Rowe, Stewart Jackson, Paul Day and Sue Milward for their help with this research. I am also indebted to Khalid Sattar for providing excellent system support. The author's address is Department of Computer Science, University of Exeter, GB Exeter EX4 4PT, EC, email richard@uk.ac.exeter.cs .

frame schemes based on the pioneering ideas of Fillmore [3]. Successful applications of these techniques include Wilks' machine translation system [4], the SAM system of Schank and others [5], BORIS [6], and the Unix Consultant [7].

However, despite the success of the above systems and many others like them, symbolic representations suffer from a number of shortcomings. In particular, it is very difficult to capture the fluidity and subtlety of language using such schemes. The possibility of representing a concept as a set of *features* was proposed as early as 1963 by Katz and Fodor [8] and has been widely used since. In a feature based representation scheme, a concept is encapsulated by simply specifying what attributes it has. For example a bird might be represented as [has_wings, has_beak, is_fluffy, is_small, flies, walks, etc], while a poodle could be [is_fluffy, is_small, walks, barks, etc]. The assumption is that many different concepts can be represented using the same set of features. Two concepts can then be compared by seeing which attributes they have in common. For example birds and poodles are both small and fluffy, and they both walk.

The term *microfeature* was first coined by Hinton [9, p172] although the first use of the term in the sense being described here occurred in an article by Waltz and Pollack [10]. There are two main differences between the microfeatures described by the latter authors and the features of earlier AI systems. Firstly, a set of microfeatures for describing a set of concepts is assumed to be much larger than a set of features intended to be used for the same purpose. Secondly, the algorithms used for comparing and processing microfeature representations tend to be much more sophisticated than those employed with schemes involving feature sets.

It should be pointed out at this stage that there are two types of microfeature discussed in the literature. There is a growing body of literature on generating distributed representations of concepts by learning to perform a task which effectively requires knowledge of those concepts. These distributed representations are considered by some workers to be couched in terms of microfeatures which are meaningful but whose meanings can not readily be determined. Meaning has come to be associated with each microfeature by virtue of the learning algorithm and not by an explicit association determined by the system's implementors. Such microfeatures may be termed *semantically uninterpretable*. The work of Hinton and Sejnowsi [11], Hinton [12], Miikkulainen and Dyer [13,14], and Elman [15] may be turned to for more information on this approach.

The second type of microfeature can be termed *semantically interpretable* because each member of a set of such features has a known attribute associated with it. Concept representations are generated by an explicit consideration of how the attributes associated with the microfeatures relate to the entity whose meaning is to be represented. Such features are closely related to the semantic features used within linguistics as exemplified by Chomsky [16], or psycholinguistics, for example the work of Schaeffer and Wallace [17], Rosch [18,19], or Smith, Shoben and Rips [20]. In addition, such microfeatures are close to the features found in traditional AI, for example as used by Brachman in semantic networks [21]. It is representations constructed using semantically interpretable microfeatures which are the subject of this paper.

The author's work in the area of distributed representations using microfeatures originated with the PARROT system [22]. PARROT was able to 'paraphrase' simple action vignettes based around stereotypical events such as getting up or having lunch. The central result of this work was that actions, objects and indeed states of the world could all be represented using a particular variant of the microfeature idea which is described in the next section.

However, while the PARROT project was an interesting initial investigation into the use of distributed representations, it left a lot of important questions unanswered. One shortcoming of particular importance was the question of which features are really important to the representation of concepts: can some features be dispensed with from all concept representations within a system without losing too much descriptive power?

There are three main reasons for being interested in this issue. Firstly, it is desirable to be able to manipulate the level of detail with which a lexicon expresses a set of concepts. In particular the amount of space used up by a lexicon may be of crucial importance especially if it defines a large number of words. Being able to balance level of detail and thus expressive power against efficiency of operation seems a very desirable property of these kinds of representations - a property, moreover, which is not shared by their symbolic counterparts.

The second reason for being interested in controlling level of detail is as follows. While distributed representations of the kind being discussed here may be used in traditional AI systems, they can also be exploited in so-called connectionist ones. In the latter they could for example serve as the inputs to a network which learned to map concepts at the lexical level into concepts at a higher level such as the propositions underlying sentences and clauses. Learning algorithms such as the *Back Propagation* of Rumelhart, Hinton and Williams [23] are capable of learning mappings of this kind from one distributed representation into another. However, the performance of such algorithms does not necessarily scale up well - small networks may learn considerably more quickly than large ones. (See [24, p264 New

Edition only] for a discussion.) Since concise representations - short vectors - lead to smaller networks of weights which can as a result learn much more quickly than larger networks, this provides a further justification for interest in controlling the size of a set of concept representations.

A third reason for interest in this issue is more general. If we can determine which features can be eliminated with least effect we have found out something interesting about the way in which the features are capturing information about objects in the world. The information gleaned from studies such as this may be able to tell us something both about the world itself and about the way in which linguistic meaning relates to the world. To take a hypothetical example, suppose it was found that features associated with the size and shape of an object always occurred in a feature set even when the features associated with colour and usefulness had been eliminated. This might license us to conclude that size and shape were more fundamental classifiers of objects in the world than colour and usefulness. A thorough study on these lines might begin to isolate the 'primitive' features which can be used to describe objects.

The rest of this paper is structured as follows. In the next section the essential techniques behind the implementation of the microfeature concept in the PARROT project are outlined. After this, two simple experiments are described which test a simple hypothesis: If the contribution to meaning of a particular feature varies widely across a set of concepts to be represented, then that feature is more significant to the representations in which it participates than another whose contribution to meaning varies to a lesser degree.

The final section discusses the results of the experiments and suggests areas for further work.

DISTRIBUTED REPRESENTATIONS IN PARROT

In the PARROT system there are three sets of microfeatures, one each for actions, objects, and contexts. Actions are considered to be the meanings underlying verbs such as Hit or Touch. Objects are the meanings of nouns such as Bicycle or Person. Finally, Contexts capture the gist of elusive concepts like Summer or Happiness - in essence states of the world in which actions can occur.

The lexicon in PARROT consists of three sections, one each for verbs, nouns and context words. We will now outline the manner in which a noun's meaning is represented.

A set of 166 microfeatures was devised for representing noun meanings. This set was constructed in an ad hoc fashion - see [25] or [22] for a description - although more systematic methods based on established psycholinguistic techniques are now being explored.

The meaning of a given noun was expressed by associating an integer *centrality* with each microfeature according to how strongly that feature applied to the noun.

For example, suppose we wish to encode the word Book. If, for simplicity, we use a set of only nine features, the result might be the following:

Book:

small_size	5
made_of_paper	5
made_of_organic_material	0
is_common	5
found_in_bookshops	10
found_in_offices	5
found_in_foodshops	0
found_in_kitchens	0
found_in_homes	5
is_functional	10
involves_information	15
is_aesthetically_pleasing	5
is_edible	0

Here we have chosen to emphasise found in bookshops and is_functional because they seem important to the concept Book. The feature involves_information is emphasised even further because it appears to be the most important one associated with the concept.

The main reason for adopting representations such as this in the PARROT system was so that representations corresponding to different concepts could be compared in order to see how similar in meaning the concepts themselves were.*
Treating a representation such as the above as a vector of numbers, comparison of meaning can conveniently be undertaken by computing the dot product of two vectors, each of which represents a distinct concept.

The dot product of two vectors P1 and P2 can be defined thus:

$$P1 \cdot P2 = \sum_{k=1}^{k=n} P1_k * P2_k \qquad (1)$$

* I am grateful to Alan Smeaton for pointing out the close relation between these ideas and the Vector Space Model of Information Retrieval [26]. This was not known to the author at the time of [25].

where both P1 and P2 have n elements, $P1_k$ is the kth element of P1, and $P2_k$ is the kth element of P2.

For the dot product method to work it is necessary to *normalise* each vector before conducting any comparisons. Normalisation simply involves scaling each vector up or down so that the dot product of that vector with itself is 1. This effectively ensures that a concept is more similar in meaning to itself than it is to any other concept.

To see these ideas in practice consider the normalised representation of Book together with normalised vectors for Magazine and Cornflakes couched in terms of the same example set of microfeatures:

	Book	Magazine	Cornflakes
small_size	0.209	0.200	0.048
made_of_paper	0.209	0.250	0
made_of_organic_material	0	0	0.095
is_common	0.209	0.250	0.048
found_in_bookshops	0.417	0.250	0
found_in_offices	0.209	0.250	0
found_in_foodshops	0	0	0.095
found_in_kitchens	0	0	0.095
found_in_homes	0.209	0.250	0.048
is_functional	0.417	0.250	0.949
involves_information	0.626	0.750	0
is_aesthetically_pleasing	0.209	0.150	0.237
is_edible	0	0	0.095

Intuitively we would expect that the meaning of Book would be more similar to that of Magazine than it is to Cornflakes, and indeed this proves to be the case: Book · Magazine is 0.966 while Book · Cornflakes is 0.474.

So far we have presented examples which use only a small set of features and not the 166 features used in PARROT. Figure one shows an example lexical entry taken from the actual system lexicon. The entry is shown in un-normalised form but would be normalised by the system prior to being used.

Note that we have only discussed the representation of nouns using object microfeatures. The meanings of verbs and context words are represented analogously. The only difference is that two different microfeature sets are used - the action microfeature set for verbs and the context microfeature set for context words. However, the study on cutting microfeatures presented in later sections was conducted only on the nouns in the lexicon.

This completes our brief exposition of distributed concept representations in PARROT. For further information on the construction and use of the original lexicon, see [25]. The next section turns to the present study whose objective was to investigate which microfeatures could be removed from a representation without substantial loss of performance. This study took as its starting point the noun definitions from PARROT's lexicon whose construction we have just been discussing.

CUTTING MICROFEATURES FROM THE LEXICON

In this section we will discuss two experiments which took as their starting point the lexicon in PARROT whose construction we have just been discussing. The purpose of these experiments was to investigate whether the representations employed in the PARROT lexicon could be thinned down by removing certain microfeatures by eliminating the centralities associated with those microfeatures from each lexical entry. The general experimental design employed here was thus as follows:

- Take the noun representations from the PARROT lexicon
- Eliminate a given set of features from each such representation
- Renormalise the lexical entries and re-assess their performance.

By 'performance' here, we mean the ability of the modified concept representations to capture the meanings of the noun concepts defined in the lexicon, and in particular how this ability compares with that of the original lexicon. Various measures of performance were used as will shortly be discussed.

The hypothesis being investigated in these experiments was as follows:

A microfeature whose centrality varies by a large amount across lexical entries expresses more useful information than a microfeature whose centrality varies by a small amount across those lexical entries.

It is quite easy provide an intuitive justification for this hypothesis. Suppose every concept in the lexicon exhibits a particular characteristic to the same degree. If there is a microfeature corresponding to this characteristic then each concept in the lexicon will have in its distributed representation the same centrality associated with that microfeature. (For example if there was a feature is_a_concept it might show this behaviour.) According to the above hypothesis, such a microfeature is not expressing any information because it is not telling us anything which enables a pair of concepts to be distinguished.

On the other hand, consider a microfeature representing a useful attribute such as involves information. The centralities associated with this microfeature in different representations are likely to vary depending on the concept being defined in each case. We have already seen this in the above example where Books and Magazines involve information while Cornflakes do not.

Experiment One - an Initial Test of the Elimination Hypothesis

The objective of this first experiment was to provide a basic test of the above hypothesis concerning which microfeatures express useful information. The method adopted may be summarised as follows:

- Compute the Standard Deviation of each microfeature's centralities across all lexical entries.

- Order the microfeatures by standard deviation putting the microfeatures with highest standard deviation first and those with lowest standard deviation last.

- Divide the ordered set of microfeatures down the middle, to produce two sets. Call the first set the *important microfeatures* and the second set the *unimportant microfeatures*.

- Construct two new sets of representations for nouns in the lexicon. In one set, called the *minus-important lexicon*, eliminate any centrality if it corresponds to a microfeature in the set of important microfeatures. Renormalise each concept representation. In the other set, called the *minus-unimportant lexicon*, eliminate any centrality if it corresponds to a microfeature in the set of unimportant microfeatures.

- Compare the performance of the minus-important and minus-unimportant lexicons with the original lexicon.

Two important points arising from the above description should be noted. Firstly, the standard deviation was used as the measure of variance in centrality of a microfeature across lexical entries. For completeness we define this here. Firstly, the mean of the set of values x_i is computed to yield $\bar{x}$. The standard deviation SD is then computed as follows:

$$SD = \sqrt{\sum \frac{(x_i - \bar{x})^2}{N - 1}} \qquad (2)$$

where N is the number of nouns in the lexicon and hence the number of centralities associated with a particular object microfeature in the lexicon.

The second point arising from the method description is that we have not as yet defined a suitable performance measure. For this experiment a simple measure was used. Recall that there are three lexicons of interest, *original (orig)*, *minus-*

important (-imp) and *minus-unimportant (-unimp)*. A comparison between the original and minus-important lexicons was made by computing the mean of all dot products of words in the original lexicon with their counterparts in the minus-important lexicon $\overline{word_i^{orig} \cdot word_i^{-imp}}$. A similar comparison was made between the original and minus-unimportant lexicons to produce $\overline{word_i^{orig} \cdot word_i^{-unimp}}$. We call these *word-word* measures because they are measuring how similar in meaning a word in one lexicon is with the same word in another lexicon.

In essence, the purpose of this first experiment was to see if there was any truth in the hypothesis about centrality variance as expressed earlier. Recall that the minus-important lexicon defines the set of noun meanings in terms of their centralities against a set of microfeatures which are not held to express much useful information. The 'important' microfeatures - those whose centrality values across all noun entries in the lexicon are high - have been taken out. By contrast, the minus-unimportant lexicon defines the noun meanings in terms of their centralities against the set of microfeatures which *are* thought to convey useful information. Here, only the 'unimportant' microfeatures - those whose centrality values across all noun entries in the lexicon are low - have been taken out. Based on these assumptions, we would expect the performance of the minus-unimportant lexicon to be much closer to that of the original lexicon than was the performance of the minus-important lexicon. The table summarises the results with lexicon performance being measured using the word-word method.

Experiment I Results		
Lexicon A	**Lexicon B**	**Mean Word-Word Dot Product**
Original	Minus-important	0.545
Original	Minus-unimportant	0.818

As can be seen, the performance of the minus-unimportant lexicon was much better than the minus-important lexicon. This result therefore lends weight to the original hypothesis. From now onwards, it was assumed that the unimportant microfeatures should be taken out of a representation if it was to be made smaller, and not the important microfeatures. The next step was to carry out a more detailed study to investigate how the performance of the lexicon degraded as unimportant microfeatures were removed.

Experiment Two - the Gradual Removal of Microfeatures from the Lexicon

In this second experiment, the objective was to investigate further the extent to which the performance of the lexicon degraded as microfeatures were removed. Following on from Experiment One, it was assumed that unimportant microfeatures - those whose centralities across all nouns in the lexicon had a low standard deviation - should be removed first. The method adopted was thus:

- From the original lexicon construct five new ones. The first lexicon was constructed by eliminating 25% of the microfeatures in increasing order of standard deviation across all lexical entries, starting with the feature with lowest standard deviation. The second lexicon was constructed by eliminating 50% of the microfeatures from the original lexicon in an analogous way. Similar lexicons were constructed by eliminating 75%, 80% and 90% of microfeatures from the original lexicon.

- The performance of the five new lexicons was then measured and compared to the performance of the original lexicon.

In this experiment three measures of performance were used. The first measure was the word-word method which we have already discussed. The results of the experiment using this method are shown in the table.

Experiment II Results (Word-Word Method)		
Lexicon A	Lexicon B	Mean Word-Word Dot Product
Original	Minus 25% of Features	0.950171
Original	Minus 50% of Features	0.818075
Original	Minus 75% of Features	0.648513
Original	Minus 80% of Features	0.599427
Original	Minus 90% of Features	0.476726

This data is also shown in Figure 1 which indicates how performance as measured using the word-word method declines as features are removed (see the end of the paper).

What these results show is that, under this performance measure at least, performance of the algorithm is good. While performance is naturally declining as features are removed, the match of a word with itself is staying as high as 0.8 even bicycle noun

{

obj_one_to_three_letters	5
obj_four_to_six_letters	5
obj_more_than_six_letters	5
obj_starts_with_vowel	5
obj_ends_with_vowel	5
obj_starts_with_consonant	5
obj_ends_with_consonant	5
obj_one_syllable	5
obj_two_syllables	5
obj_three_syllables	5
obj_first_vowel_a	5
obj_first_vowel_e	5
obj_first_vowel_i	5
obj_first_vowel_o	5
obj_first_vowel_u	5
obj_second_vowel_a	5
obj_second_vowel_e	5
obj_second_vowel_i	5
obj_second_vowel_o	5
obj_second_vowel_u	5
obj_time_early_morning	15
obj_time_morning	12
obj_time_lunchtime	10
obj_time_afternoon	5
obj_time_evening	12
obj_time_night	2
obj_lasts_several_years	5
obj_lasts_decade	10
obj_lasts_half_lifetime	10
obj_lasts_lifetime	5
obj_found_garden	10
obj_found_outside	10
obj_found_shopping_area	5
obj_found_residential_area	12
obj_found_urban_area	12
obj_found_countryside	3
obj_rubbery	1
obj_rigid_pliable	15
obj_solid	5
obj_plastic	5
obj_metal	15
obj_shiny	5
obj_smooth	5
obj_opaque	5

obj_dog_weight	5
obj_person_size	8
obj_dog_size	5
obj_specific_limits	2
obj_holey	12
obj_hollow	8
obj_sparse_filling	18
obj_virtical	5
obj_square_lamina	8
obj_rect_lamina	15
obj_round_lamina	5
obj_oval_lamina	5
obj_very_cold	5
obj_cold	5
obj_functional	5
obj_concerns_tiredness	5
obj_concerns_action	12
obj_involves_one_object	5
obj_horiz_motion	25
obj_red	1
obj_blue	1
obj_green	1
obj_orange	1
obj_yellow	1
obj_white	1
obj_black	1
obj_brown	1
obj_one_colour	10
obj_several_colours	5
obj_many_colours	1
obj_inanimate	5
obj_one_person	5

}

Figure 1. An example lexical entry from the original PARROT system [22]. Note that the features are arranged along different dimensions such as time, lifetime and colour. The entry shown is in text form and thus has centralities expressed as integers. Before use the vector generated for the word would be normalised by the system.

when 40% of the original features have been removed. The match is over 0.6 even when 60% of the original features have been removed.

However, it was not felt that the Word-Word method was in fact a particularly good measure to use for judging performance of a given lexicon. What is effectively happening using this method is that lexical entries between different lexicons are being compared. Such entries would of course never occur together in the same system. What is needed is a measure of how concept representations relate to each other *within* a given lexicon. The second performance measure was an attempt to measure exactly this relationship.

Ten pairs of words were selected from the lexicon. Four pairs were chosen such that each pair of words was similar in meaning. The dot product computed with the distributed representations of each pair was therefore high (above 0.6). Four further pairs were chosen to have medium semantic correlation and thus have medium dot products between their respective representations (0.5 to 0.6). Finally, two pairs were selected which had low correlation and thus low dot products (around 0.4).

The way in which the correlation between the various word pairs was sustained as microfeatures were removed from the lexicon was then investigated. We call this the *Dog-Cat* Method because it is measuring how correlations between pairs of words such as Dog and Cat are varying as features are removed from the lexicon. Figures 2 to 4 are graphs which show the performance of the lexicon using this method. There is one graph for high correlation words (Figure 2), one graph for medium correlation words (Figure 3) and one graph for low correlation words (Figure 4). On each graph the correlation between the chosen pairs of words is shown against the number of features removed from the lexicon.

These results appear to show the feature elimination algorithm in a better light than did the Word-Word performance measure. Take the high correlation words for instance. The level of correlation of each word pair is sustained accurately until at least 60% of the features have been removed. In fact it is only after about 75% of features have gone that break up of the correlations starts to occur. These results are mirrored in the medium correlation words although, interestingly, the results here are not so good. Even so, while there is some variation in individual correlations (mostly downward) no dramatic changes occur until after 75% of features have been removed. The fact that Desk-Room crosses over Food-Table is probably not too serious as the correlations of all words in this group were very close together in the original lexicon. Finally, the match between low correlation words tends to fall as features are removed, as can be seen from the third graph. Insofar as the Dog-Cat measure can be trusted, the results seem better than would be expected from if

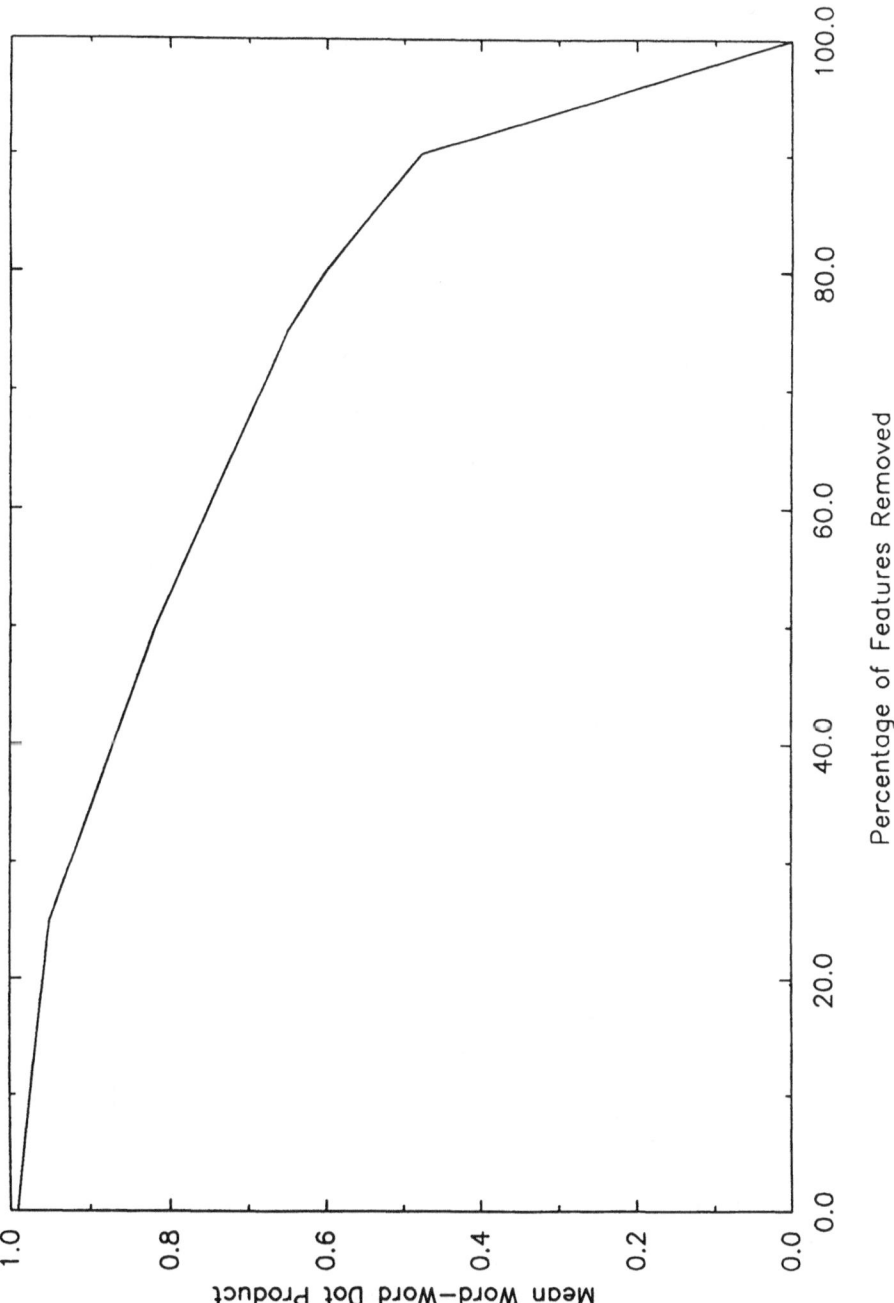

Figure 2. A graph showing how the performance of the lexicon varies as features are removed from it. The performance measure being used here is the Word-Word method.

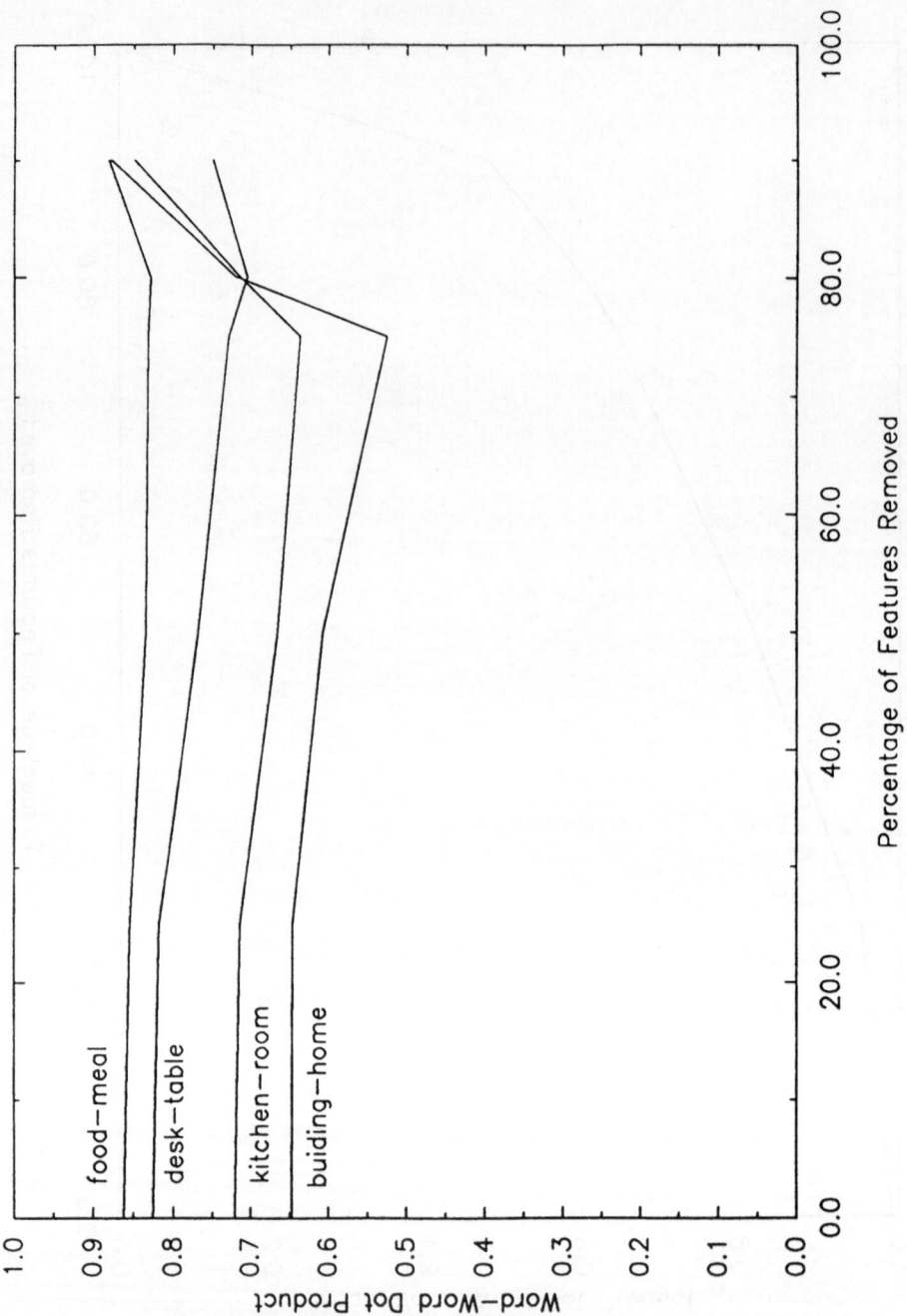

Figure 3. The performance of the lexicon measured as features are removed from it using the same algorithm as in Figure 2. The performance measure here is the Dog-Cat method being used on high correlation words.

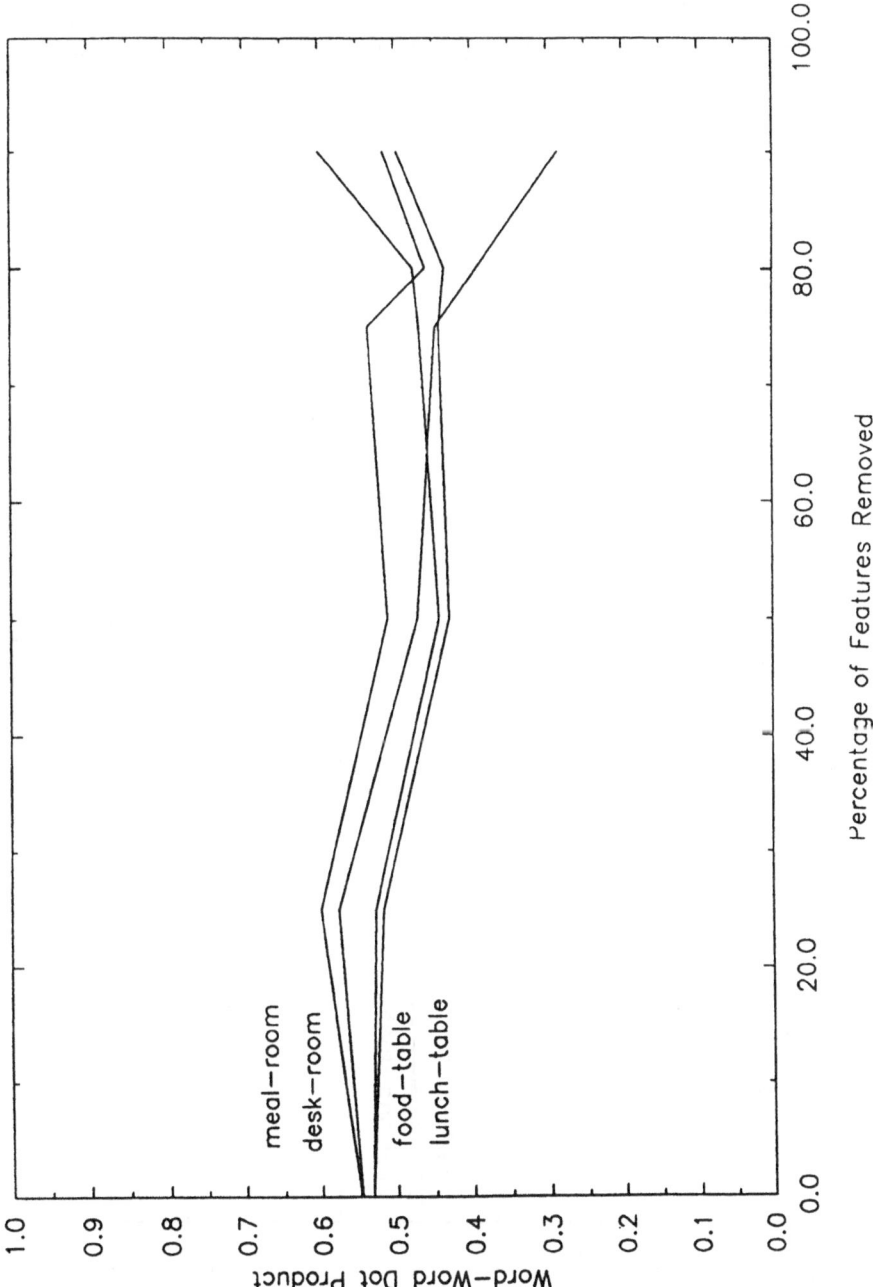

Figure 4. The performance of the lexicon measured as features are removed from it using the same algorithm as in Figure 2. The performance measure here is the Dog-Cat method being used on medium correlation words.

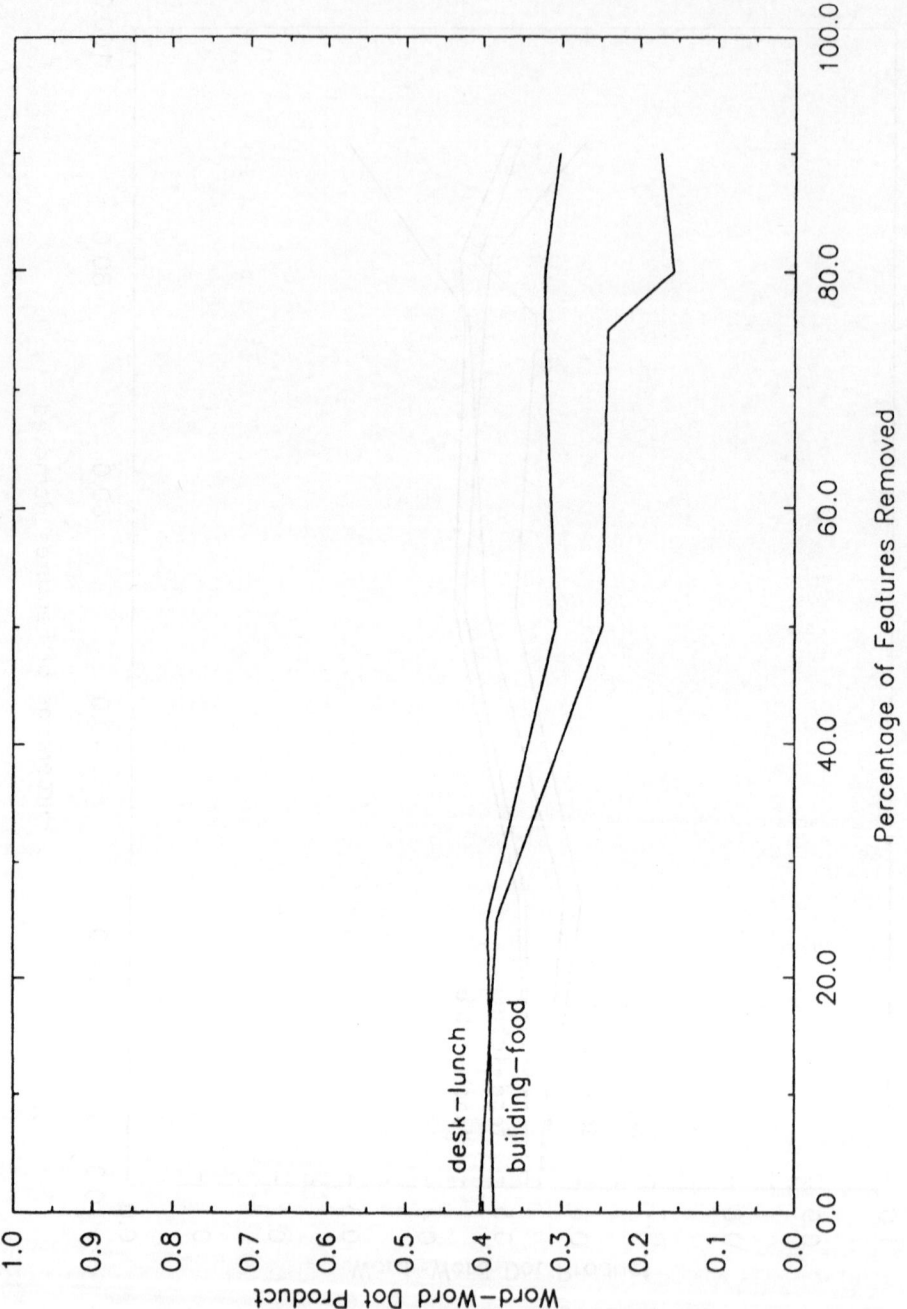

Figure 5. The performance of the lexicon measured as features are removed from it using the same algorithm as in Figure 2. The performance measure here is the Dog-Cat method being used on low correlation words.

the Word-Word measure alone had been used.

The final method for measuring performance was to perform cluster analyses on the various lexicons and to study how the clusters altered as microfeatures were removed. This gives us another measure of how the internal semantic structure of the lexicon is being altered as the representations within it become more and more sparse. The procedure adopted was as follows. The input to the algorithm was the set of distributed patterns, one for each noun in the lexicon. These patterns were normalised before any processing took place. Clustering was then performed on this set of patterns using the Ward Method [27,28]. Clustering was carried out on six distinct sets of data, namely the original lexicon and the five further lexicons derived from it with respectively 25%, 50%, 75%, 80% and 90% of the original features being removed. Some of the resulting clusters are shown in Figures 6 to 8.

In the original lexicon (see Figure 6), the words are well clustered into distinct semantic categories. Examples are washing-articles (Detergent to Bathroom), foods (A_drink to Cornflakes) culinary-objects (Fork to Toaster) clothes (Clothes to Gloves) and people (John to He).

As features are removed form the original lexicon, the original groups of semantically related words tend to remain similar despite the fact that the representation of these words altered by the loss of features. For example after 50% of the original features have been removed (see Figure 7), the foods group still contains the same words (in a different order). The washing-articles group contains the same words plus Kettle, Saucepan and Glass. Eight of the original eleven culinary-objects are still in the culinary-objects group. The clothes group is still intact. Finally the people group contains the same words plus Cashier and Queue. (in the original lexicon cluster these are grouped with other money and purchase objects (Coins to Queue).

Finally, once 90% of features have been removed, the categories of semantically linked words become merged with related categories (see Figure 8). For example some washing articles and culinary objects become grouped together. Foods and meals also become mixed into a group together with other culinary objects. However it should be noted that even after 90% of features have been removed, the groupings of words still seem to be sensible even though the groups are clearly more indistinct than they are in the original lexicon.

In summary, the results of the clustering tend to show a process of gradual decline in specificity of the lexicon as features are removed, and this is consistent with the findings of the other performance measures. However groupings in the lexicon continue to look sensible even when 90% of features have been removed.

296

Dendrogram using Ward Method: All Features

Rescaled Distance Cluster Combine

```
            C A S E          0         5        10        15        20        25
        Label              Seq  +---------+---------+---------+---------+---------+
      ┌─ detergent          30  -+---+
      │  washing_up_liquid  104 -+    +-----------------+
  W   │  apron              5   -----+                  I
  a   │  soap               87  -+                      +-----------------------------+
  s   │  tap                92  -+-----------+          I                             I
  h   │  towel              100 -+     +----+ I                                        I
  i   │  water              105 ----------+ +----+                                     I
  n   │  toothbrush         98  ---+-------+    I                                      I
  g   │  toothpaste         99  ---+        +---+                                      I
      │  basin              8   ----+-------+                                          I
      └─ bathroom           9   -----+                                                 I
      ┌─ a_drink            2   ---+-+                              +-----+
      │  milk               57  ---+ +-----------------+           I     I
  F   │  coffee             24  ------+                I           I     I
  o   │  bread              16  ---+-------+           +-----------+     I
  o   │  toast              96  ---+       +-----------+ I               I
  d   │  cornflakes         28  -----------+           +---+           I I
  s   └─ meal               55  -----------+           I   I           I I
         supper             90  -+          +---------+ I   I           I I
         food               36  -+-+       I          I   +-------+     I I
         lunch              53  -+ +-----------------+ I           +-----+ I
         breakfast          17  ---+                 I I           I       I
         dinner             31  -----------+         I I           I       I
         recipe             76  -----------+  +------+ I           I       I
         waitress           103 -----------+  +------+             I       I
         mouth              59  -----------------+    I            I       I
         nonoun             63  -----------------+    I            I       I
      ┌─ fork               37  -+-------------+       +-----------+       I
      │  knife              51  -+             I  I                        I
  C   │  kettle             49  -+-----------+ I  I                        I
  u   │  saucepan           80  -+      +----+ +----+                      I
  l   │  glass              38  -----------+ I  I   I                      I
  i   │  dish               32  -+-+      I  I  I                          I
  n   │  plate              71  -+ +-+    +-----+                          +------+
  a   │  bowl               15  ---+ +----+ I                             I      I
  r   │  spoon              89  ------+  +-+ I                             I      I
  y   │  jar                48  -----------+ +----+                        I      I
      └─ toaster            97  -----------+                               I      I
         coins              25  -+                                         I      I
         small_change       86  -+-----------------+                       I      I
         money              58  -+           I                             I      I
         price              72  ---+-----+   +---------+                    I      I
         till               95  ---+     +---+ I       I                    I      I
         charity            22  -----------+ +-------+ I                    I      I
         cashier            20  ------------+-+       I                     I      I
         queue              73  ------------+         +------+              I      I
         alarm              3   -+-----------------+  I      I              I      I
         alarm_clock        4   -+                 I  I      I              I      I
         phone              70  -+-------------+   I  I      I              I      I
         telephone          94  -+             +---------+ +----+           I      I
         chair              21  -+-----------+ I I       I I    I           I      I
         table              91  -+          +-+ I I     I I    I            I      I
         computer           27  ---+-----+ I     I I    I I    I            I      I
         desk               29  ---+     +-----+ +--+   I I    I            I      I
         office             65  -------+       I      I I    I              I      I
         list               52  -------------------+  I      I              I      I
         shopping_list      84  -+            I   I   +------+              I      I
         book               14  ---+          +------+                      I      I
         paper              67  -+-+----+      I                            I      I
```

Figure 6a. A dendrogram of the original lexicon after clustering the distributed word representations using the Ward Method. Note the clusters for washing, foods, and culinary objects and observe how these change as features are removed (Figures 7a to 8b).

```
papers            65   ----+    +-------------+                    I           I
newspaper         62   --------+                                   I           I
home              42   ------+                                     I           I
kitchen           50   -----+----+                                 I           I
workplace        106   ------+   +---------+                       I           I
tv               102   ---------+          +---------------+       I           I
face              35   ------------+-----+  I              I   I   I           I
teeth             93   --------+   +------+                I   I   I           I
sound             88   ------------+                       I   I   I           I
bed               10   -----------+---------------+         +---+   I
bedroom           11   --------+               +------+ I           I
clothes           23   ---+---+           I       I   I           I
jacket            47   ---+   +----------------+   I   I           I
gloves            39   --------+                   +---+           I
bicycle           12   +------------------------------+           I
bike              13   -+                          I              I
goods             40   -+----------------+         I              I
merchandise       56   -+           +--------+ I                  I
item              43   -+-+         I        I I                  I
items             44   -+ +--------------+     +-+                I
object            64   ---+                 I                     I
shopping_trolley  85   +--------------------+ I                   I
trolley          101   -+                +----+                   I
restaurant        77   ----+----------   I                        I
shop              83   -----+        +------+                      I
building          18   ----+-------+ I                             I
room              78   ------+     +-+                             I
basket             7   -----------+---+                            I
shelves           82   ---------+                                  I
John              46   -+                                          I
Nick              60   -+                                          I
Paul              66   -+                                          I
Bob                6   -+-+                                        I
Noel              61   -+ +---+                                    I
Edward            33   -+-+  I                                     I
Richard           74   -+       +------------------+               I
Amanda             1   -+-+  I              I                      I
Eunice            34   -+ I  I              I                      I
Mary              54   -+ +---+             I                      I
Susan             79   -+ I                 +--------------------+
Jane              45   -+-+                 I
Rosemary          75   -+                   I
caller            19   -----+---+           I
colleague         26   -----+   +---------------+
person            69   -+------+ I
she               81   -+       +-+
he                41   -------+
```

Figure 6b. The second half of Figure 6a.

Dendrogram using Ward Method 50% of Features Removed

Rescaled Distance Cluster Combine

```
          C A S E              0         5        10        15        20        25
          Label         Seq    +---------+---------+---------+---------+---------+

       ┌─detergent        30   -+-+-+
       │ washing_up_liquid 104  -+ +-----------+
       │ apron             5   ----+           +---------+
       │ kettle           49   -+--------+  I            I
  Washing saucepan        80   -+        +---+           I
       │ basin             8   ---+-----+                I
       │ glass            38   ----+                     +----------------+
       │ soap             87   -+                  I                     I
       │ towel           100   -+----+             I                     I
       │ tap              92   -+    +-----------+ I                     I
       │ water           105   ------+           +------+                I
       │ toothbrush       98   -+------+       I                         I
       │ toothpaste       99   -+      +-------+                         I
       └─bathroom          9   -+--------+                               I
       ┌─a_drink           2   -+-+-+                      +------+
       │ coffee           24   -+ +-----------+            I      I
       │ milk             57   ----+            I          I      I
  Foods bread             16   -+----+          +-----------+     I      I
       │ toast            96   -+    +----------+ I        I      I      I
       └─cornflakes       28   ------+          I I        I      I      I
       ┌─food             36   -+-+-+          +--+        I      I      I
       │ lunch            53   -+ +------+     I          I      I      I
       │ breakfast        17   ---+      +------+          +--------+    I
       │ meal             55   ---+----+-+                 I           I
       │ supper           90   ----+                       I           I
       │ dinner           31   ------+----+-+              I           I
       │ recipe           76   ------+      +------+       I           I
       │ restaurant       77   -------------+    I         I           I
       └─waitress        103   ---------+        +----------+          I
       ┌─fork             37   -+--------------+ I                     I
       │ knife            51   -+              I I                     I
       │ bowl             15   -+------+       +---+                   I
  Culinary dish           32   -+  I           I                      I
  Objects jar             48   -+----+ I       I                      +------+
       │ plate            71   -+    +-+--------+                      I    I
       │ spoon            89   ------+ I                               I    I
       │ toaster          97   --------+                               I    I
       │ alarm             3   -+-----------------+                    I    I
       │ alarm_clock       4   -+                 +------------+       I    I
       │ nonoun           63   ---------------+--+ I           I       I    I
       │ sound            88   -----------+   +---+            I       I    I
       │ mouth            59   ----------------+                I      I    I
       │ list             52   -----------------+               +---+  I    I
       │ shopping_list    84   -+               I               I   I  I    I
       │ book             14   -+-+             +--------+       I   I  I    I
       │ paper            67   -+ I             I        I       I   I  I    I
       │ newspaper        62   ---+-----------+         +------+  I   I  I    I
       │ papers           68   ----+          I              I   I  I    I
       │ phone            70   -+-------------+        I      I   I  I    I
       │ telephone        94   -+             +--------+      I   I  I    I
       │ computer         27   -+-----------+  I             I   I  I    I
       │ desk             29   -+           +---+            I   I  I    I
       │ chair            21   -+------+     I               I   I  I    I
       │ table            91   -+      +------+              +-------+   I
       │ shelves          82   ---------+                    I          I
       │ coins            25   -+                            I          I
       │ small_change     86   -+----------+                 I          I
       │ money            58   -+          +-----------------+          I
       └─price            72   ---+----+ I                I            I
```

Figure 7a. A dendrogram of a stripped down lexicon after clustering the distributed word representations using the Ward Method. Here, 50% of the original features have been removed using the Standard Deviation Method. Note that there are minor differences in the clusters for washing, foods, and culinary objects as compared with the clusters for the original lexicon with 100% of its features (Figures 6a and 6b).

```
till               95   ---+   +----+                    +------+ :          I
charity            22   --------+                I       I I :          I
bicycle            12   -+---------------------+          I       I I          I
bike               13   -+                    +-----+     I :          I
shopping_trolley   85   -+----+                I          I :          I
trolley           101   -+   +----------------+          I :          I
basket              7   -----+                           +--+          I
home               42   ---------+                    I          I
kitchen            50   -----------------+-------+          I          I
tv                102   ---------+       +------------+     I          I
shop               83   ---+--------+    I          I   I          I
workplace         106   ---+   +------+          I   I          I
building           18   -----+--+   I          I   I          I
room               78   -----+ +---+          +---+          I
office             65   --------+                 I          I
bed                10   ---+--------------+          I          I
bedroom            11   ---+              +-----+   I          I
clothes            23   -+----+          I      I   I          I
jacket             47   -+   +--------------+    +----+          I
gloves             39   -----+                   I          I
face               35   ---------------------+ I          I
teeth              93   -----------+          +--+          I
goods              40   -+-----------+          I          I
merchandise        56   -+          +----------+          I
item               43   -+--+          I          I
items              44   -+ +----------+          I
object             64   ---+          I
Paul               66   -+          I
Richard            74   -+          I
Bob                 6   -+          I
Nick               60   -+-+          I
Noel               61   -+ I          I
Edward             33   -+ I          I
John               46   -+ I          I
Rosemary           75   -+ +------+          I
Susan              79   -+ I      I          I
Amanda              1   -+ I   I          I
Jane               45   -+ I      +--------------+          I
Mary               54   -+-+   I          I          I
Eunice             34   -+   I          I          I
person             69   -+   I      +--------------------------+
he                 41   ---------+          I
cashier            20   -----+------+          I
queue              73   -----+   +------------+
caller             19   -----+----+ I
colleague          26   -----+   +--+
she                81   ---------+
```

Figure 7b. The second half of Figure 7a.

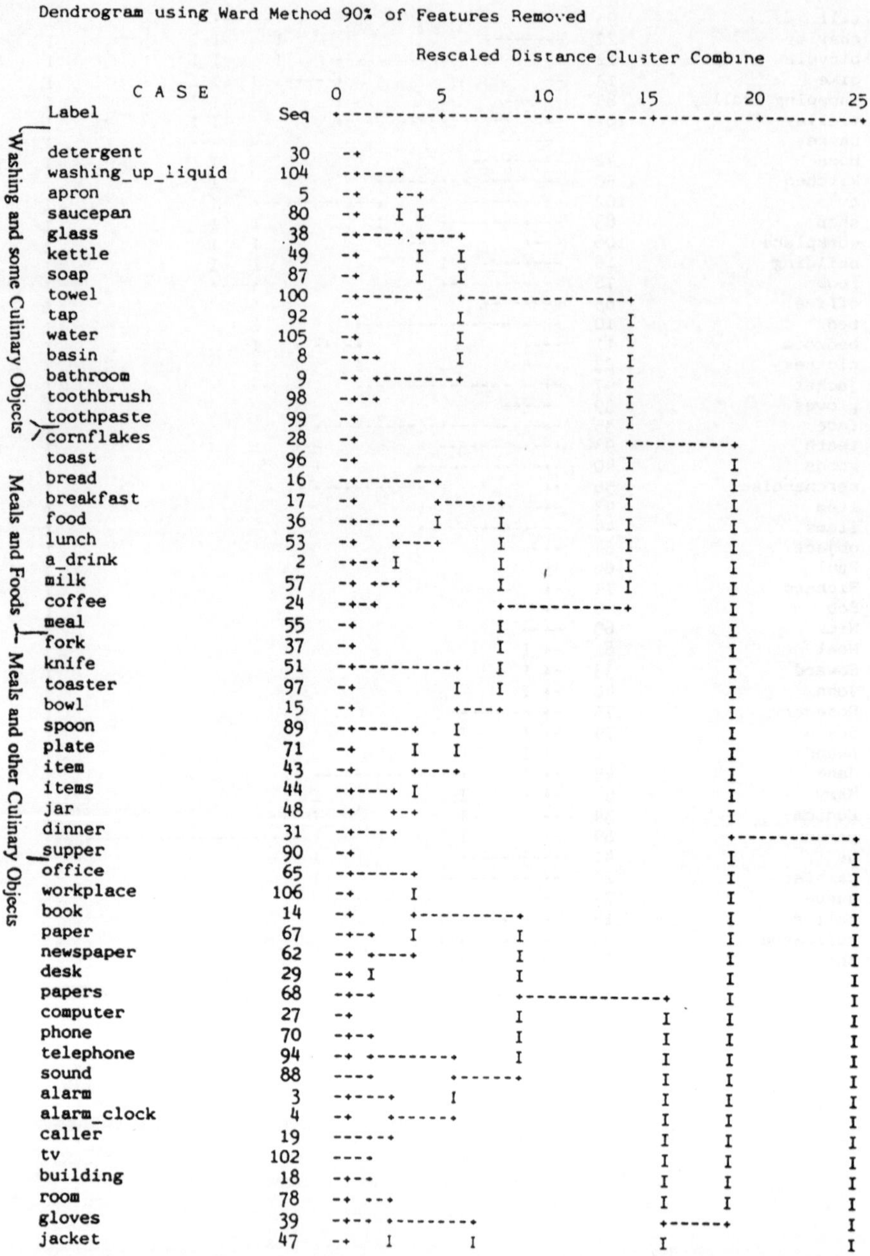

Dendrogram using Ward Method 90% of Features Removed

Rescaled Distance Cluster Combine

```
         C A S E            0       5      10      15      20      25
         Label         Seq  +-----------+-------+-------+-------+-----+
 ┌─ detergent           30  -+
 │  washing_up_liquid  104  ------+
 │  apron                5  -+    +--+
 │  saucepan            80  -+  I I
 │  glass               38  -+---+.+----+
 │  kettle              49  -+   I I
 │  soap                87  -+   I I
 │  towel              100  -+-----+  +--------------+
 │  tap                 92  -+      I               I
 │  water              105  -+      I               I
 │  basin                8  -+-+    I               I
 │  bathroom             9  -+ +-------+             I
 │  toothbrush          98  -+-+                     I
 └─ toothpaste          99  -+                       I
 ┌─ cornflakes          28  -+          +----------+
 │  toast               96  -+          I          I
 │  bread               16  -+-------+   I          I
 │  breakfast           17  -+     +-----+          I
 │  food                36  -+---+ I    I           I
 │  lunch               53  -+  +---+   I            I
 │  a_drink              2  -+-+ I     I            I
 │  milk                57  -+ +-+     I            I
 │  coffee              24  -+-+      +-----------+  I
 │  meal                55  -+        I           I  I
 │  fork                37  -+        I           I  I
 │  knife               51  -+--------+   I        I  I
 │  toaster             97  -+      I I            I  I
 │  bowl                15  -+      +----+         I  I
 │  spoon               89  -+------+  I           I  I
 │  plate               71  -+   I I                I  I
 │  item                43  -+   +----+             I  I
 │  items               44  -+---+ I                I  I
 │  jar                 48  -+ +-+                  I  I
 │  dinner              31  -+---+                   +-----------+
 └─ supper              90  -+                       I           I
 ┌─ office              65  --------+                I           I
 │  workplace          106  -+   I                   I           I
 │  book                14  -+   +----------+        I           I
 │  paper               67  --+-+ I         I        I           I
 │  newspaper           62  -+ +----+        I        I           I
 │  desk                29  -+ I            I        I           I
 │  papers              68  -+-+            +--------------+      I
 │  computer            27  -+       I               I        I  I
 │  phone               70  -+-+     I           I   I        I  I
 │  telephone           94  -+ +-------+        I   I        I  I
 │  sound               88  ----+      +------+  I   I        I  I
 │  alarm                3  -+----+    I        I   I        I  I
 │  alarm_clock          4  -+ +------+         I   I        I  I
 │  caller              19  ------+              I   I        I  I
 │  tv                 102  ----+               I   I        I  I
 │  building            18  -+--+               I   I        I  I
 │  room                78  -+ +--+             I   I        I  I
 │  gloves              39  --+-+ +----------+   +------+     I
 └─ jacket              47  --+  I          I        I           I
```

Figure 8a. A dendrogram of a stripped down lexicon after clustering the distributed word representations using the Ward Method. Here, 90% of the original features have been removed using the Standard Deviation Method. Note that there are now more major differences between clusters here and those for washing, foods, and culinary objects in original lexicon with 100% of its features (Figures 6a and 6b). Here, elements in the original cluster of culinary objects have become mixed with the washing objects (top cluster) and with meals (bottom cluster). In the middle cluster are other meals and foods.

```
home              42    -+----+                    I        I               I
kitchen           50    -+                         I        I               I
goods             40    -+--------+                I        I               I
merchandise       56    -+        I                I        I               I
shopping_trolley  85    -+        +------+         I        I               I
trolley          101    -+        I      I         +--------+               I
shop              83    -+--------+      I         I                        I
basket             7    -+               I         I                        I
list              52    -+-+             +-----+   I                        I
shopping_list     84    -+ +-+           I     I   I                        I
object            64    ----+ I          I     I   I                        I
charity           22    -----+--------+ I I     I   I                        I
recipe            76    ------+        I I I    I   I                        I
restaurant        77    -+--+          +-+ I    I   I                        I
waitress         103    -+ +------+      I I    I   I                        I
dish              32    ---+      +---+  +---+  I                            I
mouth             59    -------+-+       I                                  I
nonoun            63    -------+         I                                  I
cashier           20    ---+---+         I                                  I
queue             73    ---+   I         I                                  I
coins             25    -+     +------------+                               I
money             58    -+-+   I                                            I
small_change      86    -+ +---+                                            I
price             72    ---+                        I                        I
till              95    ---+                        I                        I
face              35    -+-----------+              I                        I
teeth             93    -+           +------+       I                        I
shelves           82    ---+         I              I                        I
table             91    ---+---------+              I                        I
chair             21    ---+                                                I
Susan             79    -+                                                  I
she               81    -+                                                  I
Amanda             1    -+                                                  I
Richard           74    -+                                                  I
Rosemary          75    -+                                                  I
Paul              66    -+                                                  I
person            69    -+                                                  I
Nick              60    -+                                                  I
Noel              61    -+                                                  I
John              46    -+                                                  I
Mary              54    -+                                                  I
he                41    -+                                                  I
Jane              45    -+                                                  I
Edward            33    -+------+                                           I
Eunice            34    -+      +-------------------------------------------+
Bob                6    -+      I
colleague         26    -------+
```

Figure 8b. The second half of Figure 8a.

Clearly it is hard to make conclusions from this measure alone as to how many features can be removed before the level of performance becomes unacceptably low, as this is likely to be dependent on the application for which the lexicon is going to be used. For example, in a stereotypical event recognition task (such as could be performed crudely by the PARROT system) it may well be that a confusion between washing and culinary objects (as found in the Minus 90% Clusters) is not too crucial. This might however mean that 'She picked up the saucepan' is recognised as an acceptable part of going to bed, rather than say 'She picked up the toothpaste'. More seriously though, 'She picked up the toothbrush' might start to activate a knowledge structure associated with cooking a meal.

SUMMARY AND CONCLUSIONS

We have introduced a simple technique for deciding which features to remove from a set of distributed representations. This involves ordering features by their standard deviation across lexicon entries and removing low deviation features first.

Three techniques for measuring lexicon performance have been discussed. These are the Word-Word, Dog-Cat and cluster analysis methods. The results from the Word-Word and Dog-Cat methods indicate that at least 70% of the original features can be removed using the standard deviation method while retaining an ostensibly acceptable level of performance. The cluster analyses show that original groups of semantically related words are resistant to change. Groups of such words become confused once 90% of features have been removed, but these groups still appear semantically well formed, though more general than the original groups.

The work reported here is a first step toward building up a corpus of techniques which will enable us to produce the optimum set of distributed concept representations for a given AI application. It may also be possible to devise reliable performance measures to use in predicting how a particular lexicon will perform.

Finally, it is our contention that thinning-out techniques of the kind proposed here can be used as one way of investigating what the important attributes of concepts in the world are. This is because we can scale down a set of concept representations using, say, the Standard Deviation method and then elicit more features in order to tease apart concepts which are now indistinguishable. This procedure can then be repeated until the feature set seems to be stable. This set can then be analysed for internal semantic structure.

Next Steps

There are several interesting areas of work which stem from the experiments reported here. First of all, the Standard Deviation measure is only one possible method for choosing which features to eliminate. Another obvious strategy is to see whether the contribution to meaning of some pair of features is similar. If so, it may be possible to merge these features into a single feature. One could do this by comparing the centralities associated with the pair of features for each lexical entry. If these are similar in each case (ie correlated to a high degree) then the features could be merged. It is not yet clear how effective a technique will turn out to be. It is currently under investigation.

A second line of work is to test further the efficacy of the lexicon performance measures discussed here and to investigate new ones. A measure which ought to have been used is to load a stripped down lexicon into the original PARROT system and compare its performance with the documented examples [22]. Unfortunately it was impossible to do this at Exeter for technical reasons.

Finally, all the experiments reported here have concerned only the nouns in the lexicon. It would be interesting to perform the same experiments on the verbs as well.

REFERENCES

1. Schank RC. Conceptual Dependency: A Theory of Natural Language Understanding. Cognitive Psychology 1972; 3(4): 552-630

2. Wilks Y. An intelligent analyser and understander of English. Communications of the Association for Computing Machinery 1975; 8(18): 264-274

3. Fillmore C. The case for Case. In Bach E and Harms RT (eds) Universals of Linguistic Theory. Holt, Rinehart and Winston New York, 1968, pp 1-88

4. Wilks Y. An Artificial Intelligence Approach to Machine Translation. In: Schank RC and Colby KM (eds) Computer Models of Thought and Language. WH Freeman San Fransciso, pp 114-151

5. Schank RC, Abelson RP. Scripts, Plans and Knowledge. In Proceedings of the 4th IJCAI, Tbilisi, USSR, 1975

6. Dyer MG. In-Depth Understanding. A computer model of integrated processing for narrative comprehension. Research Report Number 219, Department of Computer Science, Yale University, 1982

7. Wilensky R, Arens Y, Chin D. Talking to UNIX in English: An overview of UC. Communications of the ACM 1984; 27(6): pp 574-593

8. Katz JJ, Fodor JA. The structure of a semantic theory. Language 1963; 39(2): pp 170-210. Also in Fodor JA, Katz JJ (eds) The structure of Language. Prentice Hall Englewood Cliffs NJ 1964

9 Hinton GE. Implementing semantic networks in parallel hardware. In Hinton GE. and Anderson JA. (eds) Parallel models of associative memory. Lawrence Erlbaum Associates Hillsdale NJ, 1981, pp 161-187

10. Waltz DL, Pollack JB. Massively parallel parsing: A strongly interactive model of natural language interpretation. Cognitive Science 1985; 9: 51-74

11. Hinton GE, Sejnowsi TJ. Learning Semantic Features. In: Proceedings of the 6th Annual Conference of the Cognitive Science Society, 1984, pp 63-70

12. Hinton GE. Learning distributed representations of concepts. In: Proceedings of the 8th Annual Conference of the Cognitive Science Society, 1986, pp 1-12

13. Miikkulainen R, Dyer MG. Building distributed representations without microfeatures. Technical Report UCLA-AI-87-17, AI Laboratory, Computer Science Department, University of California at Los Angeles, CA, 1987

14. Miikkulainen R, Dyer MG. A Modular Neural Network Architecture for Sequential Paraphrasing of Script-Based Stories. Technical Report UCLA-AI-89-02, AI Lab, Computer Science Department, UCLA, LA, LA 90024, 1989

15. Elman JL. Finding structure in time. TR 8801, Center for Research in Language, University of California, San Diego, CA. April, 1988

16. Chomsky N. Aspects of the theory of syntax. MIT Press Cambridge MA, 1965

17. Schaeffer B, Wallace R. Semantic similarity and the comprehension of meanings. Journal of Experimental Psychology 1969; 82:343-346

18. Rosch E. Natural categories. Cognitive Psychology 1973; 4:328-350

19. Rosch E. Human categorization. In: Warren N (ed) Advances in cross-cultural psychology (Vol I). Academic Press London UK, 1977

20. Smith EE, Shoben EJ, Rips LJ. Structure and process in semantic memory: A featural model of semantic decisions. Psychological Review 1974; 81: 214-241

21 Brachman RJ. What IS-A is and isn't: An Analysis of Taxonomic Links in Semantic Networks. IEEE Computer 16(10, 1983, pp 30-36

22. Sutcliffe RFE. A Parallel Distributed Processing Approach to the Representation of Knowledge for Natural Language Understanding. Unpublished doctoral thesis, University of Essex, UK, 1988

23. Rumelhart DE, Hinton GE, Williams RJ. Learning internal representations by error propagation. In: Rumelhart DE, McClelland JL (eds) Parallel Distributed Processing: Explorations in the microstructure of cognition. Volume I: foundations. MIT Press Cambridge MA, 1986, pp 318-362

24. Minsky L, Papert, S. Perceptrons. MIT Press Cambridge MA, 1988 (1969)

25. Sutcliffe RFE. Representing Meaning using Microfeatures. In: Reilly R, Sharkey NE (eds) Connectionist Approaches to Natural Language Processing. Erlbaum Hillsdale NJ, 1991

26. Salton G (ed). The SMART Retrieval System - Experiments in Automatic Document Processing. Prentice-Hall Englewood Cliffs NJ, 1971.

27. Ward JH. Hierarchical grouping to optimize and objective function. Journal of American Statistical Association 1963; 58: 236-244

28. Everitt B. Cluster Analysis. Heinemann Halstead John Wiley London 1974

Section 7:

Explanation

Allowing Multiple Question Types to Influence the Resulting Explanation Structure

Efstratios Sarantinos and Peter Johnson

ABSTRACT

This paper presents an empirical study and analysis of natural dialogues between experts, novices and partial experts. From this analysis a theory of explanation dialogues is developed known as Extended Schema based Theory (EST). In EST questions are interpreted by combining information from different, semantically related question types which together best capture the essence and meaning of the question. This theory is then applied to the design of an architecture and computational model of interpreting questions and generating explanations. The expert system, named EXPLAIN, understands the nature of the question and is able to take account of the previous dialogue. Also, the system can tailor its responses to an individual user's characteristics, including level of expertise and depth of knowledge in the domain.

INTRODUCTION

Most existing so called expert or knowledge based systems provide inadequate explanations e.g. [1], [2], [3]. These systems can only cope with a very limited set of questions. Each question is considered in isolation and not as part of an on-going dialogue. Explanations are produced without taking into account the previous discourse and are not tailored to the individual user's knowledge of the domain, beliefs and attitudes. Furthermore, most current advice-giving systems are unable to provide multiple explanations in response to the same question or if the user is in any way dissatisfied with the response. In this paper we develop an explanation theory and discuss the extent to which it alleviates these problems. The theory interprets questions by combining information from different semantically related question types, and generates explanation schemata. Effective question interpretation is achieved by; (i) relating the information presented in the question with the preceding dialogue history, (ii) by having a powerful set of question types, and (iii) by employing a two component classification scheme; The first component provides a detailed classification of question types linked hierarchically in which the bottom of the hierarchy consists of very specialised question types, whilst the second provides a mechanism that assigns to a question multiple question types that are semantically different but together best capture the meaning of a complex question.

An investigation has been carried out to develop an initial theory of explanation. This theory has been revised and developed in the light of further empirical studies [4]. A suitable knowledge domain was chosen and a natural experiment was undertaken in which "subjects" were free to ask experts questions about any aspect of the domain and experts gave whatever response or explanation they felt was appropriate, [5]. Data, in the form of audio tape recordings, of natural dialogues between experts and subjects were collected in every-day rather than a laboratory setting. From a detailed analysis of those data some interesting conclusions about question/answer dialogues were drawn, with a particular focus on explanations. The results of this experiment are reported here and interpreted in terms of an explanation theory which is then applied to the design and implementation of a computational model - EXPLAIN - for interpreting question types/paths and generating tailored explanation schemas.

In this paper we first describe the experiment and propose an analysis of the questions asked by the subjects together with the explanations supplied by the experts. Next we summarize the results of a review of related work on question classification. We then propose an explanation theory and discuss the extend to which the question classification and explanation generation schemes developed overcome the limitations of earlier approaches to question interpretation and explanation generation. Finally, we briefly describe the implementation of EXPLAIN [5], which embodies the extended schema-based explanation theory.

THE CHOICE OF THE KNOWLEDGE DOMAIN

One criterion in the choice of a knowledge domain was the availability of experts. The domain also needed to accommodate shallow (associational) as well as deep (causal) knowledge in order to provide a sufficiently complex subject matter to allow research to have some level of generality to other domains. The knowledge domain chosen was facial skin care. This domain has the added advantage of being novel as far as studies on expert systems and explanation dialogues are concerned. There are different areas of skin care, such as cleansing or exfoliating, and classes of products that are available in each area, such as day cream or lotion. For example, exfoliating the skin can be achieved by using either a facial scrub, an abbrassive-puff or an exfoliating mask. Various attributes may be associated with a given product. The type of skin for which a product is best suited is one (and possibly the most) important factor in deciding which product to buy because there are different products for different skin types. The skin type for which the product is most suited is therefore a property of the domain. The age of a person is another factor that influences the choice of a product and facial treatment. Every product has relevant information associated with it. For example, some attributes associated with the abbrassive-puff are: it is soaked in soap and water, it is applied by massaging with circular movements (everywhere except the eye area), it is normally used once a day. Having chosen the knowledge domain an experiment to investigate explanation dialogues in that domain was devised.

DESCRIPTION OF THE EXPERIMENT

The experiment was based on skin care consultations in which an expert gave advice on facial skin care to an individual. Three experts were used to ensure that the explanations recorded were not dependant on any individual expert's style. An expert is a professional consultant who has special knowledge, skill and training in the facial care domain. Two experts used in the experiment were employed by well-known cosmetics houses (viz Estee Lauder, Christian Dior, and Chanel). The third expert was not a professional but an amateur skin care consultant. Eight subjects participated, four partial-experts and four novices. A partial-expert is someone who is familiar with the knowledge domain but does not have a thorough knowledge of it, while a novice is someone who does not have any knowledge at all of the domain. Partial experts in the facial-skin care domain were female 'subjects' that had experience in using facial care products of more than five years, and read articles in books or magazines, about the face care domain. Novices were male 'subjects' who wanted to buy a product for their female friends and/or relations and had no experience nor any knowledge of the face care domain themselves. By these criteria the subjects were classified as partial-experts or novices after the dialogues had been recorded. No consultations between experts were recorded because, in this setting, this is rare and it did not occur. The dialogues were naturally occurring dialogues without any interruption by the experimenter. Each session lasted on average 25 minutes. Some consultations were recorded at two department stores (Boots and Debenhams). The remaining consultations were recorded privately at the subject's or the experimenter's residence and they were also buying/selling situations.

QUESTION TYPE TAXONOMIES

Previous studies have attempted to identify a classification of question types, [2], [3], [6], [7], [8], [9], [10], [11], [12], [13], [14] (these are summarized in figure I) however it was found that no single one of these classifications were capable of adequately accommodating the 74 questions occuring in the experiment. Consider the following generic questions taken from the corpus collected in the experiment:

'Is approach 'A' preferable to approach 'B' under the circumstances ?' (1)

'Why is approach 'A' preferable to approach 'B' under the circumstances ?' (2)

'How is approach 'A' preferable to approach 'B' under the circumstances ?' (3)

'What are the steps involved to accomplish this process ?' (4)

'Is entity 'A' similar to entity 'B' ?' (5)

'What is entity A ?' (6)

'Why did you diagnose Y ?' (7)

'What is the difference between X and Y ?' (8)

'Do you have a different X for every Y ?' (9)

'By doing X would I influence Y ?' (10)

There are many variations in the definition of question types shown in figure I. Swartout and Cawsey define question types which partly depend on the form of the question, e.g. "why" or "what". In the collected corpus evaluation questions of type (1), were frequently asked by partial-experts. Swartout's classification [3] is unable to cope with such questions since they cannot be reduced to a 'why' or 'how' question type. Furthermore, type (1) question is not the same as type (2) as question (1) does not prejudge the outcome of the comparison while (2) does. The emphasis of (2) is not on the outcome but on the reasoning behind choosing 'A' as the preferred approach. Similarly, (1) is not the same as (3), since (3) asks for a justification or evaluation of the advantage of A over B. Therefore Swartout's classification is unable to cope with such questions as (1).

Cawsey [14] defined her question types along three orthogonal dimensions: a) Whether the subject was requesting information, suggesting information and asking for confirmation or a repeat or rephrase of the information was requested, b) What type of question is asked (What, Why, Others), c) What type of information was involved. This classification can deal with a larger number of questions than [2], [3], [4], [5], [6], [7], [8], [9]. Cawsey classifies question types according to their

Cawsey, [14]

1. Whether the subject was:
(a) Requesting information
(b) Suggesting information and asking for confirmation
(c) Repeat or rephrase information to check if it is understood
2. What type of question
(a) What
(b) Why
(c) Others (What if, Why not, How)
3. What type of information was involved
(a) Behaviour
(b) Part recognition
(c) Function
(d) Notation
(e) Structure
(f) External actions
(g) Limits and assumptions
(h) Input, output and variables
(i) Concept
(j) Information
(k) Rest

Wexelblat, [11]

How shall i do what you ask me to do?
Why do you ask me to do that?
How did you come to that question or conclusion?
By what steps did you get here?
What shall I do next?
What do you know about?

McKeown, [2]

Requests for definitions
Requests for available information
Requests about the difference between objects

Gilbert, [6]

Instantiation
Classification
Prescription
Sensitivity
Procedure
Antecedence
Justification
Rationale
Generalisation
Strategy
Comanison
Exemplification
Identification
Definition
Relevance
Implication
Capability
Existence
Warrant
Process
Hypothesis

Nicolosi, [9]

Definition
Generalisation
Exemplification
Goal Orientation
Justification
Causal Antecedent
Causal Consequent

Swartout, [3]

Why questions
How questions

Lehnert, [8]

Causal antecedent
Goal Orientation
Enablement
Causal Consequent
Verification
Disjunctive
Instrumental/Procedural
Concept Completion
Expectational
Judgmental
Quantification
Feature Specification
Request

Maybury, [12]

Define
Justify (named 'explain')
Compare

Pilkington et al, [10]

Category
(a) Elaboration
(b) Enablement
(c) Exploratory
(d) Evaluatory
Referent/Context
(a) Object
(b) Command
(c) Modality
(d) Task or task-plan
(e) Claims
(f) Counter-Claims
(g) Comments on performance

Valley, [13]

Description
Comparison
Feature analysis
(a) General
(b) Properties
(c) Activities
Relationship Analysis
(a) General
(b) Classify
(c) Component
(d) Action
(e) Other

Hughes, [7]

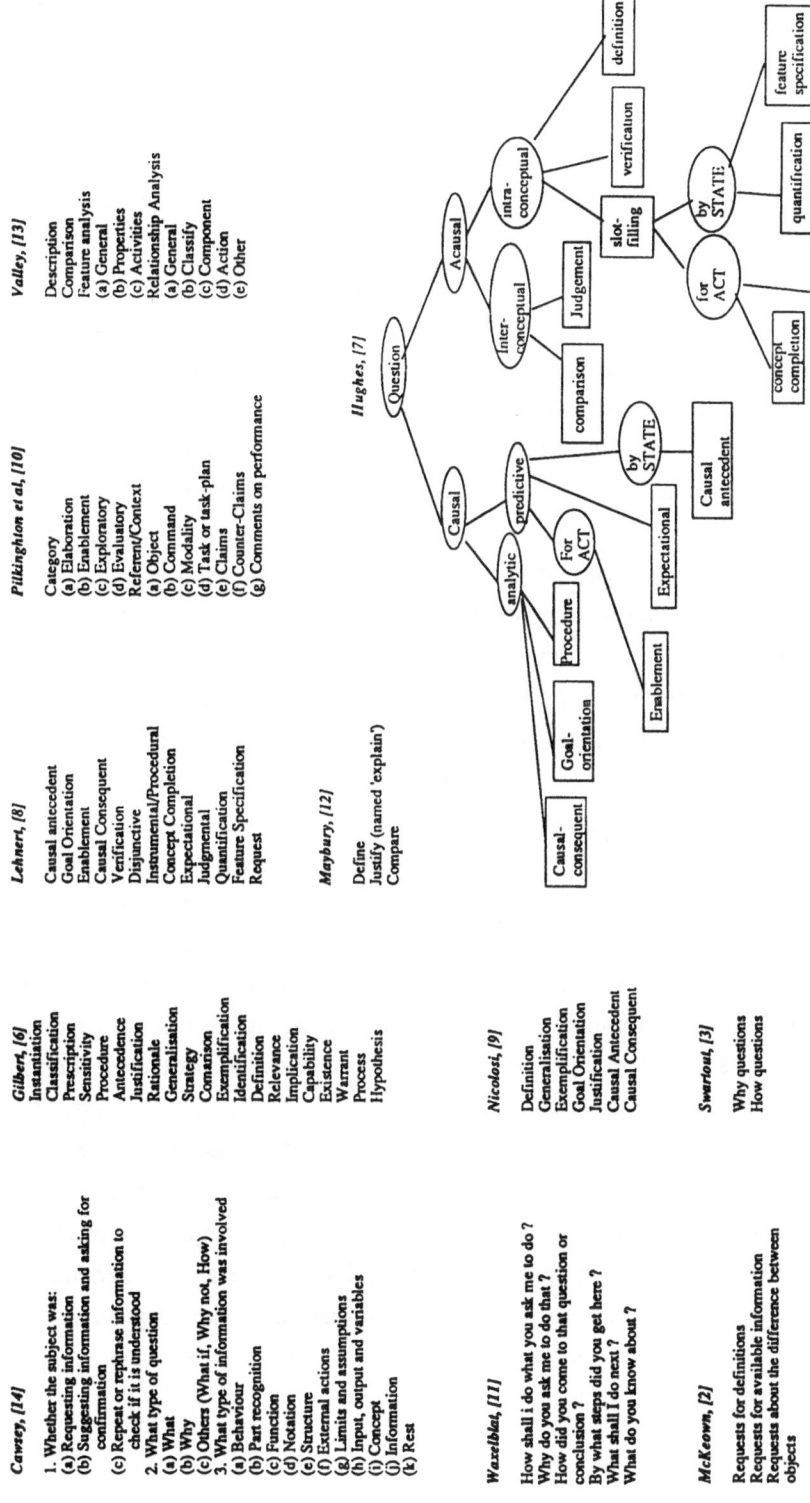

Figure 1: Summary of existing classifications of question types

syntax. Furthermore, the 'type of information' dimension does not include procedural questions of type (4). Cawsey also has a category called 'rest' in which all other questions are classified. This is inadequate since too many questions from the collected corpus would have fallen into this category.

McKeown's [2] question classification also includes a general category, namely the category 'requests for available information'. Again many questions in our study would have fallen into this category. Furthermore, her classification does not accommodate queries of type (1) and (4). Maybury [12] classified questions into three categories, examples of which are questions (6), (7) and (8). However, his classification does not accommodate any other question which does not conform with (6), (7) and (8). This makes his classification very restricted.

Waxelblat's [11] question classification cannot accommodate evaluation questions as in question (1) nor comparison questions like (5), both of which were frequently asked in the experiment. Pilkinghton et al [10] defined the question types along two dimensions, 'category' (elaboration, enablement, exploratory, evaluatory) which includes the type of question and 'referent/context' which is the referent or other contextual information. This classification excludes question (6). Nicolosi's [9] and Valley's [13] question classifications do not include evaluation (1) and procedural question types (4) both of which were found to be frequently asked in the experiment. Also, the 'general' and 'other' question sub-types of Valley's classification are too vague to be of use in generating explanations.

Lehnert's [8] classification cannot accommodate question (6). Hughes' [7] classification takes into account the relationships between the question types. However, many different questions asked in the experiment, such as (5) or (10) are not distinguished between. In addition, Hughes classification cannot accommodate questions such as (9). Consequently, this question classification is not precise enough and results in a general explanation with no identification of the question's details. It is important to know exactly what the question is so that the right points are emphasized in the explanation. Gilbert [6] decided not to classify the question types but the types of answer. His classification covers a range of non-related answer types and as such is not relevant to our present analysis of question types.

In the present analysis the question types defined in the above mentioned studies were modified in order to suit the questions asked by the subjects in the experiment. Our classification of question types adequately covers all the questions asked in the experiment.

CLASSIFYING QUESTIONS

The dialogues were recorded on a tape recorder and later transcribed. In this way it was possible to study all questions and responses/answers for their explanation requirement and content. A total of 74 questions were asked by all the subjects and each one was identified as being of a given kind. Questions asked by experts were not classified. The identified question types used to classify the subject's questions are not domain dependant. Each actual question can be considered as consisting of a path of classified question types. Figures II and III show the relations between the different question types derived from the empirical study. Figure V compresses figures II and III into a single diagram. The highlighted boxes represent questions that are further decomposed. We call these questions *root* question types as opposed to the rest of the questions which we call *leaf* question types. The links between the questions types are called *question paths*.

There are two types of links, the *multi-concept* and the *hierarchical*. The multi-concept link is shown by a line marked with a small circle in the middle, and the hierarchical link is represented using a line with an arrow at the receiving end (one level down the hierarchy). There are important conceptual differences between the two kinds of links. The hierarchical link joins question types that are semantically related with each other, e.g. the *comparison* question type has a number of sub-types such as *difference* or *similarity*. The sub-types inherit the properties of their super class but also have local properties of their own. Given an actual question that matches the super-class question type there may or may not be a match of all its sub-types depending on the question semantics. A root question type (super class) can belong in the question types path without necessarily all its sub-classes (its leaf question types) being members of the same path. For example, a given question such as "Where do I apply the sun protecting cream ?" is classified as *feature analysis* which is a root question type without all its subclasses (viz *collection, frequency, timing, contents, place*) being members of the question path; in fact in this example its only the *place* question type that is included in the path. Consequently not every *feature analysis* type question is equated with information about frequency, timing, contents etc.

The multi-concept link relates question types that are semantically and conceptually different. Each concept in a multi-concept link has a non-overlapping definition and does not inherit properties or have any related properties with other concepts. Relations between these question types come into existence because of the multiple semantics of the asked question which links together the different question types. Question types that are not related hierarchically belong to the same

316

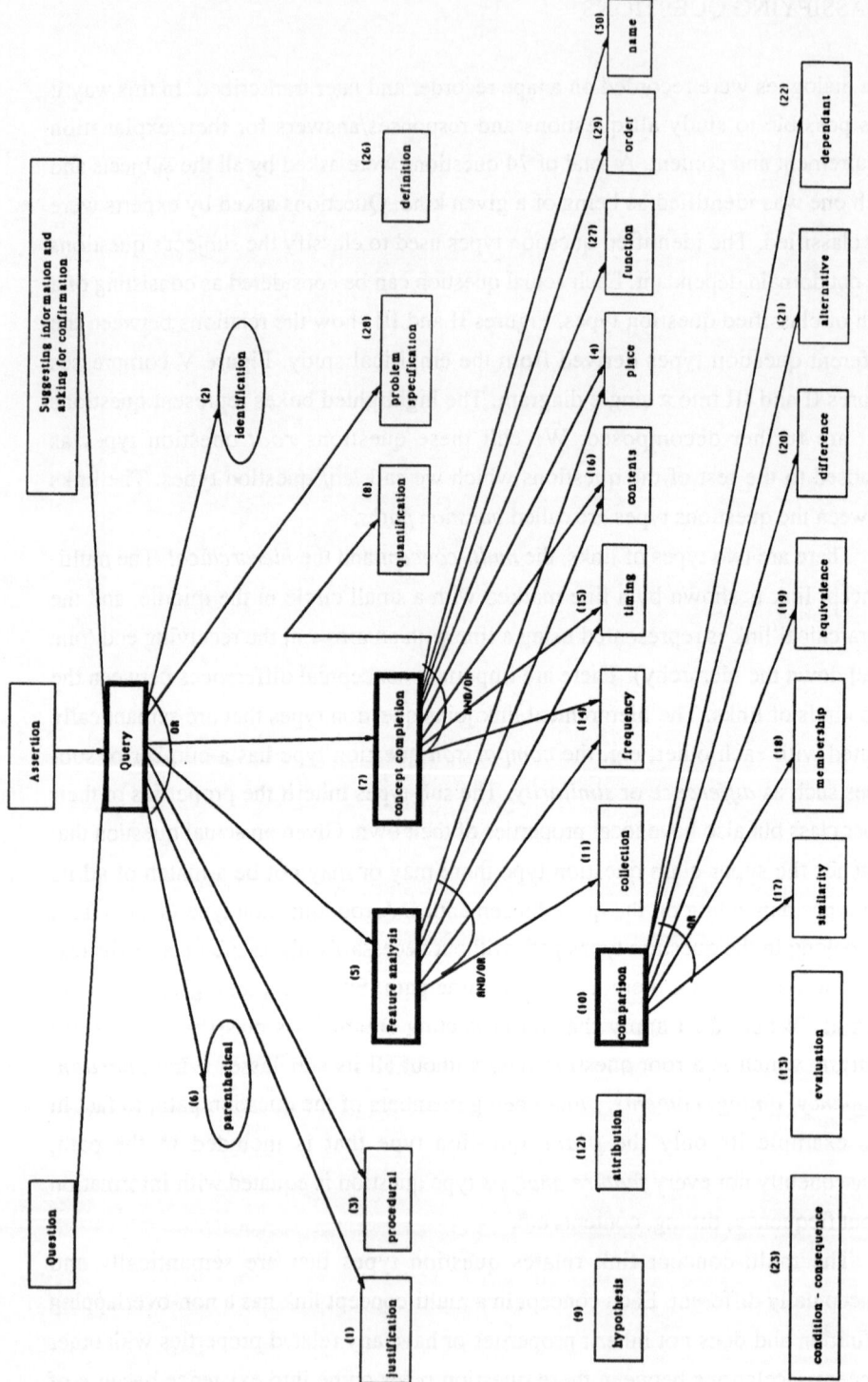

Figure 11: The hierarchical relations between the question types (the number in brackets is a question type identifier)

317

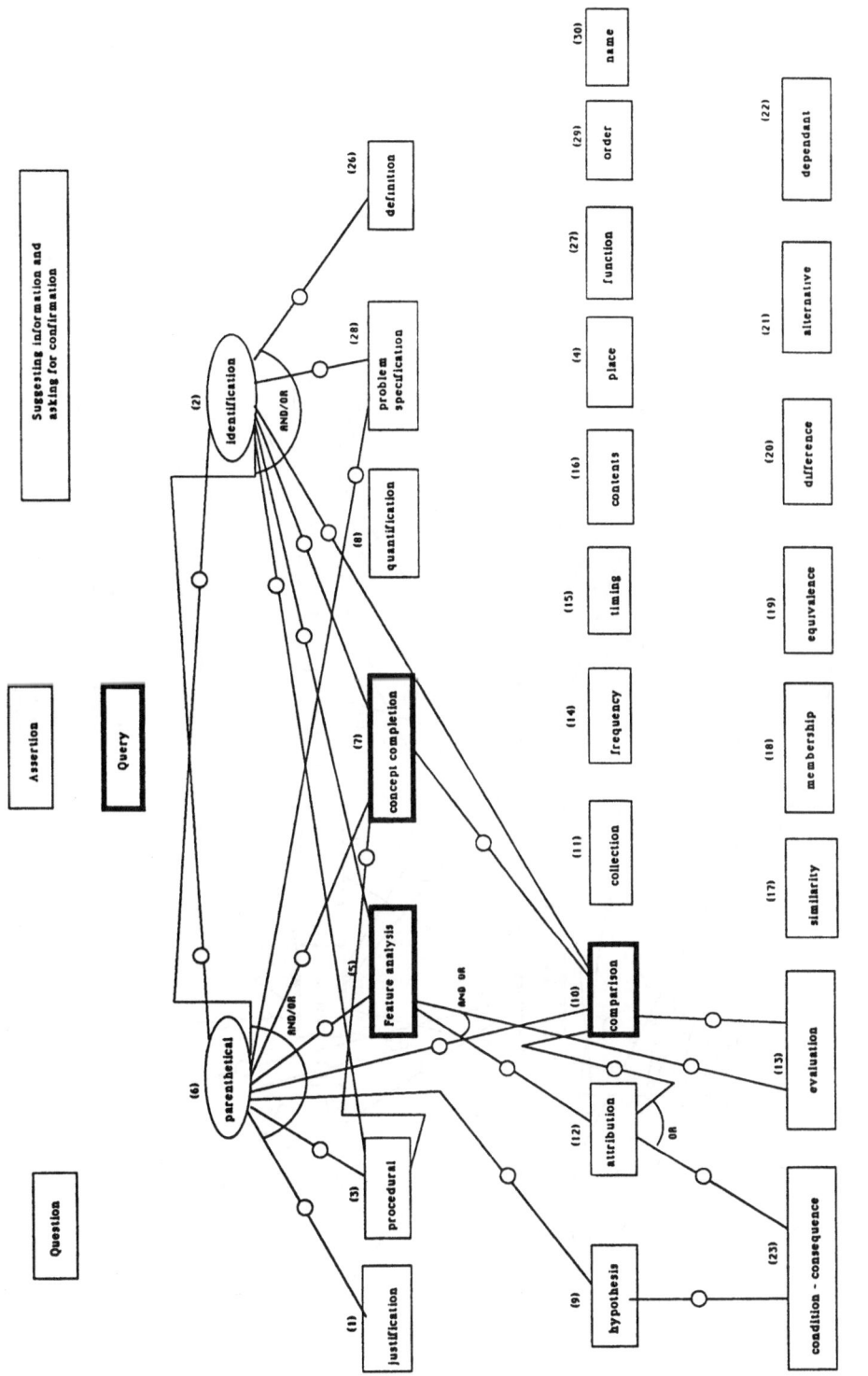

*Figure III: The multi-concept relations between the question types
(the number in brackets is a question type identifier)*

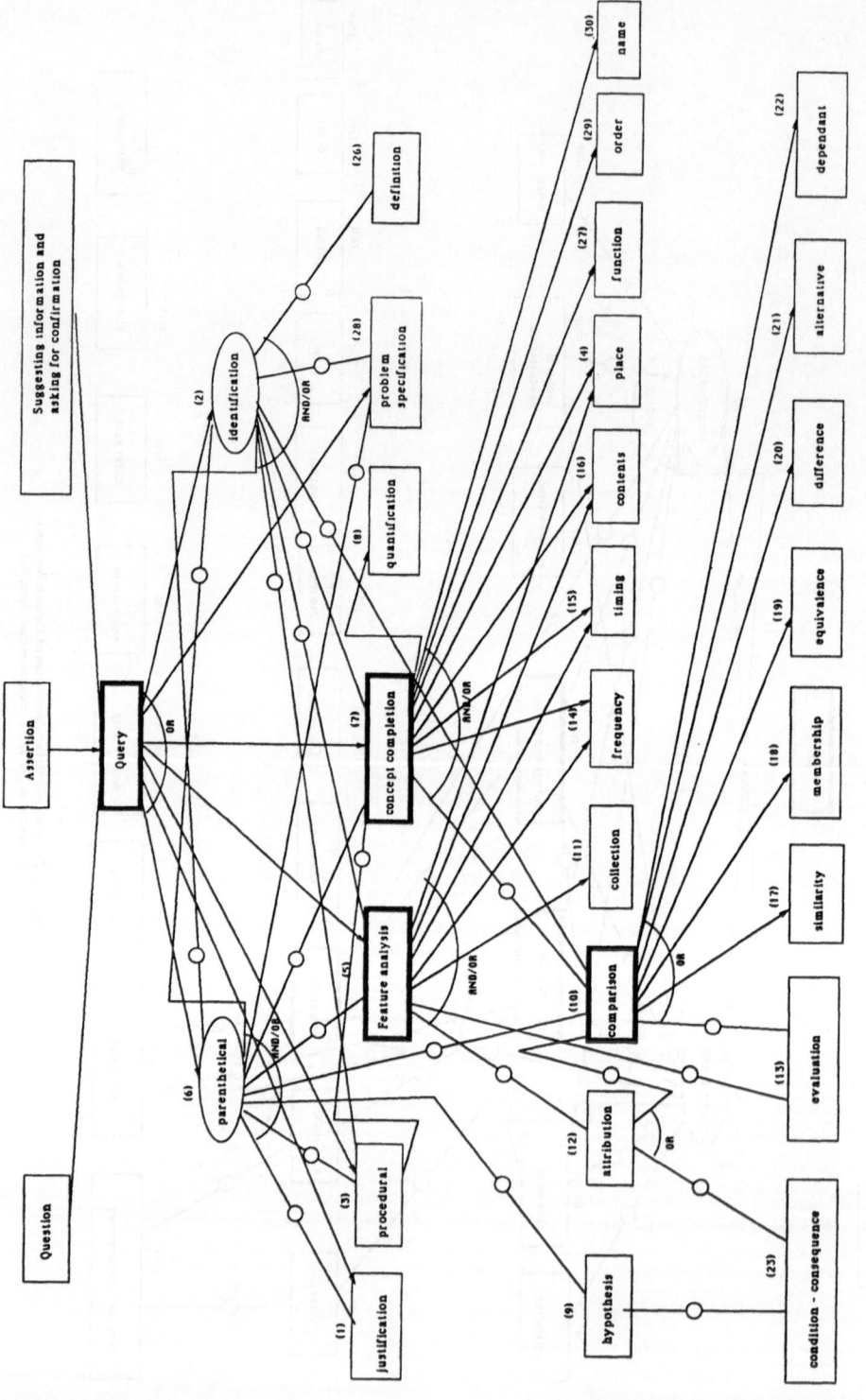

*Figure V: Prioritised relations between the question types
(the number in brackets is a question type identifier)*

question path by virtue of the asked questions. For example, a question such as "How about a cleansing mask; is this better than cream or soap ?" can be classified as *concept completion* by the phrase "How about a cleansing mask" where the subject requires some general information regarding cleansing masks; the question is also classified as *evaluation* by the phrase "Is this better than cream or soap ?" where there is an implied request to give a judgement; evaluate the usability and usefulness of cleansing masks with respect to other cleansing products notably cream and soap. The identification of multi-concept links between the question types is a major contribution of this research work. Previous research has not attempted to relate conceptually different question types to fully characterize the semantics of a complex question. During the question classification process both types of links are instantiated into a path of distinct question types that fully specify and interpret the question.

INTERPRETING A QUESTION

Question interpretation is achieved by applying the classification of question types and paths discussed above. Effective question interpretation is not a simple assignment of an individual question to a given question type; instead it requires the combination of information from different, semantically related question types to form composite structures. These structures capture the essence of a question and permit more realistic inferences that are assumed to resemble the inferences made by experts in explanation dialogues. A prerequisite for achieving such goals is that information presented in the current question must be related to relevant facts that were presented earlier in the request, via the *parenthetical* and *identification* question types. These question types are designed to capture the chronological sequence of events and their relative importance - compared to the current topic of conversation - in the preceding discourse. This ensures that the previous dialogue is considered when answering a question and that follow-up questions can be answered differently if necessary. Another prerequisite is that there is a sufficient set of question types which can accommodate the particular question. The question classification scheme described earlier has two components. One component provides a detailed classification of hierarchically linked question types - in which the bottom of the hierarchy consists of primitive question types. The second component provides a mechanism that identifies multiple question types that are semantically different and together best capture the meaning of a complex question. Satisfying both these prerequisites ensures a deeper understanding of the question.

From the empirical studies it was observed that any given question is composed of a number of different conceptual meanings. For example, a question can be simultaneously *identification, concept-completion, comparison and evaluation,* as in "How about a cleansing-mask; is this better than scrub or soap ?". It is an *identification* question because 'cleansing-mask' is introduced for the first time in the dialogue. (This is established by studying the preceding dialogue sequence). It is also a *concept-completion* question because the subject asks for general, but not detailed information about 'cleansing-masks' in the phrase "How about a cleansing mask". Also, a *comparison* is requested between a 'cleansing mask', 'scrub' and 'soap' from the phrase "Is this better than scrub or soap ?". Finally, there is an implicit request to *evaluate* this (give a judgement) with respect to the subjects' preferences and subjective properties, such as her skin-type, implied from the context of the dialogue.

From this analysis it is observed that there is a relation between the *identification, concept-completion, comparison* and *evaluation* question types with respect to the question examined. We call this relation the "question path". The question types characterize completely the semantics of the asked question. From the above analysis it was also observed that usually a question is not only just of a given question type as it is thought to be in the current research literature ([2], [3], [7], [8], [9]). A question is normally composed of a variety of conceptual interpretations which are related in a question path. Failure to detect the multiple semantics of the question path results in a lack of, or partial understanding of the intended meaning of the question; this consequentially, is reflected in an incomplete explanation which is unsatisfactory to the inquirer, for it may fail altogether, or only partially answer the query. Therefore the question interpretation process amounts to a particular instantiation of a question path. The question paths shown in figures II and III are considered to be applicable to many domains*. The question paths were identified in the experiment, and were verified by classifying various texts such as

---

* A second empirical study and analysis of natural dialogues between experts, novices and partial-experts was conducted, in the domain of European Community law, to validate and verify the set of question types and paths developed. The transcribed dialogues were analysed by two subjects and a positive correlation (0.94, 0.67) was found between our and their classification of question types and paths; no new question types were needed to classify the questions asked in the domain of European Community law nor any additional question paths than those shown in Figure V were found. Furthermore their perception of the meaning of question types was found to be in agreement with ours. This supports our view that these question types are not domain or dialogue specific.

magazine articles, personal letters, student essays, administrative reports and text-book questions. In the dialogues examined, these are an adequate set of question paths for relating the question types shown. These question paths are the most frequent semantic interpretations of questions asked in these explanation dialogues. The question paths may have some generality extending beyond the experimental data.

The theory includes a method for generating additional question paths to those shown in figures II and III. The question paths shown in figures II and III have been given priority consideration to reduce processing time during question interpretation. Multi-concept and hierarchical links join the most likely question type candidates to interpret the current question. If the most likely question types fail to match the semantics of the question, all question types are examined for a possible match until a question path is established. Question interpretation is performed in a top-down, left to right fashion (using figure V). The question paths identify specialised patterns of question type usage that can be employed to interpret questions. Question paths are not pre-compiled but are generated during the question classification process. It is therefore possible to arrive at new question paths. This makes it possible for a question to be interpreted differently depending on the preceding dialogue or the user's acceptance of a previous explanation. It enables the theory to recover from failure by planning an alternative interpretation of the same question and resulting in a different explanation.

QUESTION PARSING

To give an idea of how a given subject's statement from the experiment relates to the question types graph, shown in figure V two examples are shown below. The questions below must be interpreted bearing in mind that these are questions posed to an independent skin care consultant, and occur in the same order as shown below.

1) *I would like a cleansing product for my skin type.*
This is an assertion for it states the person's intentions. Although this utterance is not phrased as a question it is obvious from its content that the subject expects an answer as a reply. It was observed that a large majority of statements were implicit questions like the above, and as such the expert replied by giving an answer or explanation, as appropriate. The question path is the following:

identification=>concept completion=>
name, frequency, place, function, contents, order.

Here the subject introduces *cleansing products* as the area of her interest (*identification*). Further she wants to find out more about cleansing products with the aim of using them herself (*concept completion*). In order to do that she indirectly asks for information about which products to use, how to use them and if they are appropriate for her skin type. That kind of information needs to be provided so that the subject has a good idea of what a product does and if she buys it, how she can use it. Therefore the question is classified as *name* (name of the product recommended), *frequency* (frequency of use), *place* (where to apply the product and which areas to avoid), *function* (what the product does to the skin), *contents* (what the product contains so that it is not harmful to her skin type), and *order* (in which order to use the products recommended); this information is not usually included by the expert when she recommends a product because she takes this into account anyway but in this question the subject asks specifically for it ("my skin type"). The above information is usually disclosed by the expert skin care consultant for she assumes that the subject is interested in cleansing in order to buy a product at the end of the consultation.

2) *What do you think about natural products ?*
This is a question for it requests information about *natural products* directly. Added evidence is the word-order and the use of an interrogative word (What). The question path is as follows:

identification=>feature analysis=>evaluation.

Here the subject introduces *natural products* as the area of her interest. This area of interest is a member of the current focus of attention (and hence the question is not classified as *parenthetical* but as *identification*) Also in this question, the subject wishes to know more about natural products in some detail (*feature analysis*) to enable her to understand the judgment (*evaluation* - "you think") she requires from the expert as to the usefulness of such products.

EXPLANATION STRATEGIES

The next step in developing a theory of explanation is to relate our question classification to suitable explanation structures. An explanation is taken to be a declaration made with the view to make clear i.e unfold and illustrate something. To associate explanation strategies to question paths we need to choose a discourse structure theory that will be suitable for explanation generation and then map our question classification formalism to the discourse theory so that the explanations generated answer well the user's questions.

The Schema Based Approach

We adopted and extended the schema-based approach, McKeown [2], which interprets text using hierarchical structures bound together by rhetorical predicates. McKeown shows that by using an organizational framework for the text (a schema) a system can generate coherent multi-sentential texts given a discourse goal. Rhetorical predicates are the means which a speaker has for describing information. They characterize the different types of predicating acts he/she may use to delineate the structural relations between propositions in a text. Some examples are analogy (the making of an analogy), constituency (description of sub-parts or sub-types), and attributive (providing detail about an entity or event). Linguistic discussion of such predicates can be found in Williams [15], Shipherd [16] and Grimes [17]. Next we will describe two explanation predicates in more detail. The attributive predicate can be used to illustrate a particular feature about a concept. e.g. "The oily skin, in the morning, has a sheen on the surface". The "sheen on the surface" is an attribute of the concept "oily skin". The constituency explanation predicate describes an entity or event in terms of its sub-parts or sub-types. e.g. "The moisturizing products are a day cream, a night cream, an eye-cream, an eye-gel and moisturizing masks".

The schema-based approach analyses explanations in terms of hierarchically structured explanation schemata. A schema represents an explanation pattern and consists of a number of rhetorical predicates which are instantiated as propositions selectively. In this way, schemata guide the generation of explanations. Schema recursion can be achieved by allowing each predicate in a schema to expand to either a single proposition (e.g., a sentence) or to a schema (e.g. an explanation sequence). The explanation structure generated from this application of schemata will be a tree structure, with a sub-tree occurring at each point where a predicate has been expanded into a schema. Propositions occur at the leaves of the tree.

Therefore, schemata, are similar to hierarchical plans Sacerdoti [18]. Each predicate in the schema is a generation goal which can be achieved either by fulfiling a number of sub-goals (the predicate expands to a schema) or producing a single utterance (the predicate expands to a proposition). Explanation schemata are precompiled and have the advantage that guided by local focus constraints will generate coherent and cohesive explanations. Also because their structure is pre-defined explanation generation is not a time consuming process[**]. The identification and constituency schemata are shown below.

identification schema	*constituency schema*
identification	{constituency}
{analogy/constituency/	cause-effect*/attributive*
attributive/renaming/	depth-identification/
amplification}*	depth-attributive
particular-illustration/	{particular-illustration/
evidence +	evidence}
{amplification/analogy/	{comparison/analogy}}+
attributive}	{amplification/rational
{particular-illustration/	attributive/evidence}
evidence}	{amplification/rational/
	attributive/evidence}

Schemata guide the generation process about what to say next and describe the structure of text at all levels and so can be used to generate explanations, given a discourse goal.

Extending the Schema Based Approach

A total of 74 expert explanations were identified in the experiment. The explanations were given by the expert only, in response to questions asked by the subject. Any explanations given by the subjects are ignored as far as the classification is concerned. (However, these explanations, are taken into account by the dialogue modeller in order to produce tailored responses). It was observed, that the expert in her advice uses a combination of predicates when she gives an

---

[**] We are using McKeown's notation: "{}" indicate optionality, "/" indicates alternatives, "+" indicates that the item may appear 1 or more times, and "*" indicates that the item may appear 0 or more times.

explanation. We call this particular combination of explanation predicates, an explanation schema. Different question paths have different schemata associated with them. We have applied McKeown's theory to the present empirically identified natural dialogues and have defined two additional schemata for *procedural* and *cause-effect* explanations, shown below, and three additional predicates *evaluation*, *similarity* and *difference* to those identified by Williams, Shipherd, Grimes and McKeown. Those modifications increase the generative power of the schema based approach enabling it to generate a wider range of explanations.

cause-effect schema   *procedural schema*
{cause-effect/   procedural/
constituency/   constituency}+
attributive}+

Despite the above extensions the schema based approach is still unable to offer alternative explanations to achieve the same goal, repair the explanation should it fail to satisfy the user or replan it should repairing fails. The schema based approach also fails to monitor the execution of an explanation, i.e. allow the inquirer to offer a feedback as the explanation is generated. This approach cannot plan another response to answer the same question, even if the user indicates dissatisfaction with the current explanation, for a schema is the result of a compilation process where the rationale for all the steps in the process has been compiled out. What remains is the top level discourse goal and the sequence of acts that achieve that goal. Recently this has lead to the adoption of Rhetorical Structure Theory as an alternative approach to explanation generation. Next we will describe, in outline, how Extended Schema based Theory overcomes the shortcomings of the schema based approach while retaining its advantages.

Repairing Explanations

From the experimental dialogues recorded in both the facial skin care and European Law domains it appears that rarely one-shot explanations satisfy the inquirer and experts frequently have to rephrase and restructure their explanations in an attempt to convince the inquirer. Even when the inquirer cannot articulate clearly a follow-up question experts do provide alternative explanations. Clearly an explanation system needs to be able to recover and produce an alternative explanation if the explanation generated fails. Most knowledge based systems are unable to recover locally i.e re-phrase and re-structure the explanation when it fails to satisfy the user, with the exception of Moore and Swartout [19] who modified Rhetorical Structure Theory [20] by associating precondition and effect fields with operators and by devising a set of recovery heuristics to repair the explanation that failed to satisfy the inquirer. Repairing requires a theory to explicitly represent the role and purpose of each proposition that makes up the explanation so that explanation systems which embody the theory are able to recover locally by substituting or rephrasing the proposition which failed to satisfy its purpose. Here we will describe how the schema based approach is modified to account for unsuccessful explanations and generating alternative responses.

Moore and Swartout identified three kinds of local failure (i) goals that introduce new concepts, (ii) goals that required assumptions to be made about the user's knowledge, (iii) the top-level goal of the completed text plan. Analysis of the natural dialogues recorded in the experiments also show that introducing new concepts in the explanation is frequently a source of trouble for it may be that the inquirer is not familiar with the concept being introduced and further information was required, or the inquirer disagrees in the way the concept was utilised in the explanation. Clauses that require assumptions to be made may also cause an explanation to fail if the assumptions were erroneous due to incomplete or incorrect information stored in the user model. Finally, in cases where neither (i) nor (ii) appear to be reasons the explanation failed, explainers frequently attempt to answer the question differently by using a different explanation strategy. The schema based approach in its current form cannot support repairing because at any time it does not keep track of the goals each predicate is satisfying. Our extension of the schema based approach associates with each predicate an effect on the hearer which is effectively the goal the predicate is satisfying. Consequently an explanation system can recover locally by substituting the predicate that failed with another that achieves the same effect in the schema structure. Furthermore effect fields are also associated with schemata, i.e. the top level discourse goals, to enable explanation systems to recover locally by substituting a schema that failed with another that achieves the same goal.

The schema based approach can be extended to accommodate repairing by associating *precondition* and *effect* fields with each predicate/schema. The precondition field lists the conditions that need to be true if the predicate is to be instantiated to a proposition, while the effect field describes the intentional goal(s) that the predicate may be used to achieve. This extension retains that advantage of the schema based approach - computationally efficient - and enables any system that embodies EST to have an understanding of the text it generates and reason about it. Furthermore it has the advantage that a system can recover from an unsatisfactory explanation by repairing it and can reason about the explanations it generates for it knows the goal(s) each predicate achieves. The predicates and schemas identified by Williams, Shipherd, Grimes and McKeown are modified so that their definitions include the precondition and effect fields which state respectively, the constraints

---

Formal notation

Effect: (believe user (claim f))
Preconditions: (AND (not (believe user (claim f)))
(knows user (support (facts_to be_presented claim)))
(not (knows user (relation (facts_to be_presented claim)))))
Definition: (evidence based-db <entity><attr-value>)

English translation

Effect: To achieve the state where the inquirer believes the claim
Preconditions: **If** the inquirer does not believe the claim,
knows the facts that will be used to support the claim (no introduction is needed),
does not know of the relation between those facts and the claim (otherwise the evidence is obsolete for the user is already aware of it)
Definition: **Then** present information that supports the claim

Figure IV: The Evidence predicate

---

under which a predicate should be applied and the goal each predicate will achieve. For example, the *evidence* predicate is modified as shown in figure IV.

From figure IV, preconditions are associated with each predicate/schema about the user's knowledge, beliefs and attitudes, using the computational model of belief ascription described in Ballim and Wilks [21]. The ascription mechanism separates beliefs about topics into objects called "environments" which may be about individuals, classes of agents or other agents. This computational model of belief ascription is embodied in a program called ViewGen which generates a type of environment known as "believer environment" which consists of a person's beliefs about a topic. Believer environments contain "topic environments" which hold a group of propositions about the "topic". The preconditions associated with the Evidence predicate are contained in the system's believer environment which encapsulates the topic environment. The topic environment consists of a group of propositions about the user ascribing to him beliefs and knowledge about various objects.

To recover, an explanation system also needs to store (i) all the alternative predicates that could have been instantiated if it was not for the local focus constraints, so that these predicates can substitute the predicate that failed, and (ii) all the alternative schemata that could have been selected during the mapping of question types and paths to suitable explanation strategies. The untried schemata can then be used to generate new explanations for the same question. If the explanation has failed due to the fact that the user did not understand a new concept being introduced, repairing is achieved by either elaborating further on the new concept, by using schema recursion, or omitting the concept if it is not essential. If explanation failure is caused from incorrect assumptions being made about the user's knowledge, during the instantiation of predicates, then recovery can be achieved by either making true the erroneous assumptions or by selecting an alternative predicate in the schema structure that achieves the same goal without making any assumptions. Finally if the top level goal has failed then an alternative schema is selected, which achieves the same goal as the one that failed. This is not global replanning because the original question path remains the same and only the explanation plan, derived from the question path, is modified.

From figure IV, should the inquirer ask a follow-up question which indicates that (s)he is not convinced with a claim being made in an explanation then the evidence predicate can be identified as the "faulty" component of the explanation schema. Corrective action can then be taken by either substituting the evidence predicate with a single or a combination of predicates that achieve the same *effect* or making true the assumption(s) that were made about the user's knowledge from the predicate *preconditions*. This was not possible in the original schema based approach for the effect and preconditions associated with each predicate/schema were compiled out.

Even the most sophisticated knowledge based systems are unable to recover globally i.e. reinterpret a question and so replan the explanation when repairing fails to satisfy the user. Existing approaches to repair techniques e.g. Moore and Swartout [19] and Cawsey [22], make the false assumption that the system interpreted accurately the user's intentions and the goal(s) he is trying to achieve. No effort is made to re-examine the user's input and consider the case where the system misunderstood the user's question and assumed a different goal than the one the user is actually interested to fulfil. Global recovery is achieved by reinterpreting the question posed; looking at the question from a different view point. Interpreting questions using question paths provides a very detailed description of the question, captures the questions multiple semantics and enables the question classification mechanism to recover globally and generate alternative question paths for the same question [23]. EST includes a method for generating additional question paths to those shown in figure V. Question paths are not pre-compiled but are generated during the question classification process. It is therefore possible to arrive at new question paths. This makes it possible for a question to be interpreted differently depending on the preceding dialogue or the user's acceptance of a previous explanation. It enables the theory to recover from failure by planning an alternative interpretation of the same question and resulting in a different explanation.

From the experimental dialogues recorded, it was observed that experts monitor the effects of their explanations by pausing at suitable points as they utter an explanation [18], [24], [25], [26]. A truly reactive theory of explanations needs to allow users to offer a feedback while an explanation is generated and if necessary re-plan the explanation if the user indicates dissatisfaction with the portion of the explanation already produced. The difficulty with this approach is to identify suitable points where the explanation generation can halt to allow the user to react if he so chooses, while the explanation is incomplete. Our theory overcomes this difficulty due to its question interpretation mechanism which interprets a question using multiple question types. Each question type is associated with suitable explanation strategies, which together compose an explanation. Therefore explanations can be broken down into a number of constituent strategies where each strategy is directly associated with a given question type. Consequently EST includes a method where it can pause and allow the user to react after part of the explanation is generated; just before or just after an explanation strategy is executed one can interrupt while the explanation is not yet completed without violating the inferencing sequence of the explanation. Next we will describe the selection heuristics used to associate question types with appropriate explanation schemata and generate explanations for any question posed.

ASSOCIATING EXPLANATION SCHEMATA WITH QUESTION PATHS

From the experiment we observed that certain question types are associated with certain explanation schemata (strategies). Schemata describe the structure of explanations and not their content. Thus, it is possible to use the same explanation strategy (schema) to answer more than one question. Each question type has a set of schemata associated with it that restricts the choice of which explanation strategy to use to a small number of possibilities. For each question type a single schema is selected out of the set on the basis of the information available to answer the question. For example, in response to requests for *definitions* the identification schema is selected if the knowledge pool contains a lot of information about the object's sub-classes and less information about the object itself. When this is not the case the constituency schema is used. In response to a *parenthetical* question type the cause-effect schema is selected if there are specific cause-effect relations between the objects mentioned in the question and the objects currently in the focus of interest. If there are no such cause effect relations between the objects, then the contrasting schema is used provided there is a strong antithesis between the objects. If neither the cause-effect nor the contrasting schema is triggered then the comparison schema is used. In response to a *concept-completion* question type the constituency schema is selected if the relevant knowledge pool contains a "rich" description of the object's sub-classes and super-classes and less information about the object itself. When this is not the case the attributive schema is used.

When a particular schema is selected to produce the explanation, the explanation contents and structure might differ for the same type of question. Up to now, we have only associated explanation schemata with question types. We also need to associate schemata with question paths. To do that we have devised a set of heuristics derived from the experiment. For example, if the linkage type between any two question types (Q1 and Q2) is hierarchical then: The explanation strategy associated with question type Q1 creates a context and an implicit or explicit relation with what is to follow, that allows the explanation strategy associated with question type Q2 to introduce its subject matter directly without a need to link its content with the preceding explanation strategy. For example consider the question "What a cleansing cream does ?". The question path is *concept-completion=>function*. The link relating the question types (figure II) is hierarchical. The explanation associated with this question is "1) A cleansing cream cleans the skin and prepares for moisturizing because it is a cleansing product. 2) Its function in particular is to remove heavy make-up". The explanation strategy associated with the concept-completion question type is (1), while the explanation strategy associated with the function question type is (2). Note that there is no intermediate utterance linking the two strategies as they are linked implicitly. In particular explanation strategy (1) provides the context for explanation strategy (2).

If the linkage path between any two question types is multi-concept then: Since the question types are semantically disjoint no context or explicit relation exists between the two explanation strategies associated with each question type. Thus a semantic link needs to be created explicitly in the explanation by either or both explanation strategies. There are two exceptions to this rule; the identification and parenthetical question types which although are contextually different from the other question types are special cases, for their main purpose is to relate the content of past questions and explanations with the current question, as real experts do. For example consider the question "How about scrub and buff-puff, which is best for me?". The question path of this question includes *comparison=>evaluation*. The link relating the two question types (figure III) is multi-concept. The explanation strategy associated with the question is "1) The exfoliating scrub exfoliates the skin without drying, while buff-puff massages and exfoliates the skin. 2) Because you have a dry and sensitive skin type 3) you are advised to use the scrub". The explanation strategy associated with the comparison question type is (1), while the explanation strategy associated with the evaluation question type is (3). Note that explanation strategy (2) serves to link explanation strategies (1) and (3) for there is no obvious reason that (3) follows from (1). The two question types are related by virtue of the question asked and not their semantics. The two question types, mentioned above, although not related hierarchically belong to the same question path by virtue of the question's meaning.

The explanation generator developed, called EXPLAIN, utilizes the findings of the theory and generates contextualized explanations in reply to the user's questions.

AN EXPLANATION GENERATOR - EXPLAIN

On receiving a question (see figure VI) EXPLAIN determines a single question path, which reflects the questions' semantics, the preceding dialogue and the user's acceptance of a previous explanation. Using the question path EXPLAIN associates explanation predicates and schemas with each question type in the path. From the question path EXPLAIN arrives at an *explanation plan* which reflects the user's intentions. The explanation plan lists the predicates and schemas that would produce the explanation if their pre conditions and the various constraints applied after their instantiation are satisfied. The relevant knowledge pool is then constructed - this is a subset of the knowledge domain which relevant to the question posed. The resulting explanation plan consists of predicates and schemas.

Having selected the explanation plan EXPLAIN then matches each schema/predicate against the relevant knowledge pool. The matching is constrained by the context of the dialogue, the current focus of attention Sidner [27] and McKeown [2] - how the user's centre of attention shifts or remains constant over

You may ask a question: *(What is lotion ?)*

The question path is: **identificationq=>definitionq**

The explanation plan, derived from the question path, is: **identification schema**

The explanation generated is the following
(identification lotion (protecting) (restrictive (purpose protect-from-sun)))**(analogy** *similar* lotion sun-cream (different lotion (skin-type oily) (use (how apply-with-cotton-wool) (when before-going-out-in-the-sun) (where all-over-the-face)) (sun-cream (skin-type dry) (use (how apply-with-fingertips) (when before-going-out-in-the-sun) (where all-over-the-face))))) **(amplification** def lotion (old-def moisturize) (protecting))

If the above explanation was to be translated into full English the following passage would be produced:

Lotion is a protecting product its' purpose is to protect the skin from the sun. Lotion and sun-cream are similar products, both have the same purpose. They can be used everywhere in the facial area before going out in the sun. Differences are that lotion is more suitable for oily skin types whilst sun cream is best for dry skin types. You apply lotion with a cotton wool whilst, as you are already aware of, you apply sun-cream with your finger-tips. In addition to protecting, lotion also moisturizes.

Figure VI: A Generated Explanation

two consecutive sentences - and the global focus (the locus of the user's centre of attention throughout an on-going discourse). The matching is further constrained by the acquired knowledge about the user (including the assumed user's knowledge of the domain). EXPLAIN uses the relations that exist between entities to make inferences as to which entities are known by the user and which are not [28].

From the experimental dialogues it appears that people who have little or no knowledge about a part of the domain receive more elaborate explanations than people who have a higher degree of knowledge in the same part of the domain. In contrast people who have a higher degree of knowledge are given explanations which include more information from other areas of the domain. Furthermore explanations to the same question have basically the same structure and usually novices first get something illustrated, and then amplified, in contrast to partial experts who first have amplification and then illustration. Although the structure of explanations is similar their content is quite different. The emphasis in the novice explanation is on 'how to do something', while in the partial expert case it is on 'what to do'. EXPLAIN utilizes these findings to tailor the explanations it generates [29]. After an explanation is generated EXPLAIN updates the user model and the dialogue planner - whose function is to take into account the preceding dialogue.

A CONSIDERATION OF EST

EST combines five novel features:

(1) interprets questions by combining information from different, semantically related question types which together capture the essence and meaning of the question,

(2) provides a context sensitive mechanism with the ability to recover either locally by restructuring the current explanation or globally [30] by re-interpreting the question and generating a new explanation for the same question if feedback indicates the inquirer is not satisfied with the current explanation,

(3) monitors and takes account of the effects of its utterances on the inquirer by pausing at suitable points during the generation of an explanation allowing the inquirer to react as the explanation is generated [30],

(4) it answers questions taking into account past explanations [31], and

(5) has a low computational overhead made possible by the use of explanation schemata.

In consultation dialogues question interpretation and the mapping of questions to suitable explanations is of paramount importance. EST can interpret any question and generate explanations by associating information from different, semantically related question types which together capture the essence of a question and permit more realistic inferences that resemble inferences made by experts in explanation dialogues.

SUMMARY

This paper described in outline an initial version of EST - for fuller description see [5] - explanation theory. A set of question types and their relations, called question paths were defined. Questions were interpreted in terms of a set of question types which together best capture the meaning and essence of complex questions. Question paths are associated with explanation strategies composed of explanation predicates. Questions and explanations were modelled by these two components of EST. Finally, an architecture and computational processors for an implemented explanation generator EXPLAIN were presented. EST provides powerful and complete question understanding and explanation generation in an efficient computational form. EST enables questions and explanations to be more closely matched. As a consequence expert systems which embody EST could become far more convenient for the casual user and more acceptable than current systems.

ACKNOWLEDGMENTS

We are indebted to Hilary Johnson, Violetta Papaconstantinou, Stathis Gikas and Alison Cawsey for the contribution they have made to the development of the ideas and the work reported here. Also to staff and customers at Boots and Debenhams who agreed to participate in the experiments.

REFERENCES

1. Shortliffe, E. H., Computer based medical consultation: MYCIN, New York: Elsevier, 1976.

2. McKeown, K., Text Generation: Using Discourse Strategies and Focus Constraints to generate natural language text, Cambridge University Press, 1985.

3. Swartout, W., XPLAIN: a System for Creating and Explaining Expert Consulting Programs, *Artificial Intelligence* 21 p 285-325, 1983.

4. Sarantinos E. and Johnson, P. Generating explanations: There is more to it than meets the eye, *Proceedings of the International Conference on Information Technology*, Tokyo, Japan, October 1-5, 1990.

5. Sarantinos, E. and Johnson, P., Explanation Dialogues: A theory of how experts provide explanations to novices and partial experts, *Artificial Intelligence*, To Appear.

6. Gilbert, N., Question and Answer Types. D. S. Moralee (ed), Research and Development in Expert Systems IV, Cambridge: Cambridge University Press, p 162-172, 1987.

7. Hughes, S., Question Classification in Rule-based Systems. Research and development in expert systems II, Cambridge University Press, 1986, 123-131.

8. Lehnert, W., The process of Question Answering: A computer Simulation of Cognition" Chapter 3. Lawrence Erlbaum Associates, Hillsdale, N.J, 1978.

9. Nicolosi, E., Leaning, M. S, Boroujerdi, M. A., The Development of an Explanatory System Using Knowledge-Based Models, *In Proceedings of the 4th Explanations Workshop, Manchester University*, 14-16 Sept 1988.

10. Pilkinghton, R., Tattersall, C, Hartley, R., Instructional Dialogue Management. CEC ESPRIT p-280 EUROHELP, 1988.

11. Wexelblat, R., The Confidence of Their Help, *AAAI'88 Workshop on Explanation*, p 80-82, 1988.

12. Maybury, M., M.Phil Thesis: A Report Generator, Engineering Department Library, Cambridge University, 1987.

13. Valley, K., Explanation Generation in an Expert System Shell. Department of Artificial Intelligence, University of Edinburgh, 80 South Bridge, Edinburgh, EH1 1HN, 1988.

14. Cawsey, A., Explaining the behaviour of simple electronic circuits, *International Conference on Intelligent Tutoring Systems*, Montreal, June 1988.

15. Williams, W., Composition and Rhetoric, DC. Heath and Co, Boston, 1893.

16. Shipherd, H. R., The fine art of writing, The MacMillan Co, New York, N.Y., 1926.

17. Grimes, J E., The thread of discourse. Mouton, The Hague, Paris, 1975.

18. Sacerdoti, E., A structure for plans and behaviour, Elsevier North-Holland, Inc. , Amsterdam, 1977.

19. Moore, J. and Swartout, W., A reactive approach to explanation, *Proceedings of the Fourth International Workshop on Natural Language generation*, Los Angeles, 1988.

20. Mann, W. and Thompson, S., Rhetorical Structure Theory: Toward a functional theory of text organization, *TEXT b (3)* 1988 243-281.3.

21. Ballim, A., and Wilks, Y., Stereotypical Belief and Dynamic Agent Modelling, *Proceedings of the Second International Workshop on User Modelling*, Hawaii, USA, 1990.

22. Cawsey, A., Explanatory Dialogues, Interacting with Computers, 1:69-72, 1989.

23. Sarantinos E. and Johnson, P. Explanation Dialogues: A computational model of interpreting questions and generating tailored explanations, *Proceedings of the 5th UK Workshop on Explanations*, Manchester University, UK, April 25-27, 1990.

24. Wilkins, D., Domain independent planning: Representation and plan generation, Artificial Intelligence, 22:269-301, 1984.

25. Doyle, R., Atkinson, D., Doshi, R., Generating perception requests and expectations to verify the execution of plans, *Proceedings of the Fifth National Conference on Artificial Intelligence*, p 81-88, Philadelphia, Pennsylvania, 1986.

26. Broverman, C., and Croft, W., Reasoning about exceptions during plan execution monitoring, *Proceedings of the Sixth National Conference on Artificial Intelligence*, Seattle, Washinghton, 1987.

27. Sidner, C., Towards a computational theory of definite anaphora comprehension in English discourse, Ph.D dissertation, MIT, Cambridge, Mass., 1979.

28. Kass, R., Implicit Acquisition of User Models in Cooperative Advisory Systems, Technical Report MS-CIS-87-05, Department of Computer Science, University of Pennsylvania, 1987.

29. Sarantinos, E. and Johnson, P. Tailoring Explanations to the User's Level of Expertise and Domain knowledge, *Proceedings of the 2nd International Conference on Tools For Artificial Intelligence 90 (TAI 90)*, Washington D.C., USA, 1990.

30. Sarantinos, E. and Johnson, P. Consultation Dialogues, *Proceedings of the AAAI-90 Workshop on Complex Systems, Ethnomethodology and Interaction Analysis*, Boston, USA, 1990.

31. Sarantinos, E. and Johnson, P. Interpreting questions and generating explanations during consultation dialogues, *Proceedings of the Pacific Rim International Conference on Artificial Intelligence '90 (PRICAI 90)*, Nagoya, Japan, 1990.

The Construction of Explanations

Ruth M.J. Byrne

ABSTRACT

What sorts of processes are needed to construct explanations? My aim is to describe a computational model of explanatory reasoning. It implements a theory of the representations and processes that people rely on to construct explanations. I will report two experiments on the explanations people produce for unusual events. The first experiment shows that they explain a single aspect of an event even when multiple aspects are unusual, and they explain the actions carried out by an actor rather than the objects possessed by an actor, even when both are unusual. The second experiment shows that their communication goals determine what aspect of an event they explain. The results have implications for theories of explanation-based learning, and for intelligent tutoring systems.

EXPLANATORY REASONING: A PRELIMINARY SKETCH

People construct explanations to make sense of the world -- this central cognitive skill underlies their ability to learn and to communicate. Consider the following event:

The bankrobber held up the bank.

People may consider that this event is entirely usual within the context of bankrobberies, and they may explain it by retrieving a remembered explanation. Now consider the following event:

---

* I am grateful to Mark Keane for helpful comments on an earlier version of this article.

The bankrobber rang up the bank.

They may consider that this event is unusual, and they cannot explain by retrieving a remembered explanation. Instead they must construct a new explanation. How do people identify what is unusual about the event, and what unconscious mental processes do they rely on to explain it ? I will derive an initial sketch of the machinery they use to reason to an explanation based on cognitive science research on reasoning in general.

Deductive Inference

Traditionally, philosophers, linguists, psychologists, and artificial intelligence workers have characterised thinking and reasoning as the operation of formal rules of inference. For example, the premises:

> There is a triangle or there is a circle.
> There is not a triangle.

support the conclusion:

> Therefore, there is a circle.

The formal rule theory proposes that people access inference rules, such as:

> *p or q*
> *not p*
> _____
> *Therefore, q.*

But, recent experimental evidence goes against this view (see e.g. [1], [2]). Instead, the data support the view that people rely on *mental models* (e.g. [3]). They understand the premises by building a representation that corresponds to the structure of the world, not to the structure of the language in which the world is described. The first premise supports a set of models of the following sort:

Δ

    0

Δ   0

where we have adopted the convention of representing separate models on separate lines. The models represent three alternative possibilities about how the world might be when the first premise is true. The second premise rules out some of these alternatives -- those in which there is a triangle -- and leaves only the model in which there is a triangle and a circle. Hence, the deduction that there is a circle can be made by manipulating models, without recourse to formal rules of inference (see [4] for details). The experimental evidence corroborates the predictions of the model theory for inferences based on propositional connectives such as "if" and "or" ([5], [6]), spatial relations such as "in front of" ([7]), and quantifiers such as "none" and "some" (e.g. [8]).

Explanatory Inference

The model theory can be extended to account for explanatory reasoning. To explain the event:

> The bankrobber rang up the bank

people must first identify what is unusual about the event and then retrieve a relevant fact that makes sense of it. The event can be represented by a single model which contains tokens corresponding to the actor and the action:

> *b*   *r*

where "b" represents the bankrobber, and "r" represents ringing up the bank. People understand an unusual situation by comparing it to the more usual situation (see [9], [10]). For example, a person may believe that it is more usual for a bankrobber to hold up a bank, and this alternative can be represented in a second model:

> *b*   *r*
> *b*   *h*

where "h" represents holding up the bank. Each model in the set can be annotated to indicate its epistemic status, that is, whether it represents the actual situation described by the premise, or whether it represents the counterfactual, but more usual situation:

real: *b*   *r*

usual: *b*   *h*

(for a defense of propositional-like tags in models, see [11] and [12]; and for discussions of counterfactuals see e.g. Chapter 4 in [2]). A comparison of the two models will result in the identification of the unusual aspect. This aspect can then be used as a cue to access information, and the retrieved information can be integrated into the set of models by strategies designed to make sense of the unusual aspect. Before considering these processes in detail, let us turn to some of the phenomena of explanatory reasoning.

THE PHENOMENA OF EXPLANATORY REASONING: TWO EXPERIMENTS

Spontaneous Explanatory Inferences

What explanatory inferences do people make to identify that something is unusual about an event and thus requires an explanation? Past research emphases that they identify the unusual *attributes* of the situation ([9], [10], [13], [14]). Hence, we can hypothesise that people will readily identify as requiring explanation an unusual attribute such as an *object* possessed by an actor, e.g., a bankrobber who has a knife instead of a gun. Are unusual attributes the only identifiable aspect of an unusual situation? In fact, attributes may be less important than relations, as has been found, e.g., for the categorisation of information ([15]). Hence, we can also hypothesise that people will readily identify an unusual relation such as an *action* carried out by an actor, e.g., a bankrobber who rings a bank rather than holds up a bank. A third hypothesis is that people will focus on just one unusual aspect, even when more than one is unusual, because human working memory imposes limitations on the amount of explicit information that can be represented and processed.

These hypotheses were tested in an experiment involving twelve members of the subject panel of the School of Psychology, at the University of Wales College of Cardiff. They were given single sentence descriptions of various sorts of events: crimes (bankrobberies, burglaries, hi-jackings, and kidnappings), sports (mountaineering, horseriding, swimming, and tennis), and the operation of machines (computers, cars, trains, and washing machines). The events referred to an actor, who carried out an action -- either usual or unusual, and who possessed an object -- either usual or unusual. The subjects were given 12 descriptions: three contained an unusual *object* and an unusual *action*, three contained a usual object and an unusual *action*, three contained an unusual *object* and a usual action, and three contained a usual object and a usual action. Each subject took part in each of these four conditions. Their task was to explain the event, as if someone in conversation had asked "why?".

The results showed that their explanations focused on actions when they were unusual, just as readily as on objects when they were unusual: explanations that focused on the action were produced more often in the unusual-action condition (63%) than in the unusual-object condition (17%), and this difference was statistically reliable (at the 5% level, as shown by Newman-Keuls pairwise comparisons, on the main effect of usualness [$F (3,27) = 6.3$, $p < .002$] in a one-way repeated-measures analysis of variance of such explanations). For an event that contained just an unusual action, e.g.:

The bankrobber with the gun rang up the bank.

an example of an explanation focused on the action was:

He was finding out what the bank opening hours were.

(for further details, see [16]). Conversely, explanations that focused on the object were produced reliably more often in the unusual-object condition (47%) than in the unusual-action condition (the difference is reliable at the 1% level, shown by Newman-Keuls comparisons, on the main effect of usualness [$F (3,27) = 6.77$, $p < .001$] in the one-way repeated-measures analysis of variance of the explanations focused on the object). For an event that contained just an unusual object, e.g.:

The bankrobber with the knife held up the bank.

an example of an explanation focused on the object was:

He had to use physical force of some sort.

When both the action and the object were unusual, people tended to focus on just one aspect (60%) rather than on both (30%, and this difference is reliable only at the 7% level, t (9) = 1.588, one-tailed). Interestingly, when they focused on just one aspect, it tended to be the action (33%) rather than the object (17%). Finally, when the event was a usual one, most of the explanations (43%) focused on the *intentions* of the actor, more so than in any of the other conditions (and the differences are reliable at the 1% level by Newman-Keuls comparisons on the main effect of usualness [F (3,27) = 5.1, p < .006], in the one-way repeated-measures analysis of variance on the intention explanations). For example, for the situation:

> The bankrobber with the gun held up the bank.

an example of an explanation focused on the intentions of the actor is:

> He wanted lots of money.

Goal-Driven Explanatory Inferences

The second experiment examined the factors that determine what people focus on in an unusual event that contains more than one unusual aspect. Previous research suggests that goals may influence the identification of what needs to be explained ([17], [18], [19]). When people explain an event, their goal is to communicate a blueprint for the construction of a model of the unusual event and its more usual counterparts to their listener (see also [3], [20]). Hence, we can hypothesise that they will explain different aspects of an event when their goal is to communicate to different sorts of listeners, such as people close to them, or people remote from them.

This hypothesis was tested in an experiment involving sixteen members of the subject panel of the School of Psychology, at the University of Wales College of Cardiff. They were given 12 descriptions of events that contained both an unusual object and an unusual action. For half of the descriptions their task was to construct an explanation that the actor would have given to a figure close to him or her, such as an accomplice. For the other half their task was to construct an explanation that the actor would have given to a figure remote from him or her, such as a police inspector.

The results showed that their explanations were influenced by their goals. The choice of which unusual aspect to explain was affected by the two sorts of goals. People constructed explanations which focused on the action reliably more often to a close figure (45%) than to a remote figure (23%) and this difference was reliable (t (15) = -4.2, p < 0.001). For example, given the unusual situation:

> The man with the tennis shoes crawled up the mountain.

an example of an explanation to a close figure (a friend) which focused on the action, is:

> I lost my footing and slipped while mountain walking.

They constructed explanations which focused on a combination of the action, object, and actors' intention more often to a remote figure (33%) than to a close figure (23%), and this difference was reliable (t (15) = 2.44, p < 0.01). An example of an explanation for the event above but to a remote figure (a guide), which focused on a combination of the action, object and intentions of the actor is:

> I had inappropriate footwear and felt it was the safest way to proceed.

Hence, the data show that people tailor their explanations for the person to whom it is directed -- they can be flexible in their choice of what needs to be explained.

The experiments show that people can construct explanations for usual and unusual events, they focus on what is unusual in an event and try to explain it, they tend to explain just one aspect of the event even when more than one aspect is unusual, and their goals in communicating to different listeners determine which aspects of an event they explain. We can turn now to the computational model of a theory of the construction of explanations.

A COMPUTATIONAL MODEL OF EXPLANATORY REASONING

The program for explanatory inference is written in Common Lisp and it is called EXPSYCH. It is based on PROPSYCH, a program that simulates a psychological theory of propositional reasoning based on mental models, which was developed in collaboration with Phil Johnson-Laird (see Chapter 9 in [2]). EXPSYCH consists of two principal components. The first component identifies what has to be explained. It parses natural language descriptions of crimes, constructing a set of models consisting of the models for the real situation and the models for the usual situation, and it computes which aspect of the models differs. The second component explains the selected aspect. It takes the set of models and the identified unusual aspect and utilises a strategy from its repertoire to access a piece of relevant information to add to the models to explain the unusual aspect.

Identifying What to Explain

The program consists of a parser, a grammar and a lexicon for dealing with natural language input, based closely on the mechanism implemented in PROPSYCH. The program contains lexical information about the domain of crimes, committed by actors such as bankrobbers, hi-jackers, burglars, and kidnappers. The parser parses assertions that correspond, for example, to the following description:

The bankrobber with the knife rang up the bank.

It first builds a model of the actor, by accessing its lexical semantics for the noun, bankrobber. It calls a function to build an internal representation containing a token corresponding to the actor, and it keeps track of the fact that it is parsing an assertion about the real world by annotating the model with a tag:

REAL   B

When the model of the real situation is built, it builds a model of the usual situation as well, by accessing information about usual bankrobberies. The lexical entry for the noun, bankrobber, also calls a function that contains information about usual bankrobberies. This information comprises procedures that build models corresponding to the actor who usually carries out such events, i.e., a bankrobber, the usual objects of the actor, such as a gun, and the usual actions of the actor, such as holding up a bank. Because the model of the real situation is currently a model of the *actor,* the bankrobbery information is accessed solely to establish who the usual actor is in such situations. A model of the usual actor is constructed on the basis of this stored information, and the output at this point contains information about the real actor and the usual actor:

$$
\begin{array}{ll}
REAL & B \\
USUAL & B
\end{array}
$$

The tokens in the two models are compared, and because the real actor is in fact the usual actor, the usual model is eliminated -- there is no need to consider the more usual situation because it is not counterfactual. In this way the program keeps track of the more usual aspects of a situation that it needs to consider -- in this case, none.

When the parser parses the meaning of the object, it constructs a model of the real situation, containing a mental token corresponding to the object, annotating the tokens corresponding to objects with a symbol: "+". It recovers the information about usual bankrobberies to find what the usual object is, and constructs a second model consisting of this attribute. It compares the objects in the real and the usual models and because they are *not* the same both models are retained. It integrates the set of models for the objects with the set of models for the actor with the following result:

$$
\begin{array}{lll}
REAL & B & + K \\
USUAL & & + G
\end{array}
$$

where "+K" represents possessing a knife, and "+G" represents possessing a gun. Similarly, when the parser parses the meaning of the action, it constructs a model of the real situation, containing a mental token corresponding to the action, annotating the token corresponding to an action with a symbol: "*". It goes though its' model building procedures again, with the following eventual result:

$$
\begin{array}{llll}
REAL & B & + K & * R \\
USUAL & & + G & * H
\end{array}
$$

where "*R" represents ringing a bank and "*H" represents holding up a bank. The difference between the two models is computed by searching the model of the usual situation for any actor, object, or action it contains, and returning the corresponding aspect from the model of the real situation. Upon recovering such information, the output is:

> *THERE IS AN UNUSUAL OBJECT:*   *+ K*
> *THERE IS AN UNUSUAL ACTION:*   ** R*

The program deals with the situations in which there is just an unusual object, or just an unusual action in a similar way. It can also handle a description of a usual event, for example:

> The bankrobber with the gun held up the bank.

and its output is the following set of models:

> *REAL*   *B*   *+ G*   ** H*

in which there is only one model. The program thus returns:

> *THERE IS NOTHING UNUSUAL.*

Strategies to Construct Explanations

The second component of the program contains a repertoire of strategies to retrieve a piece of relevant information that will make sense of the unusual aspect. Imagine the input to the second component is the models for a burglary event containing an unusual action and an unusual object:

> The burglar with the suit knocked on the window.

The program contains information that the more usual action for a burglar is to break a window, and the more usual clothing possessed by a burglar is a balaclava. It thus constructs the models:

> *REAL*   *BG*   *+ S*   ** K*
> *USUAL*   *+ BL*   ** BR*

where "BG" represents the burglar, "+S" represents possessing a suit, "+BL" represents possessing a balaclava, "*K" represents knocking on a window and "*BR" represents breaking a window. The second component takes as input the models, and the information that the unusual aspects of the event are that the burglar possesses a suit and knocks on a window. It selects at random one of these aspects to explain, for example, the unusual action of knocking on a window. One strategy to explain an unusual aspect is to assume that what usually leads to the *more usual* aspect *failed,* and this failure led to the unusual aspect. The program accesses information about the *usual* counterpart to the unusual aspect, i.e. information about the usual action in burglaries: breaking windows. It retrieves an *antecedent* to this usual action, e.g., the burglar must have an implement with which to break the window, and it adds this information to the models:

$$REAL \quad BG \quad + S \quad * K$$
$$USUAL \quad \quad + BL \quad * BR \quad > I$$

where ">" represents an antecedent, and "I" represents having an implement. It then negates the antecedent, and adds this *negation* to the models:

$$REAL \quad BG \quad + S \quad * K \quad > \neg I$$
$$USUAL \quad \quad + BL \quad * BR \quad > I$$

where " ᄀ" represents negation. Hence, the model of the real event contains the information that what led the burglar to knock on the window was that he did not have an implement to break it with. The same strategy can be used to explain unusual objects as well as unusual actions. A related strategy accesses information not about what usually leads to the usual aspect, but what usually is a *consequence* of the more usual aspect. It assumes that the usual consequence is to be avoided and this avoidance results in the unusual aspect.

If the stored information about burglaries does not contain information about the antecedent or consequent of the usual aspect, then the second set of strategies is called. This set embody a second way to explain an event: assume that what leads to the unusual aspect *in its more usual context* is the case. The program accesses information about the unusual aspect, e.g., knocking on a window in its' more usual context: non-burglary settings. It retrieves an antecedent of the action, e.g., that people knock on windows in order to gain the attention of someone inside, and it adds this information to the models:

$$REAL \quad BG \quad + S \quad * K \quad > A$$
$$USUAL \quad \quad + BL \quad * BR$$

The model contains the information that what led the burglar to knock on the window was to get the attention of someone inside the house. The strategy can also be used to explain unusual objects. A related strategy accesses information about the usual consequence of the unusual aspect. Thus, the program currently contains four strategies for explaining unusual events:

- Assume the antecedent to the usual aspect has failed.
- Assume the consequent of the usual aspect has failed.
- Assume the antecedent of the unusual aspect in its more usual context is the case.
- Assume the consequent of the unusual aspect in its more usual context is the case.

The program contains one more strategy, used exclusively for explaining *usual* events. Recall that the data show that given a usual event people explain the actors' *intentions*. The program represents a usual event, such as:

The kidnapper with the menacing expression pointed the gun.

in the following way:

$$REAL \qquad K \qquad + E \qquad * P$$

where "K" represents the kidnapper, "+E" represents having a menacing expression, and "*P" represents pointing the gun. The program returns the information that there is nothing unusual about the situation, which triggers the strategy to identify the intentions of the actor. The procedures access information about kidnappings, to retrieve the usual intentions of a kidnapper: to get money. This information is added to the model:

$$REAL \qquad K \; ^\wedge M \qquad + E \qquad * P$$

where "^" is a tag to indicate an intention, and "M" represents getting money. The explanation for the situation is that the kidnapper wanted money.

The program constructs explanations by a model-based process of explanatory reasoning. It relies on the models to identify what is unusual, and it adds to the models the information retrieved during the execution of its strategies.

THE CENTRAL ROLE OF EXPLANATORY INFERENCE

Explanatory inferences are made when an unusual situation is encountered. I have argued that people make these inferences by constructing models of the situation and its more usual counterpart. They compare these models to identify what is unusual in the situation, and the unusual aspect serves as a cue to be used by a set of strategies to access relevant information. The data show that people can produce explanations readily for both usual and unusual events. They select for explanation both unusual actions and unusual objects, they tend to select just one aspect even when more than one is unusual, and they tend to focus on actions rather than objects even when both are unusual. Their selection is influenced by their goals: they tailor what they select to suit their listeners.

This research has implications for explanation-based learning systems, and for intelligent tutoring systems. Researchers in artificial intelligence and cognitive psychology have examined how to best capture the process of learning a concept or categorising an instance as part of a concept (e.g. [21]). One way is based on identifying the common features that classify the regularities in a set of examples, and is sometimes called similarity-based learning (SBL). An alternative way is based on using a domain theory to establish the nature of a single instance, and is sometimes called explanation-based learning (EBL). Experimental evidence suggests that humans categorise information by applying explanatory theories rather than by compiling feature lists (e.g. [15]), and EBL programs have had some success where SBL ones have failed (e.g. [22]). But, current EBL systems construct *proofs* that an example is an instantiation of a general concept (e.g. [23]). In these systems the structure of an explanation is similar to a formal derivation in logic. People, on the other hand, do not seem to construct proofs even when they make deductions, instead they construct models ([2]). I have suggested that their explanatory inferences are also model-based. If the structure of an explanation is akin to models rather than to proofs, then EBL systems that are model-based rather than proof-based may prove more fruitful. Likewise, an intelligent tutoring system must provide people with coherent explanations of, for example, their failures. It is crucial that the information they are presented with is compatible with the way they naturally represent it. If their representations are model-based then an intelligent tutoring system that presents explanations that can be readily represented in models would be fruitful.

In conclusion, cognitive science research on the construction of explanations suggests explanatory inference is a model-based process. It is built out of the basic building blocks of deductive reasoning and memory retrieval processes, and it is central to learning and to communication.

REFERENCES

1. Evans, J. St. B. T. Bias in Human Reasoning. Hillsdale, Erlbaum, 1989.
2. Johnson-Laird, P.N. and Byrne, R.M.J. Deduction. Hillsdale, Erlbaum, 1991.
3. Johnson-Laird, P.N. Mental models: towards a cognitive science of language, inference, and consciousness.Cambridge, Cambridge University Press, 1983.
4. Johnson-Laird, P. N., Byrne, R.M.J. and Schaeken, W. Reasoning by model: the case of propositional inference. Manuscript submitted for publication, Princeton University, New Jersey, 1990.
5. Byrne, R.M.J. Suppressing valid inferences with conditionals. Cognition 1989; 31: 61-83.
6. Byrne, R.M.J. Everyday reasoning with conditional sequences. Quarterly Journal of Experimental Psychology 1989; 41A: 141-166.
7. Byrne, R.M.J. and Johnson-Laird, P.N. Spatial reasoning. Journal of Memory and Language 1989; 28: 564-575
8. Johnson-Laird, P.N., Byrne, R.M.J., and Tabossi, P. Reasoning by model: the case of multiple quantification. Psychological Review 1989; 96: 658-673.
9. Hilton, D. J. and Slugoski, B.R. Knowledge-based causal attribution: the abnormal conditions focus model. Psychological Review 1986; 93: 75-88.
10. Kahneman, D. and Miller, D.T. Norm theory: comparing reality to its alternatives. Psychological Review 1986; 93: 136-153.
11. Johnson-Laird, P.N. and Byrne, R.M.J. Only reasoning. Journal of Memory and Language 1989; 28: 313-330.
12. Polk, T.A., and Newell, A. Modeling human syllogistic reasoning in Soar. In: Tenth Annual Conference of the Cognitive Science Society, Hillsdale, Erlbaum, 1988, pp 181-187.
13. Kahneman, D. and Tversky, A. The simulation heuristic. In: Kahneman, D., Slovic, P., and Tversky, A. (eds.) Judgement under uncertainty: heuristics and biases.

Cambridge, Cambridge University Press, 1982.

14. Hilton, D.J. Conversational processes and causal explanation. Psychological Bulletin 1990, 107: 65-81.

15. Murphy, G.L. and Medin, D.L. The role of theories in conceptual coherence. Psychological Review 1985; 92: 289-316.

16. Byrne, R.M.J. Cognitive Processes in the Construction of Explanations. Manuscript submitted for publication, University of Wales College of Cardiff, 1990.

17. Achinstein, P. The nature of explanation. Oxford, Oxford University Press, 1983.

18. Antaki, C. Explanations, communication and social cognition. In: Antaki, C. (ed.) Analysing everyday explanations: A casebook of methods. London, Sage, 1988.

19. Schank, R.C. Explanation patterns: Understanding mechanically and creatively. Hillsdale, Erlbaum, 1986.

20. Wittgenstein, L. Philosophical investigations. transl. Anscombe, G.E.M. Oxford, Blackwell, 1945 / 1974.

21. Eysenck, M.W. and Keane, M.T. Cognitive Psychology: A Student's Handbook. Hillsdale, Erlbaum, 1990.

22. DeJong, G. Explanation-based learning with plausible inferencing. Technical Report No. UIUCDCS-R-90-1577. Department of Computer Science, University of Illinois at Urbana-Champaign, 1990.

23. Mitchell, T., Keller, R., and Kedar-Cabelli, S. Explanation-based generalisation: A unifying view. In Collins, A. and Smith, E.E. (eds.) Readings in Cognitive Science. Calif, Morgan Kaufman, 1988.

Section 8:

Uncertainty

Incomplete Information and Uncertainty

J.M.Morrissey

ABSTRACT

In this paper we examine the problem of representing different types of incomplete data in an information system. Clausal form logic is used as a representation language and we look at the consequences of the subsequent increase in expressive power. We propose a method of improving the system's response to a query by giving an uncertainty measure based on entropy.

INTRODUCTION

In conventional database systems it is not possible to represent certain types of shallow or incomplete information. Usually we are limited to definite facts and to one type of 'null' which can mean that either the value is unknown or that the attribute is not applicable. Thus, for example, providing that we do not violate constraints and do not have a null as part of a primary key we can represent facts of the following type:

1. Supplier S3 supplies 200 of part P2 to project J5.
2. Supplier S4 supplies an unknown quantity of part P100 to project J9.
3. Part P99 has no subparts (the attribute is not applicable).

but we can't represent the following:

1. Supplier S4 supplies P3 or P4 to Project P7.
2. Supplier S10 does not supply P4.
3. There is an unknown supplier who supplies P6 and P7 to project J7
   (assuming that 'supplier' is a primary key).

Ideally an information system should be able to represent the following types of incomplete information, as summarised in [8]:

1. State that 'something' has a property without identifying 'something'.
2. State that 'everything' belonging to a certain class has a certain property without naming 'everything'.
3. State that either A or B is true without saying which is true.
4. State explicitly that something is false.
5. State whether or not two non-identical expressions name the same object.

Additionally the system should be able to distinguish between different types of null value, specifically:

1. Value 'unknown' but belonging to the set of values known to the system,
   that is part of the domain of discourse.
2. Value completely unknown.
3. Value not applicable.

Many attempts have been made to increase the expressive power of databases in relation to incomplete information. The standard treatment in relational systems [1,2] is to insert a null whenever a value is missing or not applicable. Three-valued logic is used for query evaluation, the answer being 'maybe' when ever there is a null. In [3] the treatment is extended. Nulls are replaced by two special markers which distinguish between 'value missing' and 'value not applicable'. In both cases tautologies are not recognised and thus tautologically true queries are answered incorrectly.

Indexed nulls are proposed in [9]. They enable us to know which nulls are equal, and they prevent information loss when a relation is decomposed and subsequently rejoined.

Date [4] proposes a system of default values, instead of nulls. However, the defaults may cause more problems than they solve since all operations must be allowed to distinguish between them and the real values. One advantage is that they allow tautologically true queries to be answered correctly.

Grey [5] examines the problem of treating views as first class objects. He concludes that to do so the system must be capable of storing certain types of incomplete information. He extends the relational domains to include a special value 'unknown', real number intervals and p-domains. (A p-domain is a subset of the attribute domain, the actual value is one of this subset.) The relational operations are extended to cope with these new values.

Vassiliou [10,11] is concerned with imperfect information and how to lessen its impact on the system. He is mainly concerned with missing values and inconsistent data. To each relational domain he adds the special values *missing* and *nothing*, the latter representing inconsistent data. A special four valued logic is proposed for query evaluation. The response to any query is divided to two lists: those tuples which evaluate to 'true' and those which evaluate to 'maybe'.

Lipski's work [7] is the most comprehensive treatment of incomplete information. Semantics are provided for both complete and incomplete information systems. However, the only type of incomplete information considered is the p-domain. Lipski's work will be examined again later.

Williams and Nicholson [12] show how QBE [14] can be extended to handle incomplete information. Winslett [13] examines the problem of automating the process of inserting and deleting information from an incomplete database.

In [6] each domain is augmented to include p-domains, number intervals, two special values denoting 'value not known' and value not applicable' and semantics are given for query evaluation. No other type of incomplete information is considered.

In this paper we extend the representation of incomplete information to include the different types outlined above, we examine some consequences of the increase in expressive power and we propose one method of improving the system's response to a query given that there is some incomplete data.

REPRESENTING INCOMPLETE INFORMATION

Assuming suitable descriptors and using clausal form logic as a representation language we can easily represent the various types of incomplete information.

Simple facts are represented as clauses with a single consequent. For example, 'Supplier S3 supplies 400 of part P2 to project J7' is written as

=> Supplies(S3,400,P2,J7)

Simple negative facts are represented as clauses with a single antecedent. For example, 'S1 does not supply part P9' is written as

Supply(S1,P9) =>

By introducing a Skolem Constant to represent an object we can store information about the properties of that object without identifying it. For example 'there is someone who supplies 500 of part P2 to project J5' is represented as

=> Supplier(@X)
=> Supplies(@X, 500, P2, J5)

and 'there is a part which is red and is a subpart of P100' is represented as

=> Part(@X)
=> Coloured(@X, RED)
=> Subpart_Of(@X,P100)

Skolem Constants also allow us to know whether or not two non-identical statements refer to the same object: all clauses with the same Skolem Constant refer to the same object. Thus

$$=> Part(@Y)$$
$$Coloured(@Y,RED) =>$$
$$=> Subpart\_Of(@Y, P200)$$

states that there is an unknown part which is not red and which is a subpart of part P200. But

$$=> Part(@Y)$$
$$=> Part(@Z)$$
$$Coloured(@Y,RED) =>$$
$$=> Subpart\_Of(@Z, P200)$$

says that there are two unknown parts, one of which is not red and the other is a subpart of P200.

To leave open the question of identity we use a special constant, @*NK*. Then for example,

=> Part(@NK)
=> Colour(@NK,PINK)

means that there is an unknown part, and there is an unknown object which is pink but we do not know whether or not we are talking about the same object.

Skolem constants are also used to represent information about an object or value which is unknown but part of the domain of discourse of the system. Thus for example,

=> Part(@X)
=> Supplier(@Y)
=> Supplies(@Y, 400, @X, J9)

means that there is an unknown supplier supplying 400 of an unknown part to project J9, but both the part and the supplier are part of the domain of discourse.

To represent information about an object or value which is unknown and which may or may not be part of the domain of discourse we introduce special constants. These have the same semantics as Skolem Constants except that it is not known whether or not they are part of the domain of discourse. For example,

=> Supplier(!X)
=> Supplies(!X, 400, P2, J5)

means that they is a supplier, who may not be part of the domain of discourse, who supplies 400 of P2 to J5. To leave open the question of identity, and state that the object may not be part of the domain, the special constant !*NK* is used. Thus

=> Part(!NK)
=> Colour(!NK, PINK)

says that there is an unknown part, and an unknown pink object; it is not known whether the two statements refer to the same object or not, nor is it known if the object(s) denoted by '!*NK*' are part of the domain of discourse. Thus '!*NK*' represents 'completely unknown'.

We can state the properties of a group of objects, without enumerating all the members of the group, via conditionals. For example 'every part has a part number' is represented as

Part(x) => Has_Partnumber(x,y)

meaning that every part has a partnumber which may or may not be known to the system. This can be viewed as a constraint on the database.

Disjunctive information is represented as a clause with disjunctive consequents. For example 'Supplier S20 supplies wheels or tyres' can be represented as

=> Supplies(S20, WHEELS), Supplies(S20, TYRES)

meaning that the supplier supplies wheels, tyres or both.

In conventional data modelling real world objects are modelled by selecting representative attributes to which values are assigned. But sometimes the representative attribute will not apply to all objects. For example, the attribute 'name of spouse' will not apply to any single person. To handle this exception we introduce another special value which is used to indicate that the attribute is not applicable. Thus

=> Fax(Wongs_Eatery, NA)

means that Wong's Eatery does not have a fax machine. It is important to realise that this is complete precise information and thus its semantics are the same as any regular precise value.

INCOMPLETE INFORMATION AND QUERY EVALUATION

The increase in representational power has its cost, specifically
   1. The Closed World Assumption no longer holds.
   2. Domain closure is no longer a valid assumption.
   3. The semantics of query evaluation are no longer obvious,
   4. and the 'correct' answers to queries are not informative.

All conventional database management systems either explicitly or implicitly employ the Closed World Assumption (CWA) during query evaluation. It states that if a fact is not known to be true then it can be assumed to be false. Thus if the system does not contain the fact

=> Colour(P1, GREEN)

then we can safely assume that P1 is not green. However, there is a problem when the system contains indefinite data. For example, if we have the information

=> Colour(P1, GREEN), Colour(P2, Blue)

but neither of the following facts

=> Colour(P1, GREEN)
=> Colour(P1, BLUE)

then the CWA will allow us to conclude that P1 is not green and that P1 is not blue, which is obviously incorrect. So to maintain consistency we may not use the CWA when there is indefinite data.

The domain closure assumption states that there are no objects other than those known to the system. So if asked 'is every part green ?' then the system only considers those objects known to it. However, we have introduced constants which represent objects which may not be known to the system. Theoretically this means that the system may not be able to answer such closed queries; in this example the best it can say is 'maybe' which is not very useful.

When we have incomplete information then the semantics of query evaluation are no longer obvious. Consider the following data

=> Colour(P1, RED)
=> Colour(P2, GREEN), Colour(P2, RED)
=> Colour(@X, RED)
=> Colour(P9, @Y)

and the query 'list all red objects'. Clearly P1 is red and should be part of the answer; it is possible that P2 and P9 are red but we can't be sure; and in the case of the object @X, it is red but it could also be P1, P2 or P9. It is not clear what the system's response should be. Logically the 'correct' answer consists of those objects about which we have complete information - P1 in this case - but this is not useful nor does it make use of all available information.

In the next section we look at one way of improving the system's response to make it more useful and informative.

MEASURING UNCERTAINTY

Lipski [7] handles disjunctive information by means of p-domains. Two interpretations are provided for queries, the internal and the external. The internal ignores all incomplete information and thus there is no uncertainty. The external makes an open world assumption and considers all incomplete information. Two operators are provided for querying:

Surely: this allows the user to ask for subjects which surely have a certain property. The answer is the set of objects known to belong to the external interpretation.

Possibly: this allows the user to ask for objects which possibly have a certain property. The answer is the set of objects for which we can't rule out the possibility that they belong to the external interpretation.

One problem with this treatment is that there is no measure of the certainty that an object possibly belongs to the external interpretation. Thus the answer is of limited utility since we can t distinguish between an object which is very likely to have the property and one which is very unlikely to have it. In [6] a similar treatment of disjunctive information is proposed. The answer to any query is divided into two sets of objects: those which are known to satisfy the query with absolute certainty, and those which possibly satisfy the query with varying degrees of uncertainty. The latter set are ranked by increasing uncertainty and presented to the user in that order. Different measures of uncertainty are examined, including measures based on entropy, self-information and dissimilarity. However, all measures are based on the assumption that all values were equally likely to occur. This is an unnecessary and limiting constraint.

Entropy is a direct measure of uncertainty, it is at a maximum when there is complete uncertainty and at a minimum when there is no uncertainty. It is defined as

$$- \sum_{k=1}^{N} P(k) \, ln \, P(k)$$

where $P(k)$ is the probability of event k occurring, and there are N possible events. The limits are defined as

$$0 \leq entropy \leq ln\, N$$

where N is the number of possible events.

In [6] entropy was used to measure the uncertainty that an object satisfied the query, but the assumption was made that all events are equally likely. Here we propose that subjective estimates of the likelihood of an event are supplied by the user, either by specifying a range or giving some qualitative term. These subjective estimates are converted into 'probabilities' by the system for insertion into the entropy formula. The user does not need to know anything about probability nor does he need to enumerate all possible events. At query evaluation time two sets are formed in response to open queries: the set of objects known to satisfy the query and the set which possibly satisfies the query. For each object in the latter set we estimate the uncertainty involved, and the objects are ranked and presented in order of increasing uncertainty. A simple example illustrates the key ideas.

Example

A user enters the following data about part colours: *P1 is blue; P2 is either blue or red but it is very likely to be blue and very unlikely to be red; P3 is blue, red or green but most likely it is green and the other colours are equally likely to occur; the colour of P4 is unknown but is one of the colours in the domain and it is four times as likely to be red as any other colour; there is no information about the colour of P5.* The only colours known to the system are red, orange, green, blue, indigo and violet and therefore the qualitative terms given by the user are converted into for example, the following numbers:

P(P1 is blue) = 1
P(P2 is blue) = 0.8
P(P2 is red) = 0.2
P(P3 is green) = 0.6
P(P3 is blue) = 0.2
P(P3 is red) = 0.2
P(P4 is red) = 0.4

P(P4 is orange) = 0.1
P(P4 is yellow) = 0.1
P(P4 is green) = 0.1
P(P4 is blue)   = 0.1
P(P4 is indigo) = 0.1
P(P4 is violet) = 0.1

Since we have absolutely no information about P5 then uncertainty is at the maximum possible, it is infinite and there is no need to do any further calculations. Given the open query 'list all blue parts' we form the two sets of objects: those known to be blue and those which are possibly blue.

Known Blue: P1
Possibly Blue: P2, P3, P4, P5.

We now calculate the entropy, or uncertainty about the colour of each part.

Given 7 colours the maximum uncertainty = 1.94

Uncertainty about the colour of P1 = 0

Uncertainty about the colour of P2 = -(0.8 ln 0.8 + 0.2 ln 0.2)
$$= .501$$

Uncertainty about the colour of P3 = -(0.6 ln 0.6 + 2(0.2 ln 0.2))
$$= 0.95$$

Uncertainty about the colour of P4 = -(0.4 ln 0.4 + 6(0.1 ln 0.1))
$$= 1.747$$

Uncertainty about the colour of P5 = infinity

The response to the query is to display the set of objects known to be blue, followed by the set which is possible blue. For each object in the latter set the uncertainty is displayed either graphically or numerically. The maximum possible uncertainty, given that the value is in the domain, is displayed for reference.

In a similar way, given a closed query like 'are all objects blue' the system responds with a yes, no or maybe. In the latter case both the maximum uncertainty and the uncertainty with respect to the query are displayed giving users an understanding of the amount of uncertainty in the system's reply.

CONCLUSIONS

In this paper we have shown how clausal form logic can be used to represent certain types of incomplete information. We have also shown how an uncertainty measure can be used to give a more informative answer to queries. While this work has its origins in databases the aim is to explore the applications to knowledge bases and to develop a method of reasoning with incomplete information. This paper is quite tentative and probably raises more questions than it answers.

References

[1] Codd, E.F. Understanding relations. ACM SIGMOD 7, (1975), 23-28.

[2] Codd, E.F. Extending the database relational model to capture more meaning. ACM TODS 4,4 (Dec. 1979), 394-434.

[3] Codd, E.F. Missing information (applicable and inapplicable) in relational systems. SIGMOD Record 15,4 (Dec. 1986).

[4] Date, C.J. Null values in databases. Proceedings of the 2nd British National Conference on Databases, (1982).

[5] Gray, M. Views and imprecise information in databases. Ph.D Thesis, Cambridge, England, (Nov. 1982).

[6] Morrissey, J.M. A treatment of imprecise data and uncertainty in information systems. Ph.D. Thesis, National University of Ireland, (Aug.1987)

[7] Lipski, W. On semantic issues connected with incomplete information databases. ACM TODS 4,3 (Sept. 1979) 262-296.

[8] Moore, R. Reasoning about Knowledge and Action. Technical Note191. SRI International, Menlo Park, CA. 1980.

[9] Ullman, J. Universal relation interfaces for database systems. In: Information Processing '83. Mason, R.E.A., Ed. Elsevier Science Publishers, B.V.(North Holland) (1983).

[10] Vassiliou, Y. Null values in database management: a denotational semantics approach. Proceedings of the 1979 ACM SIGMOD international conference on management of data. Boston, (1979).

[11] Vassiliou, Y. A formal treatment of imperfect information in database

management. Ph.D. Thesis. University of Toronto, (Sept. 1980).

[12] Williams, M.H.; Nicholson, K.A. An approach to handling incomplete information in databases. Computer Journal, 31(2), (1988), 133-140.

[13] Winslett, M. A model based approach to updating databases with incomplete information. Database Systems 13(2), (1988), 167-196.

[14] Zloof, M.M. Query-By-Example: a database language. IBM Systems Journal, 16, (1977), 324-343.

PROPAGATING BELIEFS AMONG FRAMES OF DISCERNMENT IN DEMPSTER-SHAFER THEORY

Weiru Liu,   Michael F. McTear,   Jun Hong

ABSTRACT

Propagating beliefs is a major problem in dealing with uncertainty. In this paper we discuss two aspects in propagating beliefs among frames of discernment in D-S theory. First of all, we extend Yen's probabilistic multi-set[1] mapping to evidential mapping in order to propagate beliefs from an evidential frame of discernment to a hypothesis frame of discernment the relations between which are uncertain. Secondly we introduce an approach to pooling beliefs in a complex frame of discernment.

1. INTRODUCTION

Uncertainty is not a new problem for academic research. But it is really a difficult problem in the area of Artificial Intelligence, especially in knowledge-based systems. There are several methods which can be used to deal with uncertainty in knowledge-based systems, such as certainty factors in MYCIN, inference nets in PROSPECTOR, fuzzy sets, Bayesian nets and Dempster-Shafer belief functions [2], among which Dempster-Shafer theory is more widely accepted than others. The main advantage of Dempster-Shafer theory over other approaches to the uncertainty problem is that belief is distributed over subsets of the possible propositional space, thereby allowing belief to be attributed to a hypothesis H without all remaining belief being attributed to ¬H[3].

There are two kinds of uncertainty problem: one is caused by evidence; another is caused by the uncertain relations between evidence and hypotheses. If there are several elements in an evidence space and belief can not be distributed to each individual element in the evidence space, it is difficult to use the approaches in MYCIN and PROSPECTOR, but D-S theory is powerful in such situations as shown in [2,4]. On the other hand, if the knowledge is in the form of heuristics, D-S theory can not be applied directly to represent the uncertain relationships between evidence and hypothesis, but the approaches in MYCIN and PROSPECTOR can.

Yen[1] proposed an extension to D-S theory by introducing a probabilistic multi-set mapping between the evidence space and the hypothesis space in which conditional probability is used to represent uncertain relations between evidence and hypothesis groups. But belief distributions in an evidence space is represented in the form of probabilities.

In this paper, we extend Yen's theory by introducing an *evidential mapping* in which we use *mass functions*, instead of conditional probabilities, to represent these uncertain relations. Mass functions are also used to distribute belief in an evidence space. We also use Shafer's partition technique to obtain the mass function in an evidence space when this space is made of several variables. The extended D-S theory has the ability to deal with uncertainties in heuristic rules.

In Section 2, we explore Yen's extension and give our evidential mapping based on Yen's extension. The detailed discussion on propagating uncertainty from evidence spaces to hypothesis spaces is presented as well. In Section 3 we introduce Shafer's partition technique at first and then give the method to get the mass distribution in a complex evidence space through using that partition technique.

2. YEN'S PROBABILISTIC MAPPING AND OUR EVIDENTIAL MAPPING

2.1 Dempster-Shafer Theory

Now we briefly review the basics of D-S theory of evidence.

Frame of discernment — the set of mutually exclusive and exhaustive propositions. It is often referred to as Θ.

Mass distribution — the distribution of a unit of belief over a frame of discernment. It is also called a mass function or basic probability assignment. It satisfies the following conditions

$$(1).\ m(\phi)=0;$$
$$(2).\ \Sigma\,m(A)=1. \tag{1}$$

Multivalued mapping — the notation $\Gamma:A\to B$ indicates that Γ is a mapping, assigning to each element a of A a set of elements $\Gamma(a)$ of B.

$$\Gamma\ A\to 2^B \tag{2}$$

If Θ_1 and Θ_2 are two frames of discernment, Γ is a multivalued mapping and m is a mass distribution in Θ_1 then the mass distribution in Θ_2 can be got by:

$$m'(B) = \frac{\sum_{\Gamma(a)=B \wedge a \in \Theta 1} m(a)}{1 - \sum_{\Gamma(a)=\phi \wedge a \in \Theta 1} m(a)} \qquad \text{where } B \subseteq \Theta_2 \qquad (3)$$

Usually there is no element in Θ_1 mapped to the empty set in Θ_2, so the formula (3) can be changed as

$$m'(B) = \sum_{\Gamma(a)=B \wedge a \in \Theta 1} m(a) \qquad (4)$$

where the subset B is called the focal element.

Fusion (or combination) — suppose m_1 and m_2 are two mass distributions in Θ from independent sources then the new mass distribution m in Θ is as

$$m(C) = \frac{\sum_{A \cap B = C} m_1(A)m_2(B)}{1 - \sum_{A \cap B = \phi} m_1(A)m_2(B)} \qquad (5)$$

2.2 Yen's Probabilistic Mapping

It is difficult for D-S theory to deal with an uncertain problem like *"IF a symptom e is observed, THEN the patient is more likely (with probability 0.8) to have a disease in {h1, h2} than to have a disease in {h3, h4} (with probability 0.2)"* described in [1] directly. Based on such motivation Yen extended the theory in two aspects: a). the multivalued mapping in the D-S theory is extended to a probabilistic one; b). D-S combination rule is modified to combine belief updates rather than absolute belief measures. Based on such an extension he implemented a system Gertis [1].

Here is an example

Rule 1: IF *e* THEN {h1, h2} with 0.8
            {h3, h4} with 0.2
        ELSE {h1, h2, h3, h4} with 1.0
where E={e, ¬e}, H={h1, h2, h3, h4}.

The frame E is called an evidence space and the frame H is called a hypothesis space. The rule given above is called a heuristic rule.

Using E' to denote the background evidence source that determines the posterior probability distribution of the evidence space E, and if $p(e/E')=0.7$, $p(\neg e/E')=0.3$ then the basic probability assignment in H will be obtained by using rule 1:

$$m(\{h1,h2\}/E')=0.7\times0.8=0.56$$
$$m(\{h3,h4\}/E')=0.7\times0.2=0.14$$
$$m(\{h1,h2,h3,h4\}/E')=0.3\times1=0.3$$

Definition 1: A probabilistic multi-set mapping from a space E to a space Θ is a function Γ^*: $E\rightarrow2^{2^{\Theta}\times[0,1]}$. The image of an element in E, denoted by $\Gamma^*(e_i)$, is a collection of subset-probability pairs

$$\Gamma^*(e_i)=\{(H_{i1}, P(H_{i1}/e_i)), ... ,(H_{im}, P(H_{im}/e_i))\}. \qquad (6)$$

that satisfies the following conditions:

a. $H_{ij}\neq\phi$, $\qquad\qquad$ $j=1...m$
b. $H_{ij}\cap H_{ik}=\phi$ $\qquad\qquad$ $j\neq k$
c. $P(H_{ij}/e_i)>0$, $\qquad\qquad$ $j=1...m$ $\qquad\qquad$ (7)
d. $\Sigma P(H_{ij}/e_i)=1$

where e_i is an element of E, $H_{i1},...,H_{im}$ are subsets of Θ.

There are two main limitations in Yen's extension: 1).the certainty degrees of heuristic rules are described using conditional probability; and 2).the belief distribution in an evidence space is measured by probability. For a given evidence space, if belief cannot be distributed on each individual element of the space as probability then the system fails to use this piece of evidence. If there is prior probability on elements in the evidence space, the system has to consider the effects of prior probability. Furthermore, it is not clear how to get the probability in an evidence space E when E has more than one variable i.e. $E=E_1\times E_2...\times E_n$.

2.3 Evidential Mapping

With respect to the first limitation of Yen's extension, we propose an evidential mapping that uses mass distributions to express these uncertain relations between the evidence space and the hypothesis space.

In order to distinguish the evidential mapping from the probabilistic mapping, we use $f(e_i\rightarrow H_{ij})$ to represent the uncertain relation between evidential element e_i with a subset H_{ij} of the hypothesis space by evidential mapping. The definition 1 can be modified as:

Definition 1': An evidential mapping from an evidential space E to a hypothesis space H is a function $\Gamma^*: E{-}{>}2^{2^H}{\times}[0,1]$. The image of an element in E, denoted by $\Gamma^*(e_i)$, is a collection of subset-mass pairs:

$$\Gamma^*(e_i)=\{(H_{i1},f(e_i{\rightarrow}H_{i1})),....,(H_{im},f(e_i{\rightarrow}H_{im}))\}$$

and the formula in (7) becomes

a.	$H_{ij}{\neq}\phi$	$j=1,...,m$
b.	$f(e_i{\rightarrow}H_{ij}){>}0$	$j=1,...,m$
c.	$\Sigma_j\ f(e_i{\rightarrow}H_{ij}){=}1$	

(8)

where e_i is an element of E, $H_{i1},...,$ H_{im} are subsets of H. The condition (b) in formula (7) is deleted and condition (c) is modified as (b) in (8) because of the evidential mapping.

In particular we establish the mapping from the whole evidence space E to the whole hypothesis space H for the pair of (E, H) and let $f(E{\rightarrow}H)$ equal to 1 no matter whether the user declares it or not, that is $\Gamma^*(E)=\{(H,1)\}$.

For the second limitation in Yen's extension, we use the mass function to distribute a piece of evidence in an evidence space rather than using probability. So we do not need to consider Yen's second extension in our approach and we simply treat the prior probability as a mass function in an evidence space.

Now we consider rule 1 here again. Suppose we get a mass distribution through a piece of evidence as $m(e)=0.7$ and $m(H)=0.3$, then we can propagate it to the hypothesis space through rule 1 and obtain

$$m'(\{h1,\ h2\})=m(e){\times}f(e{\rightarrow}\{h1,\ h2\})=0.7{\times}0.8=0.56$$
$$m'(\{h3,\ h4\})=m(e){\times}f(e{\rightarrow}\{h3,\ h4\})=0.7{\times}0.2=0.14$$

by the mapping $e{\rightarrow}H$ and

$$m'(\{h1,\ h2,\ h3,\ h4\})=m(E){\times}f(E{\rightarrow}H)=0.3{\times}1=0.3$$

by the mapping $E{\rightarrow}H$. It is easy to see that m' is a mass function in H.

Considering the more general situation, suppose there are n elements in an evidence space E and there are n' elements in a hypothesis space H, and the evidential mapping from E to H is defined as

$$\Gamma^*(e_i)=\{(H_{i1},f(e_i\rightarrow H_{i1})),....,(H_{ir},f(e_i\rightarrow H_{ir}))\}, \text{ for } i=1,...,n$$

where $E=\{e_1, e_2,..., e_n\}$ and $H_{ij}\subseteq H$. Particularly we define $h_i=\{H_{i1}, ..., H_{ir}\}$.

Given the mass distribution m in E, firstly suppose a piece of evidence related to m is distributed on each element e_i in E as $m(e_i)$, then a function m' in H can be obtained by mass function m and evidential mapping Γ^*: E$\rightarrow$ H as:

$$m'(H_{ij})=m(e_i)\times f(e_i\rightarrow H_{ij}) \quad H_{ij}\in h_i$$

define m'' as:

$$m''(H_{ij}) = \Sigma m'(H_{kt}) \text{ for all } H_{kt}=H_{ij}, H_{ij}\subseteq H, H_{kt}\in h_k \qquad (9)$$

The function m'' is again a mass function in the hypothesis space and has the following features:

1) $m''(\phi)=0$
2) $\Sigma m''(H_{ij})=1 \quad H_{ij}\subseteq H$

It can be proved by the following two steps according to definition 1$'$ and features of the mass function.

$$\Sigma_j m'(H_{ij})=\Sigma_j m(e_i)\times f(e_i\rightarrow H_{ij})$$
$$=m(e_i)\times\Sigma_j f(e_i\rightarrow H_{ij})=m(e_i) \qquad (10)$$

$$\Sigma_i \ \Sigma_j \ m'(H_{ij}) = \Sigma_i \ \Sigma_j \ m(e_i)\times f(e_i\rightarrow H_{ij})$$
$$= (\Sigma_i \ m(e_i)) \ (\Sigma_j \ f(e_i\rightarrow H_{ij})) = 1\times 1 = 1 \qquad (11)$$

Secondly, suppose m is a mass distribution, a unit of belief related to m is distributed on the subsets A_j of E as $m(A_j)$ which contains more than one element and the remaining belief is distributed to each elements of E-A_j. Furthermore suppose there are k elements in A_j.

In order to make things clear, we rearrange the element sequence in E and put the elements in A_j as the first k elements such that:

$$E=\{e_1, e_2, ..., e_k, e_{k+1}, ..., e_n\}$$
$$where\ A_j=\{e_1, e_2, ..., e_k\}.$$

Through the evidential mapping Γ^* from the evidence space to the hypothesis space we have $h_1, h_2, ..., h_k$ corresponding to each element in A$_j$. The subset h in H which is related to A$_j$ can be defined as:

$$h= \overset{k}{\underset{i=1}{\cup}}(\overset{r}{\underset{j=1}{\cup}}H_{ij})\quad where\ H_{ij}\in h_i$$

Because based on this piece of evidence we can only get $m(A_j)$ rather than getting each $m(e_i)$ for e_i belonging to A_j, we ignore those $f(e_i\text{-}\rightarrow H_{ij})$ related to $e_i\in A_j$ and establish the mapping from A$_j$ to h as:

$$\Gamma^*(A_j) = \{h,1\}$$

Define m' as a function in the hypothesis space H by propagating the mass function m to H through evidential mapping we can get

$$m'(h)=m(A_j)\times 1=m(A_j)$$
$$m'(H_{ij})=m(e_i)\times f(e_i\rightarrow H_{ij})\ for\ i=k+1, .., n, H_{ij}\in h_i$$

Using formula (9) to define m'' in H, we can also prove that m'' is a mass function in H based on formula (10) and (11):

$$\Sigma m''(X) = m'(h) + \overset{n}{\underset{i=k+1}{\Sigma}}\ \overset{r}{\underset{j=1}{\Sigma}}m'(H_{ij})$$

$$= m'(h) + \overset{n}{\underset{i=k+1}{\Sigma}}\ \overset{r}{\underset{j=1}{\Sigma}}m(e_i)\times f(e_i\rightarrow H_{ij})$$

$$= m(A_j) + (\overset{n}{\underset{i=k+1}{\Sigma}}m(e_i))\ (\overset{r}{\underset{j=1}{\Sigma}}f(e_i\rightarrow H_{ij}))$$

$$= m(A_j) + (1-m(A_j))\times 1 = 1$$

Similarly we can prove that m'' is always a mass function in H based on a mass distribution m in E and evidential mapping Γ^*: E$\rightarrow$H for other situations which are different from the above.

For two independent evidence spaces E_1, E_2 and the hypothesis space H, if there are evidential mappings Γ_1^* : $E_1 \rightarrow H$ and Γ_2^*: $E_2 \rightarrow H$, then for the given mass distributions m_1 in E_1 and m_2 in E_2, we can get two mass distributions in H: $m_1"$ and $m_2"$. Combining $m_1"$ and $m_2"$ using D-S's combination rule, we get the new mass distribution in H. We also treat the prior probability as a mass distribution in the evidence space.

3. MASS DISTRIBUTION IN A COMPLEX EVIDENCE SPACE

3.1 Shafer's Partition Technique

As defined in Section 2.1, a frame of discernment, or simply a frame, is a finite set[1] of possible answers to some question, where one and only one of these answers can be correct. The act of adopting a frame for a question formalizes a variable. The elements of the frame are the possible values of the variable. For example, when we adopt the set $\Theta_1 = \{red, white, yellow\}$ as our frame for the question *"What colour rose is Bill wearing today?"*, we formalize the variable $X_1 = $*"the colour rose Bill is wearing today"*, with possible values, *red, white,* and *yellow*. We can also give another question as *"what colour shirt is Bill wearing"*, and the possible values set is $\Theta_2 = \{white, blue\}$. The variable corresponding to Θ_2 is $X_2 = $*"the colour shirt Bill is wearing today"*. Then the conjoined question from these two questions will be *"what colour rose and what colour shirt is Bill wearing today?"*, and we can get the frame for the new question by Cartesian product

$$\Theta_1 \times \Theta_2 = \{(red, white), \quad (red, blue), \quad (white, white), \quad (white, blue), \quad (yellow, white), \\ (yellow, blue)\}$$

Ruling out the possibility of Bill wearing either a red rose on a blue shirt or a white rose on a white shirt, a smaller frame Θ will be obtained as

$$\Theta = \{(red, white), (white, blue), (yellow, white), (yellow, blue)\}$$

By establishing the multivalued mapping from Θ_1 to Θ, we can consider a variable as a function or mapping defined from its frame to a conjoined frame which is made by the variable's frame and other frames through Cartesian product.

A partition T of a frame of discernment Θ has the following features:

---

[1] T.Strat has discussed infinite situation in [7].

suppose $A_1, ..., A_n$ are subsets of Θ partitioned by T, then

(1) $A_i \neq \phi \quad i=1,...n$

(2) $A_i \cap A_j = \phi \quad i \neq j$ $\qquad\qquad$ (12)

$$(3) \quad \overset{n}{\underset{i=1}{\cup}} A_i = \Theta$$

Such a partition T itself can be regarded as a frame of discernment. More details can be found in [5,6].

3.2 Mass Distribution in A Complex Evidence Space

Another major problem in propagating a mass distribution from the evidence space to the hypothesis space is how to get a mass distribution in an evidence space which contains more than one variable. That is, if an evidence space E is made of $E_1 \times E_2 \times \times E_n$ and E_1 E_2,.., E_n are independent, then how do we get the mass distribution in E by knowing m_1 in E_1, m_2 in E_2,.. , and m_n in E_n.

In order to deal with this problem, we adopt Shafer's partition theory by dividing a larger frame into smaller frames for each variable in the larger frame.

For any rule such as "IF A and B and C THEN D with d", we firstly form a frame or space Θ which is made of $A \times B \times C$. Suppose the value set of A is $\{a_1, a_2, ..., a_n\}$, the value set of B is $\{b_1, ...,b_j\}$, the value set of C is $\{c_1, ..., c_k\}$, then there are $n \times j \times k$ possible elements in Θ. For each a_i of A there is a corresponding subset of Θ and in fact this subset is $\{a_i\} \times B \times C$. By using the value set of A, frame Θ can be divided into n subsets and this division suits the definition of partition in Section 3.1, so A, B, and C can cause three partitions in Θ.

Based on Shafer's partition, we call A, B, and C as variables in Θ, so we can get three different partitions by A, B, and C separately. Each partition itself is also a frame.

Suppose three evidential mappings from A, B, and C to Θ are $\Gamma_1^*, \Gamma_2^*, \Gamma_3^*$, then we can get

$$\Gamma_1^*(a_i) = \{(\{a_i\} \times B \times C, 1)\} \text{ for each } a_i \text{ in A}$$
$$\Gamma_2^*(b_i) = \{(\{b_i\} \times A \times C, 1)\} \text{ for each } b_i \text{ in B}$$
$$\Gamma_3^*(c_i) = \{(\{c_i\} \times A \times B, 1)\} \text{ for each } c_i \text{ in C}$$

When A, B and C are independent and each of them has a mass distribution on it, we can propagate these three mass distributions to frame Θ according to the discussion in Section 2.3 and then combine these three mass distributions using D-S's rule in frame Θ.

Frame Θ will become very large when each of A, B and C has a large value set separately, and usually a set of rules which establish the relations from Θ to a hypothesis space only specify some of the values in Θ. That means only a subset of the product of A×B×C is meaningful. So in order to decrease the computational complexity, we adopt the following method to cut off some elements in Θ. The effect of the method can also cut off the elements in A, B, and C as well.

First of all, we can form a subset of Θ based on a set of rules related to it, and suppose this subset contains l element. We define $\Theta_1=\{s_1, s_2, ...s_l\}$, in which $\Theta_1 \subseteq \Theta$ and $s_i \in \Theta$. Similarly three partitions can be made in Θ_1 based on some elements in A, some elements in B, and some elements in C to form another three smaller frames. We call these three partitions as A_1, B_1, C_1, and they suit the conditions $A_1 \subseteq A$, $B_1 \subseteq B$, $C_1 \subseteq C$.

Because A_1, B_1, C_1 are the partitions of Θ_1, so we can establish evidential mappings Γ_{A1}^*, Γ_{B1}^*, Γ_{C1}^* from A_1, B_1, C_1 to Θ_1 separately.

Secondly, we use s to represent the subset of $\Theta - \Theta_1$, and create a new frame, noted as Θ_2, by union $\{s\}$ with Θ_1.

$$\Theta_2 = \Theta_1 \cup \{s\}$$

Similarly, we use a denote $A - A_1$, b denote $B - B_1$, and c denote $C - C_1$, and obtain three new frames by union $\{a\}$ with A_1, $\{b\}$ with B_1, and $\{c\}$ with C_1 respectively.

$$A_2 = A_1 \cup \{a\} \quad B_2 = B_1 \cup \{b\} \quad C_2 = C_1 \cup \{c\}$$

Finally, we define Γ_{A2}^*, Γ_{B2}^* and Γ_{C2}^* as follows:

$$\Gamma_{A2}^* (a_i) = \Gamma_{A1}^*(a_i) \text{ and } \Gamma_{A2}^*(a) = \{(s,1)\} \quad a_i \in A1$$
$$\Gamma_{B2}^* (b_i) = \Gamma_{B1}^*(b_i) \text{ and } \Gamma_{B2}^*(b) = \{(s,1)\} \quad b_i \in B1$$
$$\Gamma_{C2}^* (c_i) = \Gamma_{C1}^*(c_i) \text{ and } \Gamma_{C2}^*(c) = \{(s,1)\} \quad c_i \in C1$$

Through the definitions of Γ_{A2}^*, Γ_{B2}^* and Γ_{C2}^* we can see that they are the evidential mappings from A_2 to Θ_2, B_2 to Θ_2, and C_2 to Θ_2.

It can be proved that A_2, B_2, C_2, and Θ_2 are all frames. By knowing the mass distributions in A_2, B_2, and C_2, the three mass distributions in Θ_2 can be obtained through propagating mass distribution from A_2, B_2, C_2 to Θ_2, and the final mass distribution in Θ_2 is the result of three mass distribution combination.

Similarly if an evidence space Θ is made of A, B, ..., G, we can use the above method to get a mass distribution in it through A, B, ..., G.

4. SUMMARY

For most domain problems the heuristic knowledge is usually much more important than the domain environment knowledge and how to deal with the uncertainty problem in both knowledge and evidence has great effects on the performance of a knowledge based system.

In this paper, based on D-S evidential theory and Yen's extension of the theory, we have proposed our approaches to representing heuristic knowledge by evidential mapping and to pooling the mass distribution in a complex frame by partitioning that frame using Shafer's partition technique. We have generalized Yen's model from Bayesian probability theory to the D-S theory of evidence.

In order to use the Dempster-Shafer theory properly there are still quite a lot of things which need to be done. We will continue to do some theoretical and practical work in this area in the future.

REFERENCE

1. J.Yen "Gertis: A Dempster-Shafer Approach to Diagnosing Hierarchical Hypotheses" Communications of the ACM, May 1989, Vol. 32. pp.573-585.

2. T.M.Strat "The Generation of Explanations within Evidential Reasoning Systems" IJCAI-87, pp.1097-1104.

3. S.Carberry and K.D.Cebulka "Capturing Rational Behavior in Natural language Information Systems" Proc. of 7th conference of AISB. London, pp.153-163.

4. J.D.Lowrance, T. D. Garvey and T.M.Strat "A Framework for Evidential Reasoning Systems" AAAI-86, pp.896-903.

5. G. Shafer, P.P.Shenoy and K.Mellouli "Propagating Belief Functions in Qualitative Markov Trees" International Journal of Approximate Reasoning 1987, Vol.1, pp.349-400.

6. G. Shafer and R. Logan "Implementing Dempster's Rule for Hierarchical Evidence" Artificial Intelligence 1987, (33), pp. 271-298.

7. T.M.Strat "Continuous Belief Functions for Evidential Reasoning". AAAI-84, pp.308-313.

EVIDENTIAL REASONING AND RULE STRENGTHS IN EXPERT SYSTEMS

Jiwen Guan , David A. Bell , Victor R. Lesser

ABSTRACT

The management of uncertainty is at the heart of many knowledge-based systems. The Dempster-Shafer (D-S) theory of evidence generalizes Bayesian probability theory, by providing a coherent representation for ignorance (lack of evidence). However, uncertain relationships between evidence and hypotheses bearing on this evidence are difficult to represent in applications of the theory. Such uncertain relationships are sometimes called "rule strengths" in expert systems and are essential in a rule-based system. Yen [1] extended the theory by introducing a probabilistic mapping that uses conditional probabilities to express these uncertain relationships, and developed a method for combining evidence from different evidential sources. We have extended the theory by introducing an evidential mapping that uses mass functions to express the uncertain relationships, and we have developed a method for combining evidence based on the extended D-S theory. It is a generalization of Yen's model from Bayesian probability theory to the D-S theory of evidence.

1. INTRODUCTION

Rules in an expert system sometimes associate evidence with a group of mutually exclusive hypotheses but say nothing about its individual members of the group. Therefore, an appropriate formalism for representing evidence should have the ability to assign beliefs to (not necessarily disjoint) sets of hypotheses and to combine these beliefs in a consistent way, when they represent evidence from different sources. The Dempster-Shafer (D-S) Theory of Evidence ([2], [3], [10]) is one such formalism. In this theory, beliefs are assigned to subsets of a set of mutually exclusive and exhaustive hypotheses, and Dempster's rule is used to combine beliefs coming from different sources of evidence. Unfortunately, an uncertain relationship between evidence and hypotheses bearing on this evidence (sometimes called "rule strength" in expert systems) is difficult to represent in applications of the theory.

Yen [1] proposed an extension to D-S theory by introducing a probabilistic mapping between the evidence domain and the hypothesis domain. Conditional probabilities are used to represent uncertain relationships between evidence and hypothesis groups bearing on this evidence.

In this paper we extend the theory by introducing an evidential mapping that uses mass functions as used in the D-S theory to express these uncertain relationships. It is a generalization of Yen's model from Bayesian probability theory to the D-S theory of evidence.

As well as using mass functions to represent evidence instead of the Bayesian probabilistic density functions in Yen's model, we also use a mass function to represent the prior belief over the hypothesis space instead of a Bayesian probabilistic function in Yen's model. This means that we can simply use Dempster-Shafer's rule to combine evidence with the prior mass function instead of Yen's three-step method for combining evidence with prior probabilities.

The sections below describe our procedure for combining evidence using an example of its use. Section 2 describes the knowledge base and rule strengths we are dealing with in the example, and Section 3 describes the evidence and data. Section 4 presents the procedure to get from evidence and rule strength to a mass distribution for each rule; i.e., the first half of a single step reasoning process. Section 5 presents the procedure for combining prior mass and different rules; i.e., the second half of the single step reasoning process. Section 6 establishes the belief interval for the final result after the single step reasoning process.

2. A KNOWLEDGE BASE AND RULE STRENGTHS

To make the description of our method clear, we trace through the analysis of the following simplified problem in the context of fault diagnosis in a distributed vehicle monitoring system [9]. The system consists of a set of nodes. A node's local sensors receive signals generated by nearby vehicles. The goal of the system is generate a dynamic map of vehicles based on this signal data.

Consider a frame of discernment with two hypotheses $\Theta = \{h, \bar{h}\}$ where

$$h = \text{There are vehicles in the node's monitoring domain}$$

and $\bar{h}$ is the complementary hypothesis

$$\bar{h} = \text{There are no vehicles in the node's monitoring domain.}$$

Consider one type of evidence e_1 which comes from a node's sensors. This evidence, when present, indicated by $\{e_1\}$, strongly supports $\{h\}$ and refutes $\{\bar{h}\}$. When the evidence is not present, indicated by $\{\bar{e}_1\}$, the support strengths are divided between $\{\bar{h}\}$ and Θ. More specifically, there is an evidence space $\Xi_1 = \{e_1, \bar{e}_1\}$ and mass functions $s_{11}, s_{12}, s_{13} : 2^\Theta \rightarrow [0, 1]$ such that

$$s_{11}(\{h\}|\{e_1\}) = 0.8, s_{11}(\{\bar{h}\}|\{e_1\}) = 0, s_{11}(\Theta|\{e_1\}) = 0.2;$$

$$s_{12}(\{h\}|\{\bar{e}_1\}) = 0, s_{12}(\{\bar{h}\}|\{\bar{e}_1\}) = 0.5, s_{12}(\Theta|\{\bar{e}_1\}) = 0.5;$$

$$s_{13}(\{h\})|\Xi_1) = 0.40, s_{13}(\{\bar{h}\}|\Xi_1) = 0.25, s_{13}(\Theta|\Xi_1) = 0.35.$$

Another source of evidence e_2 is another node with completely reliable sensors, and whose monitoring domain overlaps the first node's monitoring domain. When this evidence is present with certainty, h is true. The absence of e_2 strongly supports $\bar{h}$ and leaves Θ in part unknown. Thus, we have mass functions $s_{21}, s_{22} : 2^\Theta \to [0, 1]$ such that

$$s_{21}(\{h\}|\{e_2\}) = 1, s_{21}(\{\bar{h}\}|\{e_2\}) = 0, s_{21}(\Theta|\{e_2\}) = 0;$$

and

$$s_{22}(\{h\}|\{\bar{e}_2\}) = 0, s_{22}(\{\bar{h}\}|\{\bar{e}_2\}) = 0.8, s_{22}(\Theta|\bar{e}_2) = 0.2.$$

Summarizing, the knowledge base is the following:

RULE-1

IF EVIDENCE $\{e_1\}$ THEN

HYPOTHESIS $\{h\}$ WITH STRENGTH $s_{11}(\{h\}|\{e_1\}) = 0.8$

HYPOTHESIS $\{\bar{h}\}$ WITH STRENGTH $s_{11}(\{\bar{h}\}|\{e_1\}) = 0$

HYPOTHESIS Θ WITH STRENGTH $s_{11}(\Theta|\{e_1\}) = 0.2$

ELSE IF EVIDENCE $\{\bar{e}_1\}$ THEN

HYPOTHESIS $\{h\}$ WITH STRENGTH $s_{12}(\{h\}|\{\bar{e}_1\}) = 0$

HYPOTHESIS $\{\bar{h}\}$ WITH STRENGTH $s_{12}(\{\bar{h}\}|\{\bar{e}_1\}) = 0.5$

HYPOTHESIS Θ WITH STRENGTH $s_{12}(\Theta|\{\bar{e}_1\}) = 0.5$

ELSE IF EVIDENCE Ξ_1 THEN

HYPOTHESIS $\{h\}$ WITH STRENGTH $s_{13}(\{h\}|\Xi_1) = 0.4$

HYPOTHESIS $\{\bar{h}\}$ WITH STRENGTH $s_{13}(\{\bar{h}\}|\Xi_1) = 0.25$

HYPOTHESIS Θ WITH STRENGTH $s_{13}(\Theta|\Xi_1) = 0.35$

Here $\{e_1, \bar{e}_1\} = \Xi_1$ is an evidence space and

$$m_{11}(X) = s_{11}(X|\{e_1\}),$$

$$m_{12}(X) = s_{12}(X|\{\bar{e}_1\}),$$

$$m_{13}(X) = s_{13}(X|\Xi_1)$$

are mass functions $2^{\Theta} \to [0,1]$; i.e., they are the functions $m : 2^{\Theta} \to [0,1]$ such that

$$m(\emptyset) = 0, \sum_{X \subseteq \Theta} m(\Theta) = 1.$$

RULE-2

IF EVIDENCE $\{e_2\}$ THEN

HYPOTHESIS $\{h\}$ WITH STRENGTH $s_{21}(\{h\}|\{e_2\}) = 1$

HYPOTHESIS $\{\bar{h}\}$ WITH STRENGTH $s_{21}(\{\bar{h}\}|\{e_2\}) = 0$

HYPOTHESIS Θ WITH STRENGTH $s_{21}(\Theta|\{e_2\}) = 0$

ELSE IF EVIDENCE $\{\bar{e}_2\}$ THEN

HYPOTHESIS $\{h\}$ WITH STRENGTH $s_{22}(\{h\}|\{\bar{e}_2\}) = 0$

HYPOTHESIS $\{\bar{h}\}$ WITH STRENGTH $s_{22}(\{\bar{h}\}|\{\bar{e}_2\}) = 0.8$

HYPOTHESIS Θ WITH STRENGTH $s_{22}(\Theta|\{\bar{e}_2\}) = 0.2$

Here $\{e_2, \bar{e}_2\} = \Xi_2$ is an evidence space and

$$m_{21}(X) = s_{21}(X|\{e_2\}),$$

$$m_{22}(X) = s_{22}(X|\{\bar{e}_2\})$$

are mass functions $2^{\Theta} \to [0,1]$.

Generally, a rule can be represented in the form of

RULE-i; $i = 1, 2, ..., n$.

IF EVIDENCE E_{i1} THEN

HYPOTHESIS X_{i11} WITH STRENGTH $s_{i1}(X_{i11}|E_{i1})$

HYPOTHESIS X_{i12} WITH STRENGTH $s_{i1}(X_{i12}|E_{i1})$

...

ELSE IF EVIDENCE E_{i2} THEN

HYPOTHESIS X_{i21} WITH STRENGTH $s_{i2}(X_{i21}|E_{i2})$

HYPOTHESIS X_{i22} WITH STRENGTH $s_{i2}(X_{i22}|E_{i2})$

...

ELSE IF EVIDENCE E_{im} THEN

HYPOTHESIS X_{im1} WITH STRENGTH $s_{im}(X_{im1}|E_{im})$

HYPOTHESIS X_{im2} WITH STRENGTH $s_{im}(X_{im2}|E_{im})$

Here Ξ_i is an evidence space,

$$E_{i1}, E_{i2}, ..., E_{im} \subseteq \Xi_i$$

and

$$m_{i1}(X) = s_{i1}(X|E_{i1}),$$

$$m_{i2}(X) = s_{i2}(X|E_{i2}),$$

$$m_{im}(X) = s_{im}(X|E_{im})$$

are mass functions $2^\Theta \to [0,1]$.

Note: Items

" HYPOTHESIS X_{ijk} WITH STRENGTH $s_{ij}(X_{ijk}|E_{ij})$ "

in RULE-i can be eliminated whenever $s_{ij}(X_{ijk}|E_{ij}) = 0$.

3. DATA AND EVIDENCE

Now suppose that the first node's sensors report signals which are very weak (do not appear to belong to any vehicles). Specifically, we have data:

$$c_1(\{e_1\}) = 0.2, c_1(\{\bar{e}_1\}) = 0.5, c_1(\Xi_1) = 0.3.$$

Here c_1 is a mass function over Ξ_1, intuitively representing the confidence we have that e_1 is present.

The other node's sensor detects no signals, so it is completely certain that evidence e_2 is absent. Thus we have data:

$$c_2(\{e_2\}) = 0, c_2(\{\bar{e}_2\}) = 1$$

Here c_2 is again a mass function over Ξ_2.

Generally, there is a mass function $c_i : 2^{\Xi_i} \to [0,1]$ over the evidence space Ξ_i for $i = 1, 2, ..., n$.

We also suppose that there is a prior mass function $c : 2^\Theta \to [0,1]$

$$c(\{h\}) = 0.7, c(\Theta) = 0.3.$$

Generally, there is a such prior mass function c.

Note: data "$c_i(E), E \in 2^{\Xi_i}$" or "$c(X), X \in 2^\Theta$" can be eliminated whenever c_i or c is a vacuous mass function, i.e., whenever

$$c_i(\Theta) = 1 \text{ or } c(\Theta) = 1.$$

Notice that here we use n mass functions

$$c_1 : 2^{\Xi_1} \to [0,1],$$

$$c_2 : 2^{\Xi_2} \to [0,1],$$

$$......,$$

$$c_n : 2^{\Xi_n} \to [0,1]$$

to represent evidence instead of Bayesian probabilistic density functions $\Xi \to [0,1]$ in Yen's model, where n is the total number of rules in the knowledge base.

We also use a mass function $c : 2^\Theta \to [0,1]$ to represent the prior belief over the hypothesis space Θ instead of a Bayesian probabilistic function $2^\Theta \to [0,1]$ in Yen's model. Hence, we can simply use Dempster-Shafer's rule to combine evidence with the prior mass function instead of Yen's three-step method for combining evidence with the prior probabilities.

4. FROM EVIDENCE STRENGTHS AND RULE STRENGTHS TO HYPOTHESIS STRENGTH

Now, for each rule we can get a mass from the evidence strength and the rule strength.

For RULE-1; i.e., for rule strengths s_{11}, s_{12}, s_{13} and from evidence c_1, the node's mass distribution $r_1 : 2^\Theta \to [0, 1]$ is obtained as follows.

$$r_1(\{h\}) = s_{11}(\{h\}|\{e_1\})c_1(\{e_1\})$$

$$+s_{12}(\{h\}|\{\bar{e}_1\})c_1(\{\bar{e}_1\}) + s_{13}(\{h\}|\Xi_1)c_1(\Xi_1)$$

$$= 0.8 \times 0.2 + 0 \times 0.5 + 0.4 \times 0.3 = 0.28,$$

$$r_1(\{\bar{h}\}) = s_{11}(\{\bar{h}\}|\{e_1\})c_1(\{e_1\})$$

$$+s_{12}(\{\bar{h}\}|\{\bar{e}_1\})c_1(\{\bar{e}_1\}) + s_{13}(\{\bar{h}\}|\Xi_1)c_1(\Xi_1)$$

$$= 0 \times 0.2 + 0.5 \times 0.5 + 0.25 \times 0.3 = 0.325,$$

$$r_1(\Theta) = s_{11}(\Theta|\{e_1\})c_1(\{e_1\})$$

$$+s_{12}(\Theta|\{\bar{e}_1\})c_1(\{\bar{e}_1\}) + s_{13}(\Theta|\Xi_1)c_1(\Xi_1)$$

$$= 0.2 \times 0.2 + 0.5 \times 0.5 + 0.35 \times 0.3 = 0.395.$$

By RULE-2; i.e., for rule strengths s_{21}, s_{22} and from evidence c_2 we get the following mass distribution $r_2 : 2^\Theta \to [0, 1]$ for the other node:

$$r_2(\{h\}) = s_{21}(\{h\}|\{e_2\})c_2(\{e_2\})$$

$$+s_{22}(\{h\}|\{\bar{e}_2\})c_2(\{\bar{e}_2\}) = 1 \times 0 + 0 \times 1 = 0,$$

$$r_2(\{\bar{h}\}) = s_{21}(\{\bar{h}\}|\{e_2\})c_2(\{e_2\})$$

$$+s_{22}(\{\bar{h}\}|\{\bar{e}_2\})c_2(\{\bar{e}_2\}) = 0 \times 0 + 0.8 \times 1 = 0.8,$$

$$r_2(\Theta) = s_{21}(\Theta|\{e_2\})c_2(\{e_2\})$$

$$+s_{22}(\Theta|\{\bar{e}_2\})c_2(\{\bar{e}_2\}) = 0 \times 0 + 0.2 \times 1 = 0.2.$$

Generally, for RULE-i the mass distribution is obtained using the following equation:

$$r_i(X) = s_{i1}(X|E_{i1})c_i(E_{i1}) + s_{i2}(X|E_{i2})c_i(E_{i2})$$

$$+... + s_{im}(X|E_{im})c_i(E_{im}).$$

And the function r_i is again a mass function $2^\Theta \to [0, 1]$:

1) $r_i(\emptyset) = 0$,

2) $\sum_{X \subseteq \Theta} r_i(X) = 1$:

$$\sum_{X \subseteq \Theta} r_i(X) = \sum_{X \subseteq \Theta} \sum_{j=1}^{m} s_{ij}(X|E_{ij}) c_i(E_{ij})$$

$$= \sum_j \sum_X s_{ij}(X|E_{ij}) c_i(E_{ij}) = \sum_j c_i(E_{ij})(\sum_X s_{ij}(X|E_{ij}))$$

$$= \sum_j c_i(E_{ij}) = 1.$$

This is the first half of a single step reasoning process. It is said to be a *single step reasoning process* because it is an " atomic " step to combine all evidence strengths, rule strengths , and prior hypothesis strength.

5. COMBINING EVIDENCE FROM PRIOR MASS AND DIFFERENT RULES

Now, let us discuss the second half of the single step reasoning process.

If m_1 and m_2 are two mass functions corresponding to two independent evidential sources, then the combined mass function $m_1 \oplus m_2$ is calculated according to Dempster-Shafer's rule of combination:

1. $(m_1 \oplus m_2)(\emptyset) = 0$;

2. For every $C \subseteq \Theta$, $C \neq \emptyset$,

$$(m_1 \oplus m_2)(C) = \frac{\sum_{X \cap Y = C} m_1(X) m_2(Y)}{\sum_{X \cap Y \neq \emptyset} m_1(X) m_2(Y)}. \tag{1}$$

For illustrative purposes, the *intersection table* is a helpful device. Each entry contains two items: a subset and a value. The first row and the first column contain the subsets and the values assigned by m_1 and m_2 respectively. The subset in entry i, j in the table is the intersection of subsets in row i and column j. The value next to the subset is the product of m_1 and m_2, as shown in the following table .

$m_1 \oplus m_2$	$Y(m_2(Y))$
$X(m_1(X))$	$X \cap Y(m_1(X)m_2(Y))$

A given subset of Θ may occur in more than one location of the table. The value of $(m_1 \oplus m_2)(A)$ for a subset A is computed by summing the products adjacent to each occurrence of A.

For our example, the intersection table of $r_1 \oplus r_2$ for RULE-1 and RULE-2 is shown in the following table .

$r_1 \oplus r_2$	$\{\bar{h}\}0.8$	$\Theta 0.2$
$\{h\}.280$	$\emptyset.280 \times .8 = .224$	$\{h\}.280 \times .2 = .056$
$\{\bar{h}\}.325$	$\{\bar{h}\}.325 \times .8 = .260$	$\{\bar{h}\}.325 \times .2 = .065$
$\Theta.395$	$\{\bar{h}\}.395 \times .8 = .316$	$\Theta.395 \times .2 = .079$

We get the normalization constant (required to discount for mass committed to $\emptyset$, the empty set)

$$N = \sum_{X \cap Y \neq \emptyset} m_1(X)m_2(Y) = 1 - \sum_{X \cap Y = \emptyset} m_1(X)m_2(Y) = 1 - 0.224 = 0.776$$

and

$$(r_1 \oplus r_2)(\{h\}) = 0.056/0.776 = 0.0722,$$

$$(r_1 \oplus r_2)(\{\bar{h}\}) = (0.065 + 0.260 + 0.316)/0.776$$

$$= 0.641/0.776 = 0.8260,$$

$$(r_1 \oplus r_2)(\Theta) = 0.079/0.776 = 0.1018.$$

Let us now combine $r_1 \oplus r_2$ and prior mass c. The intersection table now has the form shown in the following table .

$(r_1 \oplus r_2) \oplus c$	$\{h\}0.7$	$\Theta 0.3$
$\{h\}.0722$	$\{h\}.0722 \times .7 = .05054$	$\{h\}.0722 \times .3 = .02166$
$\{\bar{h}\}.8260$	$\emptyset.8260 \times .7 = .57820$	$\{\bar{h}\}.8260 \times .3 = .24780$
$\Theta.1018$	$\{h\}.1018 \times .7 = .07126$	$\Theta.1018 \times .3 = .03054$

The normalization constant is

$$N_{(r_1 \oplus r_2) \oplus c} = 1 - \sum_{X \cap Y = \emptyset} (r_1 \oplus r_2)(X)c(Y) = 1 - 0.57820 = 0.42180,$$

and the combined mass distribution $(r_1 \oplus r_2) \oplus c$ becomes

$$((r_1 \oplus r_2) \oplus c)(\{h\}) = (0.05054 + 0.07126 + 0.02166)/0.42180$$

$$= 0.14346/0.42180 = 0.34011,$$

$$((r_1 \ominus r_2) \ominus c)(\{\bar{h}\}) = 0.24780/0.42180 = 0.58748,$$

$$((r_1 \ominus r_2) \ominus c)(\Theta) = 0.03054/0.42180 = 0.07241.$$

Generally, the prior mass function is

$$c : 2^\Theta \rightarrow [0,1]$$

before reasoning.

After the single step reasoning process it changes to the mass function $m : 2^\Theta \rightarrow [0,1]$

$$m = r_1 \ominus r_2 \ominus \dots \oplus r_n \oplus c.$$

6. BELIEF INTERVAL

Finally, we can determine the belief interval.

$$bel(\{\bar{h}\}) = \sum_{X \subseteq \{\bar{h}\}} m(X) = m(\{\bar{h}\}) = 0.58748,$$

$$pls(\{\bar{h}\}) = 1 - bel(\{\bar{\bar{h}}\}) = 1 - bel(\{h\})$$

$$= 1 - \sum_{X \subseteq \{h\}} m(X) = 1 - m(\{h\})$$

$$= 1 - 0.34011 = 0.65989.$$

So in our example we have the belief interval

$$[bel(A), pls(A)] = [0.58748, 0.65989]$$

for $A = \{\bar{h}\}$, and

$$ignorance(A) = 0.65989 - 0.58748 = 0.07241.$$

The lower bound $bel(A)$ gives the total amount of belief committed to the subset A, i.e., the extent to which the current evidence specifically supports A. The upper bound $pls(A)$ expresses the extent to which the current evidence fails to refute A, i.e., the extent to which remains plausible. The difference between the two represents the residual *ignorance*. The advantage of this representation of belief is that it directly accounts for what remains unknown. It represents exactly what is known, no more and no less.

7. SUMMARY

The management of uncertainty is at the heart of many knowledge-based systems. The Dempster-Shafer (D-S) theory of evidence generalizes Bayesian probability theory, by providing a coherent representation for ignorance (lack of evidence).

However, uncertain relationships between evidence and hypotheses bearing on this evidence are difficult to represent in applications of the theory. We have extended the theory by introducing an evidential mapping that uses mass functions to express these uncertain relationships, and developed a method for combining evidence based on the extended D-S theory.

As well as the extension of the D-S theory to handle the uncertainty associated with rules and evidence, the Dempster rule of combination has been used to combine belief updates rather than absolute beliefs.

References

[1] Yen, J., "A reasoning model based on an extended Dempster-Shafer theory", *Proceedings aaai-86* 125-131.

[2] Dempster, A. P., "Upper and lower probabilities induced by a multivalued mapping", *Annals of mathematical statistics*, 38(1967) 325-339.

[3] Dempster, A. P., "A generalization of Bayesian inference", *J. Roy. Statist. Soc. B*, 30(1968) 205-247.

[4] Duda, R. O.,P. E. Hart, and N. J. Nilsson ,"Subjective Bayesian methods for rule-based inference systems", *Proceedings 1976 National Computer Conference*, AFIPS, 45(1976) 1075-1082.

[5] Gordon, J. and E. H. Shortliffe ,"A method for managing evidential reasoning in a hierarchical hypothesis space", *Artificial Intelligence*, 26(1985) 323-357.

[6] Grosof, B. N.,"Evidential confirmation as transformed probability", *Proceedings of the AAAI/IEEE Workshop on Uncertainty and Probability in Artificial Intelligence*,1985, pp.185-192.

[7] Heckerman, D., "A probabilistic interpretation for MYCIN's certainty factors", *Proceedings of the AAAI/IEEE Workshop on Uncertainty and Probability in Artificial Intelligence*,1985, pp.9-20.

[8] Loui, R., J. Feldman, H. Kyburg ,"Interval-based decisions for reasoning systems", *Proceedings of the AAAI/IEEE Workshop on Uncertainty and Probability in Artificial Intelligence*,1985, pp.193-200.

[9] Hudlicka, E., Lesser, V. R., Pavlin, J., Rewari, A., "Design of a Distributed Fault Diagnosis System", *Technical Report 86-63, COINS Department, University of Massachusetts, Amherst, Massachusetts*.

[10] Shafer, G., "A Mathematical Theory of Evidence", *Princeton University Press, Princeton, New Jersey, 1976*.

[11] Yen, J., "GERTIS: A Dempster-Shafer Approach to Diagnosing Hierarchical Hypotheses", *Communications of the ACM* 5 vol. 32, 1989, pp.573-585.

[12] Guan, J., Pavlin, J., Lesser, V. R., "Combining Evidence in the Extended Dempster-Shafer Theory",*Proceedings of the 2nd Irish Conference on Artificial Intelligence and Cognitive Science*, Dublin, 1989.

Author Index